Frustrating Flowers and Puzzling Plants

Frustrating Flowers and Puzzling Plants

Identifying the difficult species of Britain and Ireland

JOHN M. WARREN
with
Jonathan Mitchley and Henry Ford

Illustrations by John M. Warren

PELAGIC PUBLISHING

Published in 2024 by
Pelagic Publishing
20–22 Wenlock Road
London N1 7GU, UK

www.pelagicpublishing.com

Frustrating Flowers and Puzzling Plants: Identifying the Difficult Species of Britain and Ireland

https://doi.org/10.53061/QYEE4043

A CIP record for this book is available from the British Library

ISBN 978-1-78427-331-6 Pbk
ISBN 978-1-78427-332-3 ePub
ISBN 978-1-78427-333-0 PDF

Designed by BBR Design, Sheffield
Typeset in Adobe Caslon Pro by S4Carlisle Publishing Services, Chennai, India
Printed by Finidr in the Czech Republic

Contents

Foreword

Alastair Fitter

This really is the book I have always wanted. Conventional floras and flower guides are either fully comprehensive – they include every species in the flora – or highly selective – often concentrating on common and charismatic species. The first option is of course fail-safe, but it means you have to carry around, and possibly wade through, information on Lady's-slipper orchid and Daisy, neither of which any botanist is going to misidentify – or in some cases ever find.

For the enthusiast starting out on plant identification, that is fine: at that stage you need to have all the information to hand. You quickly reach a stage, though, where you know all the really common and all the spectacular plants but there are genera and families where we all struggle. Most of us never get beyond that stage.

This book is made for those of us who are competent but far from omniscient botanists. It focuses entirely on those difficult groups that perpetually challenge. For those with an amazing ability to recall obscure diagnostic characters, perhaps this book will be superfluous, but such people are few in my experience.

Most botanists work to a large extent on 'jizz', the appearance of a plant. That jizz will take into account a host of subtle characters such as size, form, shade of green, locality, habitat, associates and so on. But when asked to explain to someone else what makes that plant, say, Wood Dock *Rumex sanguineus* rather than Clustered Dock *R. conglomeratus*, I may struggle to recall the diagnostic characters, or more usually which way around they go (alright, it's the leafy flowering shoots and spreading branches of Clustered Dock!).

John Warren's species accounts here dissect that jizz very expertly. There are many features that will help identification: I particularly like the tables showing the combinations of characters that each species displays, creating a type of single-access key that works visually. The other neat thing about this book is that it tells you why these plants are so confusing, whether that be because of their method of reproduction (such as apomixis), their genetic system (e.g. polyploidy) or something else, so that you will understand better why they are frustrating you.

There is a growing problem of plant blindness. Even though we depend absolutely on them for our survival, whether that is as food or as regulators of our environment (let alone any aesthetic or other dependences), plants feature all too weakly in public arenas. Even wildlife programmes, which cannot fail to show plants, usually fail to say anything about them. Fewer people have the skills to identify plants, for a variety of reasons: an increasingly urban society, a lack of prominence and a vicious cycle whereby educators are themselves educated by those who have not imbibed those skills.

This book will play its part in helping those who want to become more confident in plant identification and, if we are serious about attacking the biodiversity crisis, that is a really important step. I salute the author and look forward to taking the book out with me and using it often. I hope the binding is strong!

Preface

If you have ever tried to identify wildflowers, the chances are you already know exactly what is meant by 'frustrating flowers and puzzling plants'. Although many plants are relatively distinctive and easy to recognise, numerous others are most definitely not. Standard wildflower books tend to provide as much help and guidance with identifying the easy and distinctive species as they do with the complex, frustrating species. This book is designed to come to the rescue of the exasperated.

From the beginning it is worth pointing out that it is not illegal to misidentify a plant. Misidentification should not even be mildly embarrassing. Experienced botanists still make mistakes; they routinely compare opinions, and debate at length the identity of especially tricky specimens. There are occasions when correct identification is important, such as reporting new records, but the Botanical Society of Britain and Ireland (the BSBI) has is a network of local experts who will verify these and be happy to help hone your field skills. The BSBI also has a list of national referees who are the recognised experts in each of the plant groups covered here. They can come to your aid on particularly challenging specimens.

There are also many special-interest groups on social media that are dedicated to complex plant groups. These can be great at helping identify tricky plants and in putting you in touch with other enthusiasts who share your fascination. However, be aware that sometimes on social media the correct identification you are seeking may be drowned out by 'fake views'. This book explains why some groups of plants are complex and genuinely perplexing. Fortunately, being able to correctly identify them is immensely rewarding and well worth the effort.

This volume does not cover all British plants. It is assumed that you already have a wildflower book and can identify most of the easier species. Rather, the purpose here is to provide you with bespoke sets of tips to help identify the plants that are members of each complex group. Individual chapters focus in turn on a different group. The causes of complexity in each case are explained to help you understand why they can be so challenging to identify; after all, it is always easier to remember something once you understand it, rather than just trying to rote-learn obscure lists of facts. Furthermore, the explanations of what causes these plants to be so exasperating are often interesting stories in their own right. These stories are frequently related to the plants' ecology, evolutionary history or unusual sexual practices.

This guide will help you grow in confidence as a botanist, by starting with the species that are easiest, most distinctive or commonest within each group of frustrating flowers. It is worth remembering that we all have remarkable abilities to instantly differentiate between almost indistinguishable individuals. We might know a set of identical twins and we are usually able to tell them apart and only occasionally make embarrassing mistakes. However, if you were asked to describe these twins so that someone being introduced to them for the first time could also tell them apart, then that is a more difficult task. This second sort of challenge is the one that has been faced in writing this book. This task has been completed after consulting the parents of each set of twins in turn – or, in our case, by working with experts on each group of complex plants to develop new and simple ways of differentiating the apparently almost identical.

It is expected that users of this guide will not typically want to consult more than one chapter at a time. For this reason, there is some repetition of explanations when biological phenomena are found in more than one group of plants. Most advanced floras use long, involved keys consisting of a series of dichotomous questions to help in identification. This approach has been avoided here, because in reality most competent botanists typically jump many of these questions and focus on the one or two key characteristics that are important in a given instance. This book attempts to identify these essential characteristics for each complex group and simplifies them into tabular form and into accessible short descriptions and illustrations.

The tabular keys used throughout are designed to help you assemble a picture of what species your unknown specimen is most like, by considering several characteristics at once. This is arguably a biologically more meaningful approach and one that may help you identify specimens that are intermediate in form and might be hybrids. Traditional dichotomous keys implicitly assume that species evolve like the branches of a tree. As we will see throughout this book, hybridisation resulting in the reuniting of previously separated evolutionary lineages is a common feature in the plant world.

Example of a tabular key as used in this book

	Leaf shape		Flower colour		
Species	**round**	**elongated**	**red**	**white**	**blue**
Species A					
Species B					
Species C					
Your specimen		X		X	

FUDGE WORDS

The task of identifying plants can be infuriating because field guides are full of 'fudge words'. Species are frequently described as '*typically* hairy' or '*occasionally* toothed' or '*sometimes* erect' or '*may be* bifurcated' and so on. Although unhelpful, this reflects the complexity of within-species variation and plasticity. In the multi-access keys used here, such uncertainty is represented using pale green shading:

Boxes shaded thus indicate that the species may be variable for this character.

When attempting to identify more challenging species, it is usually wise not to rely on a single diagnostic feature. It can be frustrating if you need both flowers and fruit to be certain of an identification when your specimen lacks one or other. Sometimes this may be rectified by a little patience. On other occasions you will need to accept that you have missed flowering this year, and so it may be impossible to assign identity with 100% confidence. It is also worth being aware that some features are more reliable as identification tools than others. For example, plant height can be highly variable and is often more a reflection of local soil conditions than species. Hence, plant height has sometimes been included as a guide to identification, but generally other features are more reliably diagnostic.

The question of how many species of plant are native to Britain and Ireland sounds trivial, but actually this is far from the case. No two wildflower books will have the same content, even when they claim to cover all Britain's plant species. This is in part because of how we define what we consider as native. The most significant cause of this problem, however, is the lack of agreement about what makes a species. If you really want to start a fight in a room full of botanists (who are not renowned as violent people) show them a list of plant species that are found in the British Isles and nowhere else. Such a list will almost certainly contain plants that some of them don't consider to be true species and will exclude plants that others will think worthy of inclusion. This is a problem to keep in mind before you start using the book.

Here a species is defined as groups of individual plants that are recognisably morphologically similar to each other while often being distinctly different from other populations of individuals. For this situation to persist requires that there is little or no exchange of genetic material between the species in question and its close relatives. These conditions may arise in the first place through a number of different evolutionary mechanisms. In this book, individual chapters are grouped by these processes, which are explained in more detail in the introductions to each section. Many plants have evolved ways of reproducing that do not involve a male parent. This includes vegetative spread, self-pollination and the production of asexual seeds. Each of these methods of reproduction, along with selection and chance, generates different levels of genetic differentiation within and between populations.

In contrast, some taxonomic complexity in plants results from the hybridising and sharing of more genetic material between species, when interspecies crossing is a fairly common occurrence or when hybrid progeny are long-lived. Subsequent back-crossing of these hybrids with the parental species leads to a blurring of the boundaries between species.

As a result of the above different biological processes dividing or blurring species boundaries, all species are not equally distinct. Such complexity is why this book is necessary. Sometimes it may be perfectly legitimate to consider all these species equal, whatever their origin and however subtly they are defined. On other occasions it may be helpful to consider species that reproduce by producing asexual seeds differently from those that reproduce sexually. All these categories have fuzzy boundaries. In our own species we are sometimes happy to describe people simply as people and on other occasions we classify them as young or old, male or female, knowing that these boundaries are not absolute. So it is with plant species. Sometimes it may be helpful to consider all entities as species, while in other cases we may wish to divide them – for instance as hybrids or inbreeders. In doing so, we don't have to assign value to them, we simply recognise that they are different.

Taxonomic botanists use a curious classification system to describe themselves, employing the terms 'splitters' and 'lumpers'. The first of these, 'splitters', see the smallest morphological differences as being potentially important in describing and cataloguing species. In contrast, 'lumpers' are happy to overlook much of this variation and classify individuals into larger fuzzier boxes. This book is not intended to be a 'splitter's' guide to complexity. But each chapter in turn looks at the question of how much splitting is appropriate in the context of a particular group of plants. You will be encouraged to decide for yourself how much splitting and lumping you feel comfortable with in each case. Generally, this will depend on why you are trying to identify plants in the first place. In ecology, lumping is usually (but not always) more helpful than splitting, while questions of evolutionary genetics often require splitting to obtain the resolution required to find an answer. Thus, you are not invited to venture deep into the land of the botanical geek; some may find, however, that here is a gateway drug for this activity. For these people, appropriate support in the way of more specialised guidance is identified. Essentially though, this book is a guide to splitting for lumpers.

The penultimate section of each chapter considers whether historically people have recognised and exploited the finer levels of morphological variation that modern botanists describe. It asks if there is any cultural heritage that supports high-resolution splitting within each group of plants. By inference, a species can be regarded as a 'real species' if you can make a distinctive jam out of it.

For those struggling to identify plants, there are several very helpful apps available. However, these apps fulfil a rather different role than this book. Plant ID apps have become increasingly adept at recognising species from photographs. Yet our task here is to tackle specimens that frequently don't fall neatly within the simple definition of a species. Hopefully, this book will enable you to develop an understanding of why some plants are more difficult to identify than others. Such knowledge will allow you to decide for yourself what level of resolution is appropriate for the task before you.

It is also hoped that this volume will be helpful to the many field naturalists and botanists who find identifying certain flowers frustrating. This book is designed for those who have struggled with the standard wildflower books that gloss over or frequently omit the difficult species covered here. At the same time, it should also come to the aid of more experienced botanists, who are already comfortable with many of these groups, but who have for some reason never got around to other complex groups because they seem too daunting, fiddly or time consuming.

Acknowledgments

Firstly, I must express gratitude to my long-suffering wife Cathryn who over the years has been subjected to more undistinguished plants – often in unremarkable locations – than one can reasonably expect a non-botanist to endure. I hope that the more attractive species have been some compensation, even if they do also grow in our garden. She has kindly and carefully checked every chapter as they have emerged from the word processor. A big thank you.

I am also grateful to Sue Draper, whose 'Plant of the Day' social media posts did much to convince me that there are people out there keen to learn more about frustrating flowers.

There are many botanists and normal people, at all levels of experience and expertise, who have been kind enough to offer feedback, field-test and provide me with essential information, and whom I must thank. These include: Sarah Whild, Helen Wallace, Eleanor Brant, Mark Morgan, Paul Kenton, Richard Carter, Mike Porter, Michael Foley, Mark Duffell, Mara Morris, David Morris, Ed Rowe, Emyr Philips, Alex Prendergast, Chris Metherell, Debby Seddon, Colin Clubbe, Paul Kenton, Ian Denholm, Andy Jones, Eoin O Maoileoin and Nick Stewart for his online Pondweed course. Each of these good people has made invaluable contributions.

Finally, I must thank David Hawkins and the team at Pelagic Publishing for their immense patience and professionalism.

SECTION I

Apomictic Species

(species that produce seeds without sex)

Probably the most frustratingly difficult plants to identify are to be found in this first grouping. The ability to make asexual seeds which are genetic copies of the 'mother' plant seems to have evolved on several occasions. The plant families in this first grouping employ subtly different mechanisms to obviate the need for sexual reproduction. However, the evolutionary driver for asexuality always seems to be similar: these plants are typically of hybrid origin and contain uneven sets of chromosomes. Having asymmetrical chromosomes makes normal sexual cell division (meiosis) tricky, if not impossible. Only those plants lucky enough to acquire mutations that promote asexuality can bypass this hybrid sterility and persist in the population. It is worth noting that all these plants are long-lived perennials, and many are also able to reproduce asexually by vegetative means, such as having runners. This longevity increases the probability of them acquiring the ability to produce asexual seeds at some point during their lifetime. Nonetheless, there are very few hard and fast rules in botany and there are one or two annual species that produce seeds asexually.

Not all the species in these families produce seeds without sex. There is an amazing diversity of sexual activities found within the plant kingdom. Some members of these families are 'normal' sexually reproducing species. Others only rarely engage in asexuality. Some routinely or only ever produce asexual seeds, and a few sneaky ones seem to have the ability to alter their sexuality depending on the environment. All this diversity in modes of reproducing does not make identifying them any easier – primarily because not always being asexual allows some of these species also to engage in the activities found in other plant families, such as hybridisation and inbreeding. Thus, the groupings used here to subdivide plant families are all fuzzy in their boundaries.

The reason that producing asexual seeds causes problems for plant taxonomists relates to the definition of a species. If all the seeds produced are genetically virtually identical, and the plants they grow into are morphologically similar to their 'parent', plus because they are asexual they are unable to cross with other species, then technically they can be considered a new species. These clonal species are often described as microspecies or 'agamospecies' rather than deemed as being 'true' species. Yet enthusiasts for these groups will passionately argue that such microspecies are taxonomically equal to all other species, and that they should be recognised as true species. Unfortunately for 'outsiders', there appears to be some disparity between enthusiasts in how they have applied the definition of a species within their favoured group. Thus, we will look at this again on a case-by-case basis.

CHAPTER 1

Brambles

Most people's interaction with brambles involves picking blackberries or trying to control them as invasive weeds. Both these activities quickly reveal that brambles are a highly diverse group of plants. Some produce the most delicious fruit, while others disintegrate into a watery mush. Some grow in impenetrable thickets which are covered in vicious thorns that tear the skin, while others have a spindly, almost delicate low-growing form. Having made these observations, most people are still shocked to learn that there are more than 350 named species of bramble in the British Isles. Furthermore, the fact that the study of brambles has its own name, 'batology', can excite the geek and frighten the novice botanist to equal extents.

WHY IS THIS GROUP OF PLANTS COMPLEX?

Many of the taxonomically challenging groups of plants in this book reproduce by way of asexual seeds, which is technically referred to as 'agamospermy'. The seeds produced in this manner are genetically identical copies of the female parent; they are clones. Taxonomic complexity always arises once a plant has abandoned sexual reproduction: because all the offspring are effectively genetically identical, morphologically very similar to their 'parent' and are not able to cross with other species, then by definition they are a new species.

Blackberries have taken agamospermy to a new level of complexity. It is believed that just a single species (*Rubus ulmifolius*, the Elm-leaved Bramble) has two sets of chromosomes (one from each parent) and produces sexual seeds following cross-pollination. Such sexual brambles are thought to be ancestral to the others. Some species of brambles have three, five or six sets of chromosomes and always produce asexual seeds. Curiously, in order to be able to produce asexual seeds, they still require cross-pollination. The majority of brambles have four sets of chromosomes, and these may produce either sexual or asexual seeds depending on environmental conditions. Such facultative apomicts complicate things further: again, these species require cross-pollination before they can produce sexual or asexual seeds. Viable pollen produced by asexual brambles may sometimes cross-pollinate facultatively apomictic brambles, producing new hybrid clones with the ability to produce asexual seeds.

The complex breeding system of brambles is not yet fully understood and is thought still to be evolving. Historically, there appears to have been two periods when most bramble species arose. As a consequence of this, some regions have very few, while in other areas you may find hotspots of diversity with many different species of brambles within a single hedge. Some of these local variants appear to have very limited distributions and can subsequently be difficult to refind.

It is worth being aware that sexual brambles are more variable within a species than are the asexual species, in which genetic variation is only derived from rare mutation. Nonetheless, it does remain possible that apomictic brambles of hybrid origin may have evolved on more than one occasion and may be morphologically variable because of this.

HOW CAN I TELL THEM APART?

Many experienced botanists are wary of trying to identify brambles. To be able to reliably identify some requires a degree of commitment that may include you visiting the same plant on several occasions, to observe different features of its morphology and to reassure yourself of the stability of form. If you get in a jam, it is a good idea to befriend an experienced local batologist and ideally also an expert from mainland Europe. They will be quick to tell you it is a prickly subject.

Unfortunately, because of their size (and spiky nature!), brambles are not the easiest of species to collect, but it is worth assembling a herbarium collection to be verified by an expert (e.g. the BSBI referee). A bramble herbarium specimen should consist of a central section of the current year's vegetative stem (the primocane) roughly a metre back from the growing tip, plus at least two leaves (bramble leaves frequently comprise three to five leaflets which are usually joined by short stalks) and flowering stems with a complete inflorescence (ideally with at least one flower with fallen petals showing the developing fruit). Photographs attached to the specimen are helpful for recording colours of leaves, flowers and fruits and it is worth making notes on colours, the number, location and shape of the prickles and growth form of the plant.

The key features to look out for include growth habit (is the plant upright, does the plant arch and root again at the tips, or does it spread below ground like raspberries?) – this is a fundamental separation between the lower-growing suberect brambles and other more typical bushy brambles. To aid their identification, brambles are split into various artificial 'series'. These series are largely based on the nature of the prickles along their new (this year's) stems. You will need to look closely at the relative size and abundance of ridges, prickles, pricklets and micro-prickles (which are like stiff hairs, and technically termed acicles), hairs and resinous glands (which may be stalked or not). The presence and distribution of these features on stems, leaves and flowers is critical. The use of a hand lens is therefore essential. You will also need to look at leaves on new stems: first record the number of leaflets per leaf and whether they are overlapping; then check the underside of a leaf for the presence of downy white hairs. Then check if these hairs lie flat and close to the underside of the leaf, appearing chalky white. Finally, look at flower colour, petals, filaments (the stalks on which the anthers are held), styles and whether the very young developing fruit is hairy.

For info: technically, the terms 'thorn' and 'prickle' are used by botanists to describe different structures. Thorns are modified stems, while prickles are protuberances of the outer layer of the plant's epidermis (skin). This is not something we need to worry about here. But if you are concerned about correctness, brambles are covered in prickles not thorns.

There are a few bramble relatives and introduced species with red or orange fruit that you may wish to consider first.

Rubus chamaemorus (Cloudberry)
This is a species of the peaty uplands of northern England and Scotland, with a few sites in north Wales. **It only grows to 20 cm tall.** The stems are round and densely covered in red prickles. **The flowers are solitary**, large and white, measuring up to 3 cm across. Leaves have **5 to 7 lobes. Fruit are distinctively orange with 4–20 large druplets.**

R. saxatilis (Stone Bramble)
Occurs in woodland and mountains in Scotland, and is scattered across Ireland, Wales and northern England. It grows to a maximum height of 40 cm. **Stems are round and covered in hairs with a few weak prickles.** Flowers are white and occur in terminal clusters of 4 or 5, measuring 1.5 cm across. Leaves are usually three lobed, the central leaflet may have a short stalk. **Fruit are distinctively bright red, with 1–6 druplets.**

R. phoenicolasius (Japanese Wineberry)
This introduced species is widely grown in gardens and now has a scattered distribution in the wild. It grows to a height of 2 m, its round stems densely covered in red glandular hairs with few weak prickles. The leaves are usually divided into 3 leaflets which appear white below. **The flowers have much smaller petals than sepals. The sepals are covered in many red glandular hairs.** Fruit are deep red, with many druplets.

R. spectabilis (Salmonberry)
An introduced species widely grown as an ornamental but now commonly found in woods and hedgerows particularly in Northern Ireland, northwest England and southern Scotland. It grows to a height of 2 m. Its round stems have few weak spines. **The leaves lack hairs**, have 3 lobes, with the terminal lobe having a long stalk. **The flowers are a distinctive deep pink colour** measuring up to 3 cm across, double flowers are known. **Fruit may be an orange to rich red colour, with many druplets and hairs.**

R. tricolor (Chinese Bramble)

This introduced species is commonly grown as ground cover in many public spaces and is now widely naturalised. **The stems may be several metres long but only 50 cm high, rooting along their length.** The stems are round and densely covered in brown bristles. **The leaves are not divided into leaflets; they are glossy dark green above and white below.** The flowers are white and found in small clusters, they measure up to 2.5 cm across. **Fruit are orange red** in colour with many druplets.

Characteristics of the most frequently encountered bramble sections

					Leaflets				
Section	**Stem cross-section round**	**Prickles only on ridges**	**Pricklets**	**Micro-prickles**	**3**	**5**	**Overlap**	**Serrated**	**Leaf undersurface white-grey**
Caesii (dewberries)									
Discolores									
Vestiti									
Rhamnifolii									
Radulae									
Hystrices									
Corylifolii (hazel-leaved brambles)									
Sylvatici									
Your specimen									

Let's look at the more widespread bramble sections and species

Section *Caesii* (dewberries)

Many wildflower books divide brambles into two, the blackberries (usually lumped as *Rubus fruticosus* agg.) and Dewberry *R. caesius*. If there is one species of bramble that people can identify it is *R. caesius*. These plants are widespread, particularly on limestone. They are thin stemmed and tend to be low growing. **Leaves have three leaflets.** Flowers are large and white. **Fruit are irregular with a few large druplets covered in a distinctive whitish misty bloom.**

R. caesius (Dewberry)

Dewberry is common across the British Isles but less so in Scotland. It grows in a wide range of habitats, particularly on limestone. The stems are thin and grow along the ground, and they may root at the tip (any bushiness may indicate that the plant is a hybrid). **Stems are round in cross-section and appear white, being covered in a waxy powder.** There are a few short slender prickles which are found all around the stem. **Leaves have three leaflets** and may have a few hairs, they are green and pale green below. The petals are large, white and overlapping. **Fruit are irregular, comprising a few large druplets covered in a distinctive white bloom, and tend to fall apart when picked.**

Section *Discolores*

These are scrambling or bushy brambles, their stems may or may not be hairy, but don't have stalked glands or micro-prickles. The flowers also lack glands and micro-prickles. **Stems are angular, with large prickles which are only found on the ridges. Leaves are characteristically white on the undersurface.**

R. ulmifolius **(Elm-leaved Bramble)**
A variable species widespread across the British Isles, especially on chalk and clay, where it is often the most abundant bramble. **Its stems have a distinctive white bloom.** Prickles are large and similar in size. The palmate leaves comprise 3 or 5 leaflets (usually 5). The lower surface of the leaves is white in appearance. Petals are pink and rounded. The fruits lack a bloom, and are regarded by some as the most flavoursome of the blackberries.

R. armeniacus **(Himalayan Giant Bramble)**
Our only garden escape that can be invasive. This highly vigorous species grows up to 4–10 m along the ground and to a height of 4 m. It roots whenever it reaches the ground. **The stems are robust, measuring up to 2–3 cm diameter at the base, and are usually hexagonal in cross-section. The new stems have a distinctive orange-green colour, contrasting with the red bases of the prickles.** The vicious prickles, which are found along the stem ridges, measure up to 1.5 cm. The palmate leaves almost always comprise 5 leaflets. The lower surface of the leaves is white in appearance. Petals are white to pale pink in colour. Flowers are the largest of any bramble in Britain, occurring in inflorescences with many flowers per cluster. The fruits measure 1.2–2 cm and ripen to a purple-black colour.

Section *Vestiti*

These are arching brambles that are **noticeably hairy.** These species have **stalked glands and micro-prickles.** Their stems are angular, with prickles only found on the ridges. The prickles are mixed with smaller pricklets. The leaves are often white on the undersurface.

R. vestitus **(European Bramble)**
One of the commonest species in Britain. A spiny shrub which grows to 2 m tall. The two-year-old stems are a dull purple colour and root at the tips. Prickles are slender, straight and deep purple in colour. It has palmate leaves which are usually divided into **5 separate leaflets, each of which has a distinctive pointed tip and a white undersurface, while the terminal leaflet is rounded.** Pink- and white-flowered forms of *R. vestitus* may rarely grow together. The fruit is round,

measuring up to 15 × 15 mm and comprising up to 40 single-seeded druplets. These are very dark, nearly black, when ripe and lack any bloom.

Section *Rhamnifolii*

These brambles frequently have low, arching stems growing close to the ground. The stems may or may not be hairy and are **usually without stalked glands.** However, the flowers may contain stalked glands. Their stems are angular, with prickles only found on the ridges. The leaves are characteristically greyish-white on the undersurface.

R. polyanthemus

One of the most widespread of bramble species. It grows to about 1.2 m tall and is semi-deciduous. Arching stems root where they touch the ground, forming dense thickets. The stems are sparsely hairy. **Prickles are long and slender, red with yellow tips on year-one stems and darker purple on year-two stems.** There are 3–5 leaflets (**the terminal leaflet is often subdivided, so there appears to be 7 leaflets**) which are **markedly corrugated**, round with a distinct leaf tip. The petals are pale pink. Flowers occur in elongated inflorescences in clusters of up to 30.

Section *Radulae*

These brambles have arching stems which may or may not be hairy but have stalked glands and micro-prickles. True prickles are only found on stem ridges. **The pricklets, micro-prickles and stalked glands are all of similar lengths,** giving a downy appearance to the stems. However, **stroking the stem between prickles, they feel rough to the touch like sandpaper – hence the series name (the Latin word *radula* meaning rake).** Glands and micro-prickles are also found on the inflorescences. The leaves are white on the undersurface and green above. **The leaf margins are highly serrated, with teeth on the serrations.**

R. echinatus

Found in most of central and southern England. A medium to tall-growing species of heathlands, hedgerows, woodlands and scrub. The stems are typical of brambles in this section – being covered by hairs and short gland-tipped micro-prickles of about the same length, giving it a rough feel to the touch. Stems are often deeply grooved and purple in colour. The purple prickles have yellow tips, are much larger than the micro-prickles and are mainly found on stem ridges. **The leaves are highly distinctive, being deeply and irregularly toothed and also strongly wavy at the margins.** Flowers are typically pale pink; **the sepals are covered in red glandular hairs.** The stamens are longer than the yellowish-green styles.

R. insectifolius

A common species in southern England, where it grows on heaths and in acid woodlands. The stems trail over the ground, through other vegetation. The main prickles are slender and located on the stem ridges, while pricklets and micro-prickles occur in between. The leaves usually have five leaflets which can be **sharply toothed and have a conspicuous long leaf tip.** The leaves are frequently slightly bronzed. **The flowers have distinctive white petals and contrasting red stigmas which are visible because they have short anthers.**

Section *Hystrices*

These brambles are variable in growth form, from arching to low growing. Stems may or may not be hairy; they have stalked glands and micro-prickles which are less obvious than the pricklets. **Prickles are found all around their stem and are of variable length, so it is difficult to distinguish prickles from pricklets or micro-prickles.** Stalked grands and micro-prickles are also found on the inflorescences. Leaves are green on the upper surface and hairy to slightly hairy and grey on the undersurface.

R. dasyphyllus

One of the commonest brambles of the British Isles, frequently encountered in woodlands and hedges. The first-year stems and prickles are usually green, but become purple to brown in their second year. The stems are flat sided and sometimes have deep groves. The larger prickles are long and fine with yellow tips, and are slightly curved along their entire length. The stems appear hairy, being covered in many pricklets, long stalked glands and micro-prickles. The leaves can be very broad, sometimes with a flattish end, except for the leaf tip which may be curved to one side. Usually there are 5 leaflets, but often the basal leaflets may fuse on one or both sides. **The leaf margins are toothed, with teeth that tend to point outwards and sometimes backwards.** The undersurface is paler green than the upper. The petals are narrow and pink. **The sepals are also narrow and turn red at their base once the petals have fallen.**

Section *Corylifolii* (hazel-leaved brambles)

These are low-growing brambles which may root at the tips. They usually have few prickles, and fewer pricklets. The leaves have 3–5 leaflets, **distinctively the lower leaflets tend to lack stalks and the leaflets overlap** and are often domed/concave. The inflorescences are branched at 90 degrees from the stem. In most members of this section, the petals touch when the flower is open. **Fruit are often irregular in form and comprise a few large druplets, which usually lack any bloom.**

R. conjungens

This widespread species occurs in lowland woods and hedges. The arching stems are green in the first year, acquiring reddish streaks with age. The **stems have blunt angles and lack pricklets and micro-prickles.** Stem prickles are uniform in size, straight and usually red in colour, sometimes with a green tip. Typical of a *Corylifolii* bramble, the leaflets overlap. **The flowers are pink and have a distinctive**

crumpled appearance. Unusually for species in this section, the petals in the open flowers don't overlap.

R. tuberculatus
A widespread and common bramble that grows in a range of habitats. It is noted for having very robust and well-armed new stems. The stems are usually densely covered with both prickles and pricklets. New stems are frequently purple in colour, as are the pricklets. Its prickles tend to be straight and paler in colour than the pricklets. **The leaves are somewhat hazel like, being corrugated and rough to the touch. The leaflets overlap and the lower leaflets may fuse.** The petals are large, white and round.

R. pruinosus
A widespread and common bramble of wet and shady spots. **Young stems are characteristically circular in cross-section, green in colour with a distinct strong white bloom. The few prickles are dark red and straight.** There are no pricklets or micro-prickles. **While the lower leaflets overlap, the terminal leaflet often has two lobes.** Flowers are white, round and large – measuring up to 3.5 cm across.

Section *Sylvatici*

These are frequently tall brambles, which may or may not be hairy. Stems typically lack stalked glands and micro-prickles. Prickles only occur on stem ridges. **The leaves are green on the undersurface. There may be stalked glands but not micro-prickles within the inflorescence.**

R. laciniatus **(Cut-leaved Bramble)**
This is a widespread garden escape, although less common in the wetter western regions and in the Scottish Highlands. It may originally have been native around the Norfolk/Suffolk border. The stems are much branched, ridged and/or furrowed, and purple in colour. Stems have stout curved prickles along their ridges. Prickles are red with yellow tips. **It has highly distinctive dissected leaves** that cannot be mistaken for any other bramble. The leaves are matt green above, the lower surface whitish-green. **The petals are large 2.5–3 cm, pink or white and usually divided into three lobes.** Mature fruit can be up to 2.5 cm in length and comprise a few large, sweet druplets.

HAVE OTHERS RECOGNISED THIS LEVEL OF VARIATION?

There is archaeological evidence from bog burials that Europeans have been eating blackberries for more than 2,500 years. Of course, the custom of blackberry-picking is still alive and strong today. Given these facts, it is remarkable that there are so few common names for these plants, except those recently contrived by botanists. The Dewberry was first recorded as a separate species by Carl Linnaeus in his work *Species Plantarum* published in 1753. This is perhaps not a surprise, as its fruit and leaves are well distinct from the other brambles. With this exception,

Leaves and this year's stems of a selection of widespread brambles

Caesii
R. caesius (Dewberry)

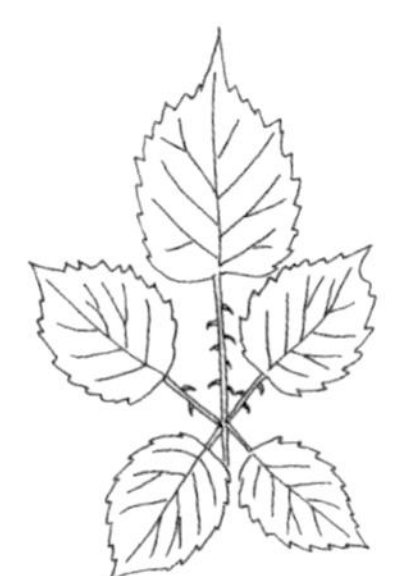

Discolores
R. ulmifolius (Elm-leaved Bramble)

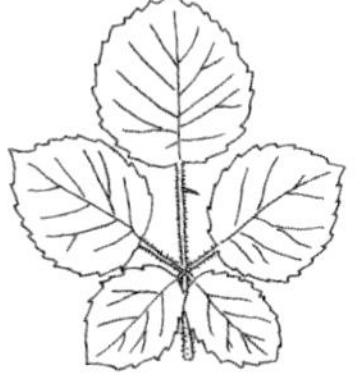
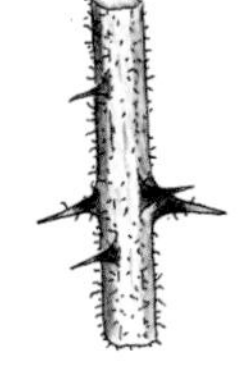

Vestiti
R. vestitus

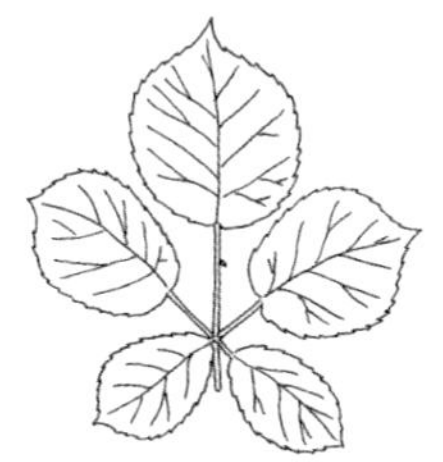

Rhamnifolii
R. polyanthemus

Radulae
R. echinatus

Hystrices
R. dasyphyllus

Corylifolii
R. conjungens

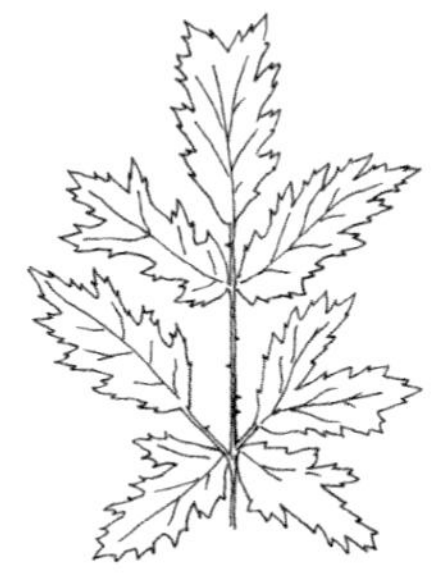

Sylvatici
R. laciniatus (Cut-leaved Bramble)

there is little evidence to suggest that any of the other *Rubus* species described by modern batologists have ever historically been recognised in Britain – although in Spanish the word 'mora' translates as blackberry, while 'zarzamora' may have referred specifically to *Rubus ulmifolius*. Its English name, Elm-leaved Bramble, is a modern invention and has never been part of the vernacular.

The absence from medieval herbals, and the lack of common names, indicates that our ancestors did not concern themselves with the diversity of brambles they encountered. However, there is one intriguing piece of evidence that suggests they may have been inadvertently aware of some aspects of the complex sex-life of these plants. According to mythology, blackberries should not be eaten after 29 September, Michaelmas Day – or Devil's Blackberry Day. It was believed that on this night the devil spat upon any unpicked blackberries and that anyone subsequently eating them would be cursed. Late September does approximately coincide with the time when many facultatively asexual brambles swap from producing sexual seeds, to producing asexual clonal seeds. It has been put forward by some that these late-season asexual blackberries contain larger seeds and are less palatable, whether the devil has spat on them or not.

HOW FAR SHOULD I GO?

There are a number of reasons for dipping your toe into the mysterious waters of batology. Firstly, you may find it rewarding to be able to put a scientific name to your favourite bramble patch, the one you consider makes the world's finest crumble or jam. You will quickly find that each batologist has their own favourites, and there is fun to be had in testing their recommendations. From a conservation standpoint, there is an argument that all field botanists should be able to correctly identify *Rubus armeniacus* as an invasive garden escape. At the other extreme, you may wish to be able to identify unusual brambles that are endemic to your local area, for example the highly distinctive *R. iceniensis* which is knee-high and only found growing under sparse bracken on heaths around Norwich.

Enthusiastic batologists use the same argument as do taraxicologists (dandelion fanatics) – claiming that identification to species level provides them with a better understanding of plant-community ecology. Both also confess to using habitat as a guide to identification. So be wary, as there may be some circularity here, and it may be problematic to use the habitat to help identify the species and subsequently use the presence of the species to provide you with information about the habitat.

The information provided above should be sufficient to allow you to identify 60% or more of brambles that you are likely to encounter to species level. Once you start examining brambles, you will begin to notice things you may have overlooked before – for instance the fact that some patches are deciduous while others are evergreen. It is fair to warn you that many experienced botanists consider batology to be something of a dark art. However, if you do wish to venture further the standard work is: Edees and Newton 1988 *Brambles of the British Isles* (available as an affordable eBook).

CHAPTER 2

Dandelions

Most people would probably not think that dandelions are difficult to identify. After all, everyone knows what a dandelion looks like. So there is no reason to look them up in your wildflower book. However, if you do turn to the section that covers dandelions you will find a short paragraph that says something like: 'a highly variable and difficult group, divided into nine ill-defined sections containing more than 250 microspecies'. Some books might even contain pictures of two or three of the most common types. Surprisingly, dandelions are probably the most challenging to identify of all our plants.

WHY IS THIS GROUP OF PLANTS COMPLEX?

The majority of plants contain equal numbers of chromosomes from each of their parents. But dandelions do not. Most British and Irish dandelions have three sets of chromosomes: 16 chromosomes (two sets of eight) from one progenitor parent species and another set of eight chromosomes from their other parent species (in genetics this is referred to as being 'triploid'). It is thought that this is the result of hybridisations that occurred during the last ice age, when northern species were driven south by the climatic conditions. This migration allowed the interbreeding of dandelions from the north and south of Europe. It is not uncommon for northern species of plants to have more chromosomes than do their southern relatives; this extra genetic material seems to provide them with the flexibility needed to survive in colder conditions. The resulting hybrid plants produced from these liaisons have three sets of chromosomes.

Under normal circumstances, such hybrids would be sterile, as it is difficult to successfully divide three sets of genes during the process of making pollen and egg cells. To avoid this problem of hybrid sterility, these new hybrid dandelions evolved the ability to make seeds without the need for sex. As a consequence, most dandelions now produce offspring that are identical copies of themselves. Technically, this is termed 'agamospermy', but we will call them asexual seeds. Each dandelion seed is an identical genetic copy of its mother, a clone. Asexual dandelions do not have male parents, they rely on virgin birth to reproduce; in fact, some dandelions are unable to produce pollen. If you want proof, it is easy to slice the end off a developing dandelion flower bud, removing its female stigmas and male anthers. Even so, a few days later it will still form a full dandelion clock of seeds without ever having been pollinated.

Once a plant has abandoned sexual reproduction, complexity is not far behind. If all the offspring are genetically identical, morphologically very similar to their 'parent' and not able to cross with individuals of another species, then technically they have become a new species. Most botanists regard these clones as microspecies or 'agamospecies'. However, some dandelion enthusiasts will passionately argue that such microspecies are taxonomically equal to all other species and that they should be recognised as true species. We can admire such enthusiasm without necessarily sharing it or agreeing.

HOW CAN I TELL THEM APART?

The first thing to be aware of when trying to identify dandelions is that they are highly plastic. Dandelions are rather unusual in growing – and occasionally flowering – all year round. By doing so, individual plants may change their appearance dramatically throughout the year. Similarly, the same plant may look very different when growing in a meadow rather than in woodland. Being able to alter their appearance is a function of dandelions' hybrid origin – sometimes they look more like one progenitor, sometimes like the other. This may confer an ecological advantage and help the plant survive in many different environments, but it can be a pain in the neck for a botanist. Even experts are reluctant to identify dandelions from extreme environments, such as deep shade or tightly mown lawns. For this reason, they are a really challenging group for the beginner.

Probably the single-most helpful characteristic in identifying dandelions is their ecology. Many sections are associated with a limited number of restricted habitat types. This can be a useful first place to start in identifying them.

Dandelion capitulum

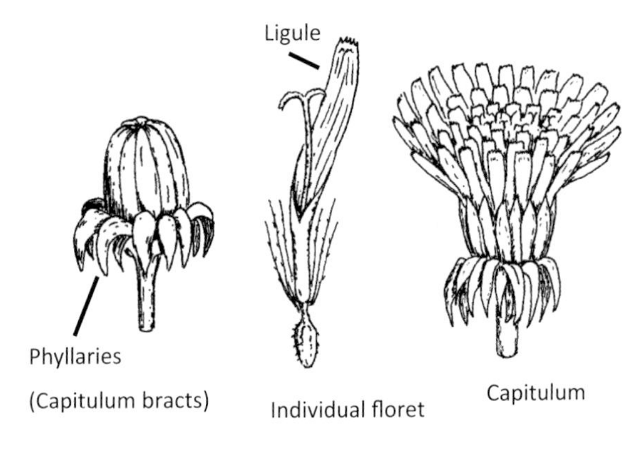

Habitat associations of dandelion sections

Section	Garden weeds and lawns	Dry grasslands	Road verges	Wet acid grasslands	Wet base-rich grasslands	Sand dunes and coastal grasslands	Uplands and mountains	Heaths
Ruderalia								
Hamata								
Erythrosperma								
Celtica								
Taraxacum								
Obliqua								
Palustria								
Spectabilia								
Naevosa								
Your specimen								

Strongly associated with — Sometimes found in

Although leaf shape can be plastic in dandelions, it is still a useful character to help identification. They are best collected in April and May when plants are growing and flowering most actively.

Another important feature is the exterior bracts that surround the flowerhead. Technically, in the dandelion family (the Asteraceae) the flowerhead is called a capitulum, the bracts that surround the capitulum are called the phyllaries, and the inner part of the capitulum is composed of many small individual flowers, called florets. These florets have single long petals (actually five fused petals) called ligules (see illustration opposite). Flower colour is mentioned in guides to identifying dandelions, but these descriptions are initially of limited help as all the species are yellow and the differences between them are very slight. In contrast, the shape of the exterior bracts (the phyllaries) that surround the flowerhead can be useful in identification.

Let's start with some common garden dandelion sections

Section *Ruderalia* (weedy dandelions)

This is by far the commonest section in lowland areas. Plants are robust, with leaves that are fairly lobed to deeply lobed. Their ligules are rich yellow, usually with grey-violet stripes on the outside. The exterior bracts that surround the capitulum measure 9–16 mm and are strongly curved. Seeds measure 2.5–4 mm and are brown to straw-coloured.

Section *Hamata* (hook-leaved dandelions)

Common in damp and dry meadows, on roadside verges and rough grasslands. They have distinctive hook-lobed leaves – the front of which is convex and the rear edge concave – with a purple mid-vein. Their ligules are deep yellow, usually with grey-violet stripes on the outside. The exterior bracts measure 8–13 mm and are curved. Seeds measure 3–4 mm and are brown to straw-coloured.

Dandelions less commonly found in gardens

Section *Erythrosperma* (lesser dandelions)

These smaller, more delicate dandelions are found growing in drier well-drained short grasslands, sand dunes and heaths. Their leaves are very deeply dissected, with a purplish mid-vein. The ligules are pale to deep yellow, usually with red-purple stripes on the outside. The exterior bracts measure 5–9 mm and are curved. Seeds measure 2.5–3.5 mm and may be red, purple, brown or straw-coloured.

Section *Celtica* (western dandelions)

These are medium-sized plants encountered in wet places such as woodland edges and shady banks. In the uplands they may occur by springs or on rocky ledges. Commonest in Scotland, Wales and the South West. Leaves are deeply lobed, with a purplish mid-vein. Their ligules are pale to deep yellow, usually with purple or grey-violet stripes on the outside. The exterior bracts measure 7–12 mm and are strongly curved outwards. Seeds measure 2.8–4 mm and are brown to straw-coloured.

Dandelion sections of restricted ecological niches with few species

Section *Taraxacum* (mountain dandelions)

Medium-sized plants that grow in upland areas. The leaves typically have small lobes with a green mid-vein. The ligules are orange-yellow with purple stripes on the outside. The exterior bracts measure 7–9 mm and typically clasp the capitulum. Seeds measure 3.5–5.3 mm and are brown to straw-coloured.

Section *Obliqua* (coastal dandelions)

These are small plants of open, sandy coastal grasslands. Leaves are delicate and deeply dissected, with green mid-veins. The ligules are deep yellow to orange with red stripes on the outside. The exterior bracts measure 6–7 mm and typically clasp the capitulum. Seeds measure 3 mm and are greyish-brown.

Section *Palustria* (marsh dandelions)

Medium-sized plants of wet, usually base-rich meadows and fens. The leaves are narrow, with very shallow lobes or not lobed at all. Leaves are unspotted and have a purple mid-vein. The ligules are yellow to deep yellow, with purple or grey

Examples of leaves and capitulum from the nine dandelion sections found in the British Isles

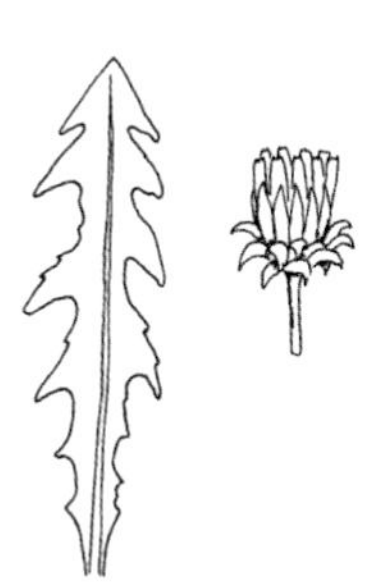

Ruderalia
(weedy dandelions)

Hamata
(hook-leaved dandelions)

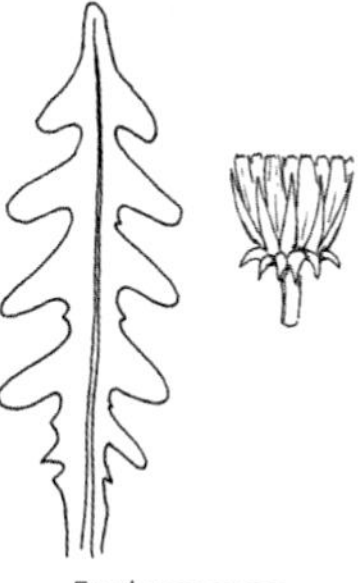

Erythrosperma
(lesser dandelions)

Celtica
(western dandelions)

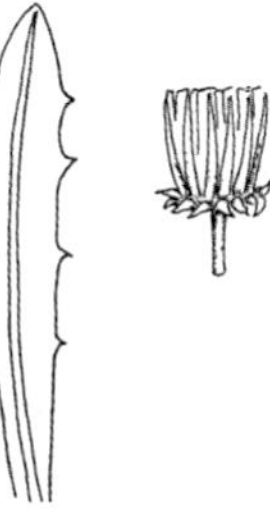

Taraxacum
(mountain dandelions)

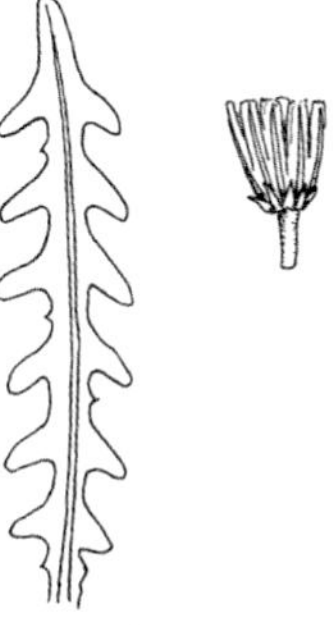

Obliqua
(coastal dandelions)

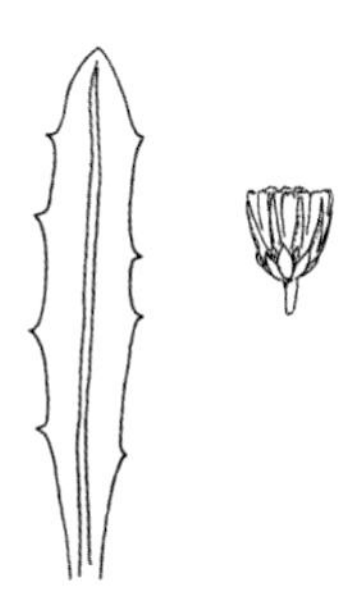

Palustria
(marsh dandelions)

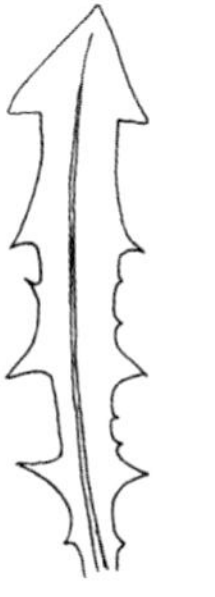

Spectabilia
(red-veined dandelions)

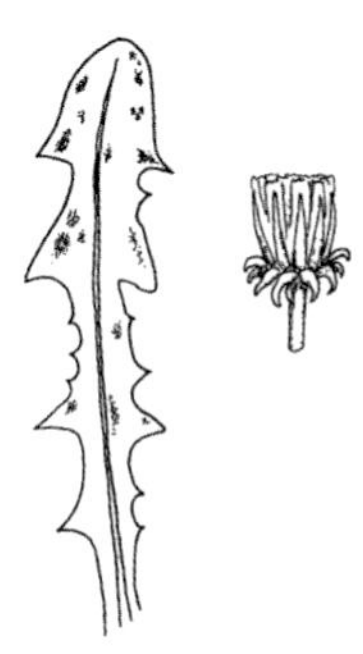

Naevosa
(spotty-leaved dandelions)

stripes on the outside. The exterior bracts tightly surround the capitulum and measure 6–7 mm. Seeds measure 3.5–4.3 mm and are brown to straw-coloured.

Other sections

Section *Spectabilia* (red-veined dandelions)

This section only contains three microspecies, but one of these is very widespread and frustratingly plastic. They grow in wet acid grasslands and on road verges. These are medium-sized plants, with narrow scarcely lobed leaves. The leaves are usually spotty with a distinctive purple mid-vein. Their ligules are bright to deep yellow with purple stripes on the outside. The exterior bracts tightly surround the capitulum and measure 8–9 mm. Seeds measure 4–5 mm and are straw-coloured.

Section *Naevosa* (spotty-leaved dandelions)

Medium-sized dandelions that are most common in the north and west, they are found in wet, acid grasslands often in the uplands. The leaves are distinctively spotted, with deep lobes and a purple mid-vein. Their ligules are mid- to deep yellow, usually with purple or grey stripes on the outside. The florets produce no pollen. The exterior bracts that surround the capitulum measure 9–14 mm and curve outwards. Seeds measure 3–4 mm and are red-brown to straw-coloured.

HAVE OTHERS RECOGNISED THIS LEVEL OF VARIATION?

Historically, dandelions have been used in a wide variety of ways. Their leaves have been boiled as a green vegetable or blanched for use in salads, their flowers have been used to make wine and the roots roasted as a coffee substitute. The plant has been employed medicinally, most noticeably as a diuretic (it is said to cause bed-wetting and one of the old folk names was 'piss-a-bed'), while the white latex it produces has been harvested to manufacture rubber. This vast array of uses provides many opportunities to exploit the taxonomic variation known within dandelions. However, with the exception of a few French salad cultivars and a Russian species identified for its high latex yield, this potential seems to have been unrecognised. This may be because the taxonomic complexity of dandelions was not understood when these uses were being developed. Furthermore, the traits that divide dandelion microspecies are unlikely to influence their palatability as a salad leaf or as a coffee substitute.

HOW FAR SHOULD I GO?

Dandelions are native to Eurasia and North America, but they are now globally distributed and often very abundant. They can be economically significant weeds of agriculture and amenity plantings. For this reason, plant scientists have needed to better understand their biology. There is also a hardcore group of botanists who delight in identifying and recording every minor dandelion variant. These taraxacologists are perhaps akin to stamp collectors who specialise in acquiring examples of otherwise common stamps with slight printing errors. Over years, taraxacologists develop amazing abilities to differentiate hundreds of dandelion microspecies. They obtain great pleasure from this pursuit, but it is of little practical value. Ardent taraxacologists themselves contend that the identification of dandelion microspecies provides them with a better understanding of the vegetation community and its ecology, without the need to record other challenging groups.

Given the fact that different dandelion sections (let alone microspecies) can only be reliably differentiated for a few months of the year – and even then, some microspecies can legitimately be placed into either of the closely related sections *Hamata* or *Celtica* – there is a strong argument for only identifying dandelions as precisely as *Taraxacum* agg. This is the approach used by most field botanists. However, if you are feeling more adventurous, and the season is appropriate, then the above descriptions may help you to identify the nine sections found in the British Isles. Alternatively, there are a few common microspecies that are more distinctive and may allow you to make a foray into the subject. Those wishing to take on the challenge of identifying all 250+ microspecies may wish to consult the online BSBI *Taraxacum* crib, while BSBI Handbook No. 23, the *Field Handbook to British and Irish Dandelions* by A. J. Richards is an essential volume, and Sell and Murrell's *Flora of Britain and Ireland* Vol. 4 is useful… before seeking medical advice for their addiction.

CHAPTER 3

Lady's-mantles

A few of these low-growing perennial plants are widely cultivated in gardens as ornamentals for their bright green foliage. From here they have escaped and are now found in the wild, along with several widespread native species. Our native species can be encountered in a range of habitats – some in lowland grasslands, while others are associated with upland mountainous conditions. What makes lady's-mantles tricky to identify is the fact that some of them are only subtly different from each other, including several rare species which are restricted to hilly areas of northern England and Scotland.

WHY IS THIS GROUP OF PLANTS COMPLEX?

Many frustrating plants are difficult to identify as a consequence of being 'apomictic', meaning that they reproduce by producing asexual seeds. Their asexual offspring can be regarded as new species because they cannot hybridise with other species and are morphologically distinct. Although many taxonomically complex plant groups are similar in that they are apomictic, the reasons why they have abandoned sexual reproduction differ.

The production of asexual seeds in Alchemillas (lady's-mantles) has enabled them to escape from hybrid sterility. When two species hybridise, their offspring are often sterile. This is because they have inherited two different sets of chromosomes – one from each of their parents. These unbalanced chromosomes are unable to neatly divide into two to produce the egg cells and pollen cells required for sexual reproduction. This problem is neatly circumvented by avoiding sexual reproduction.

Many species of lady's-mantle grow in subalpine grasslands. Here, environmental conditions are less than ideal for sex. The harsh weather often limits the number of available pollinating insects, while low temperatures can damage the sensitive reproductive organs within the flower. It is not clear if this has been a significant factor driving the evolution of asexual seed production in this group or if abandoning sexual reproduction gives these plants an advantage in such extreme conditions.

Characteristics of the widespread species of lady's-mantle as well as a few with more restricted distributions

Species	Plant height cm			Hairs present		No. leaf lobes			
	<30	30–50	>50	upper leaf	lower leaf	5–7	7–9	7–11	9–11
A. alpina (Alpine Lady's-mantle)						unfused			
A. conjuncta (Silver Lady's-mantle)						partly fused			
A. mollis (Garden Lady's-mantle)									
A. glabra (Smooth Lady's-mantle)									
A. xanthochlora (Pale Lady's-mantle)									
A. filicaulis (Thin-stemmed Lady's-mantle)									
A. acutiloba (Starry Lady's-mantle)				a few					
A. wichurae (Rock Lady's-mantle)					only veins				
A. glomerulans (Clustered Lady's-mantle)							9	9	9
A. glaucescens (Silky Lady's-mantle)									
Your specimen									

HOW CAN I TELL THEM APART?

As a group of plants that apparently only ever reproduce asexually, the differences between species can be slight. However, experiments suggest that these distinctions are maintained in different environments. The key features you will need to look for are hairs on the leaf and leaf stalk, the shape of the rosette leaves, and the number of teeth on the middle lobe of the leaf. The leaves are round and formed of 5–11 fused leaflets. You will need to count the number of lobes formed from these fused leaflets. The extent to which the leaflets are fused to become lobes is another key character used to separate species.

Let's start with the more widespread examples

Lady's-mantles can be crudely divided into two: those species with discrete leaflets (the alpine lady's-mantles *A. alpina* and *A. conjuncta*) and all the other species, which historically were lumped as *Alchemilla vulgaris* agg. or the common lady's-mantle, in which the leaflets are almost completely fused to form a round leaf with a scalloped edge.

A. alpina **(Alpine Lady's-mantle)**
Found in high-altitude pastures and rocky areas in the Lake District and northern Scotland. Grows to 20 cm tall. **Its leaves measure up to 3 cm across and are formed of 5–7 separate leaflets, so that the leaves have distinct fingers.** The central leaflet has about 7 teeth, which are only found at its tip. **The upper surface of the leaf is a rich green and appears to be outlined in white; the undersurface is covered in silky hairs and appears almost silver.**

A. conjuncta **(Silver Lady's-mantle)**
A garden escape which can be encountered in scattered locations across the British Isles. Gardeners sometimes confusingly call this 'alpine lady's-mantle' and unhelpfully even use *A. alpina* and *A. conjuncta* interchangeably. It is a larger plant than the true Alpine Lady's-mantle, growing to 30–40 cm tall. **Its leaves are similar to that species, but the 5–7 leaflets are more oval in shape, their lower thirds fused.** The upper surface of the leaf is a rich green and appears to be outlined in white; the undersurface is covered in silky hairs and appears almost silver.

A. mollis **(Garden Lady's-mantle)**
A common garden escape now found throughout the British Isles. Easy to distinguish from the other species because it is much larger, growing up to 80 cm in height. **The leaves are bright, light green, measuring up to 15 cm across and comprising 9–11 rounded lobes.** The central lobe of the leaf has 15–19 pointed teeth. The entire plant except the flowers is covered in dense hairs.

A. glabra **(Smooth Lady's-mantle)**
A species of meadows and open woods in Wales, northern England, Northern Ireland and Scotland. It is uncommon south of a line between Bristol and Hull and absent from much of southern Ireland. Plants are often large, growing to over 50 cm tall. The leaves measure up to 10 cm across and comprise 7–9 rounded lobes. The central lobe has 11–19 teeth with distinct white tips. **Only the lower part of the stem and occasionally leaf stalks have hairs, the rest of the plant lacks hairs.**

A. xanthochlora **(Pale Lady's-mantle)**
This widespread species thrives in grassy places from high ground to low altitudes in Wales, Cornwall, northern England, Northern Ireland and Scotland (except the far north). Grows to about 40 cm. The leaves measure less than 5 cm across and are made up of 7–11 rounded lobes. The central lobe has between 11 and 15 sharp teeth. **The upper surface of the leaf and flowers are hairless; the leaf undersurface and stems are covered in dense hairs.**

A. filicaulis **(Thin-stemmed Lady's-mantle)**
Grows in most of the British Isles except southeast England. Plants are around 30 cm tall. The leaves generally measure less than 5 cm across and are formed of 5–9 lobes. The central lobe has 9–11 teeth. **The upper and lower surfaces of the leaf are hairy, as are the backs of the flowers.**

Leaves of most widespread species of lady's-mantles

Alchemilla alpina
(Alpine Lady's-mantle)

A. conjuncta
(Silver Lady's-mantle)

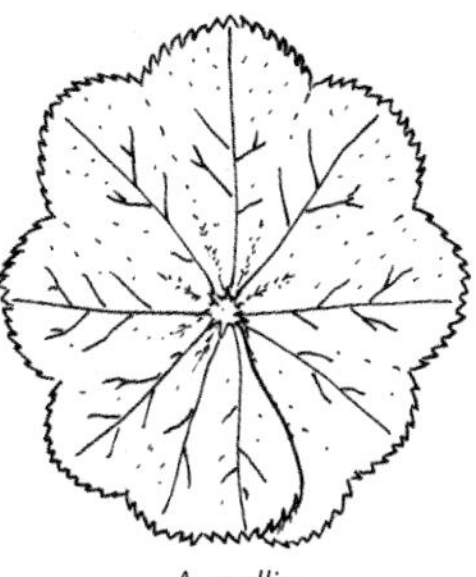

A. mollis
(Garden Lady's-mantle)

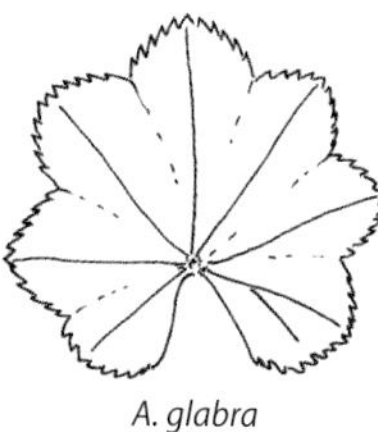

A. glabra
(Smooth Lady's-mantle)

A. xanthochlora
(Pale Lady's-mantle)

A. filicaulis
(Thin-stemmed Lady's-mantle)

Now let's look at species with more restricted distributions

A. acutiloba (Starry Lady's-mantle)

Fairly widespread in the hill pastures of Teesdale, county Durham. A large species, which grows to about 50 cm tall. The leaves measure up to 5 cm across and are formed of 9–11 lobes which are fused for three-quarters of their length. The central lobe has 15–19 teeth. **The flowers are free from hairs. The upper surface of the leaf is more or less hairless, with just a few sparse hairs. The undersurface of the leaf is covered in dense hairs.**

A. wichurae (Rock Lady's-mantle)

Found on base-rich soils from northern England to northern Scotland. Grows to about 20 cm. The leaves measure less than 3 cm and are made up of 7–11 round-ended lobes. The central lobe has 15–19 sharp teeth. **The upper surface of the leaf is hairless, as is the undersurface except along the veins.** The lower sections of stems are covered in hairs, while the top sections and flowers are hairless.

A. glomerulans (Clustered Lady's-mantle)

A species of rocky areas in the wet mountains of northern England and northern Scotland. Grows to about 40 cm. The leaves measure less than 5 cm across and are made up of 9 lobes. The central lobe has 13–19 sharp teeth. **The upper and lower surfaces of the leaf are hairy, as are the stems and leaf stalks; however, the flowers are hairless.**

Leaves of less widespread species of lady's-mantles

Alchemilla acutiloba (Starry Lady's-mantle)

A. wichurae (Rock Lady's-mantle)

A. glomerulans (Clustered Lady's-mantle)

A. glaucescens (Silky Lady's-mantle)

A. glaucescens (Silky Lady's-mantle)

Rather a rare plant that grows in calcareous grasslands in northern England, southeast Scotland and northwest Ireland. Grows to about 20 cm. The leaves measure less than 3 cm across and are made up of 7–11 rounded lobes. The central lobe has between 9–11 teeth. **The entire plant is densely covered in hairs.**

Finally, there are some species with highly restricted distributions

A. minima (Least Lady's-mantle)

This species may just be a dwarf form of *Alchemilla filicaulis* (Thin-stemmed Lady's-mantle). Only found near Ingleborough in Yorkshire and because of this sometimes considered a British endemic. Grows to less than 20 cm tall. **The leaves generally measure less than 3 cm across and are formed of 5 lobes.** The central lobe has 9–11 teeth. The entire plant is moderately hairy, although flower stalks may be free of hairs.

A. micans (Shining Lady's-mantle)

This rare species is only found in south Northumberland. The plants are around 30 cm tall. Its leaves generally measure less than 5 cm across and are usually formed of 9 rounded lobes. The central lobe has 11–15 sharp teeth. **The entire plant is covered in erect hairs, with the exception of the terminal sections of the flower stalks.**

A. monticola **(Velvet Lady's-mantle)**
Occurs only in northwest Yorkshire and County Durham. Grows to about 40 cm. The leaves measure up to 5 cm across and are made up of 9–11 rounded lobes. The central lobe has 15–19 sharp teeth. **The entire plant is covered in dense erect hairs, with the exception of the final sections of branches and flower stalks which are free from all but the occasional hair.**

A. subcrenata **(Large-toothed Lady's-mantle)**
A rare species that is only found in northwest Yorkshire and County Durham. Grows to about 40 cm. The leaves measure up to 5 cm and are made up of 7–9 rounded lobes. The central lobe has 13–17 teeth which appear almost hooked. **The upper and lower surfaces of the leaf are hairy (although the upper surface is only sparsely hairy), as are the lower stems and leaf stalks; however, the flowers and upper stems are hairless.**

HAVE OTHERS RECOGNISED THIS LEVEL OF VARIATION?

The name *Alchemilla* is derived from the historical use of these plants by herbalists and alchemists. The droplets of water that collect in the centre of their leaves were considered of the highest purity and thought to be essential in turning base metals into gold. Given the importance of these plants, it is perhaps surprising that the diversity of their forms was not recognised until recently.

Bentham and Hooker's *British Flora* of 1858 lists only three species of *Alchemilla*. These are Alpine Lady's-mantle, Common Lady's-mantle and Parsley-piert. At the time, Parsley-piert (*Aphanes arvensis/australis*) was considered to be an *Alchemilla*, but because it is an annual this tiny plant is now regarded as being in a separate family.

HOW FAR SHOULD I GO?

The introduced Garden Lady's-mantle is so distinctive, as are the alpine species, that there is no issue in being able to identify them. Beyond this there are a few widespread species that are generally relatively easy to recognise. The real frustration of *Alchemilla* identification is restricted to a few mountainous areas that you are unlikely to encounter unless on a quest to do so. However, it remains possible that some of these species have such apparently limited distributions purely as a result of recorder effort. For those wanting more information, the new BSBI Handbook No. 24 by Mark Lynes is the go-to guide.

CHAPTER 4

Sea-lavenders

You are most likely to encounter sea-lavenders as a purple haze spreading across a saltmarsh, or in a florist's bouquet. The genus has a wide global distribution and is particularly diverse around the Mediterranean. There are 14 native species of *Limonium* in the British Isles, of which eight are found nowhere else. Given that there are fewer than 50 species of plants that only grow in Britain and Ireland (i.e. are endemic to these isles), if you are botanically patriotic or interested in the unique, then this group may be for you.

WHY IS THIS GROUP COMPLEX?

Darwin was one of the first people to study why some species have different types of flowers on different individual plants. Sea-lavenders are an example of this trait. The flowers of some sea-lavender plants have smooth stigmas and produce pollen grains with a pitted surface. These are called cob plants. A second type of plant have stigmas which are covered in lots of small protuberances called papillae, and these flowers produce smooth pollen grains. The pitted pollen grains adhere easily to the surface of the papillate stigma, and vice versa. These differences ensure that only cob pollen grains can pollinate papillate plants, and only papillate pollen can pollinate cob plants. The two forms of plant (both of which are hermaphrodite) thus act as secondary genders, which promote outcrossing.

The possession of secondary genders is not without risk. If the genes for stigma type, pollen type and cross-compatibility type become unlinked by genetic recombination, the resulting plants may be unable to cross with any others. In some species of *Limonium* this problem has been sidestepped: only a single type of plant is found, and these are able to self-pollinate.

To complicate things further, sea-lavenders – like many plants – often have duplicate sets of chromosomes and genes. They also gain and lose chromosomes with ease. Under these conditions it is unlikely that the copies of the genes controlling stigmas, pollen and cross-compatibility will all become unlinked. Rather, the probable outcome is a complex mix of multiple copies of all possible combinations of these genes. When this occurs, only genotypes that are lucky enough to be able to produce asexual seeds will be able to persist. Reproduction via asexual seeds is a certain route to taxonomic complexity. Every subsequent mutation has the potential to be preserved and may result in the production of a distinctive new form of plant. All eight British endemic *Limoniums* fall into this category.

If this is not complicated enough, the sexual species are also known to hybridise, with hybrids and back-crosses being reported, as well as plants varying with local environmental conditions.

HOW CAN I TELL THEM APART?

A purist would argue that you can only definitively differentiate some species of *Limonium* by looking at their sexual parts down a microscope. Knowing this, you may wish to give up at this point. But let's remind ourselves that it is not a capital offence to misidentify a plant. If you want to have a stab at recognising these species, you will need to look at their leaves, the way their flowering stems branch, and the amount of sub-branching that occurs within their flower spikes. Only if you get really engrossed will you need to peep at their privates.

Let's start with the species that reproduce sexually

***L. vulgare* (Common Sea-lavender)**

This species is found growing in saltmarshes around much of England and Wales but is absent from Ireland. The plants are self-incompatible and either cob or papillate. **The anthers are creamy yellow.** Stems are erect and grow to 40–60 cm, with branching occurring at an angle of roughly 40°. The leaves have branching veins and die off in the autumn. **Flower spikes measure 1–2 cm and have between 5 and 8 sub-spikelets.**

***L. humile* (Lax-flowered Sea-lavender)**

Grows in saltmarshes around England, Wales and southeast Scotland; the commonest species in Ireland. May be growing with *L. vulgare*. The plants contain variable numbers of chromosomes, most commonly six sets. Able to self-pollinate,

Characteristics of sea-lavender species

Species	Branching angle			Highly branched	Anthers brown	Parallel leaf veins	Evergreen
	20	30	40				
L. vulgare (Common Sea-lavender)			■				
L. humile (Lax-flowered Sea-lavender)	■				■		
L. bellidifolium (Matted Sea-lavender)		■		■		■	
L. binervosum (Rock Sea-lavender)			■		empty	■	■
L. platyphyllum (Florist's Sea-lavender)		■	■	■			
Your specimen							

produces papillate stigma and pitted pollen which can stick to the flower's own stigma. **The anthers are brown-red in colour.** Stems are erect and grow to about 40 cm, with branching occurring at roughly an angle of 20°. The leaves have branching veins and die off in the autumn. **Flower spikes measure 2–5 cm but with only 2–3 sub-spikelets, resulting in the distinctive lax-flowered appearance.**

L. bellidifolium **(Matted Sea-lavender)**
This is a rare plant in the British Isles, only occurring in drier parts of saltmarshes in north Norfolk and Lincolnshire. All plants are self-incompatible and either cob or papillate. Its anthers are creamy white. **Stems are low growing and highly branched, giving a distinctive zig-zag appearance, growing to 30 cm.** Branching occurs at roughly an angle of 30°. The leaves usually have three parallel veins and die off during flowering. Flower spikes are short, measuring less than 1 cm but with many sub-spikelets.

And now the species that produce asexual seeds

L. binervosum **(Rock Sea-lavender)**
This highly variable asexual species is found in saltmarshes, sand dunes and growing on maritime rocks around the coast or England, Wales and Ireland. **The plants are morphologically cob, while producing no pollen** (very rarely, papillate plants are found). The empty anthers are white in colour. Different taxonomists have divided these populations into 9 or 45 subspecies, which include all our endemic *Limonium*. Stems are usually erect and vary in height from 30 to 70 cm tall. Stem branching patterns are also highly variable. **Leaves are evergreen and have parallel veins.** Flower spikes measure about 3 cm. Spikelet number is highly variable, so flowers can be densely to loosely arranged.

There are a further two species of sea-lavender that also reproduce by asexual seeds: *L. auriculae-ursifolium* (Broad-leaved Sea-lavender) and *L. normannicum* (Alderney Sea-lavender). In Britain these are both rare and restricted to the Channel Islands.

Introduced species

L. platyphyllum **(Florist's Sea-lavender)**
Grown ornamentally in gardens and, as its name suggests, frequently used in florists' bouquets. Occasionally occurs as a garden escape. This is taller and more robust than our native species, growing to 100 cm. Has creamy white anthers. **It has distinctive transparent outer bracts** and flowers that are densely packed on 2-cm-long spikes.

HAVE OTHERS RECOGNISED THIS VARIATION?

Until relatively recently, much of the variation observed within this group was thought to result from differences in local environmental conditions. The classic

Characteristics of leaves and inflorescences of native sea-lavenders

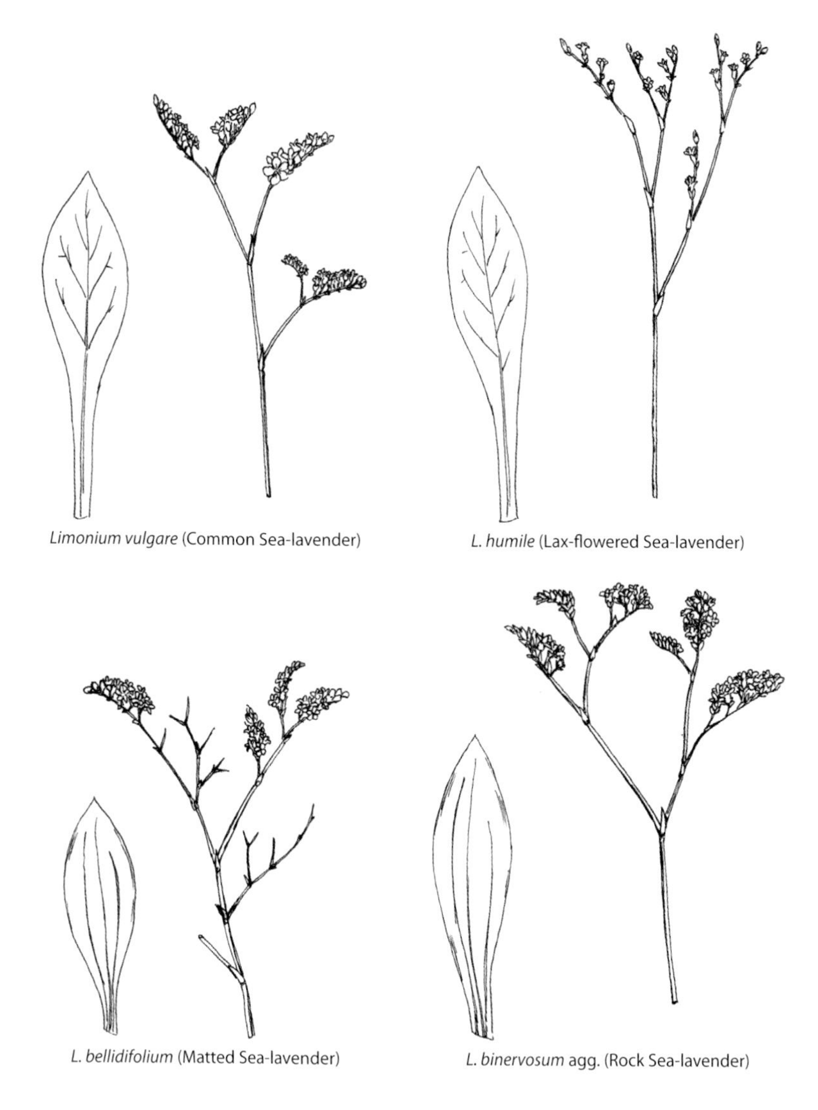

Limonium vulgare (Common Sea-lavender)

L. humile (Lax-flowered Sea-lavender)

L. bellidifolium (Matted Sea-lavender)

L. binervosum agg. (Rock Sea-lavender)

Victorian *Handbook of the British Flora* by Bentham and Hooker, first published in 1858, listed only Sea-lavender, Rock Sea-lavender and Matted Sea-lavender. The eight British endemic *Limoniums* were not recognised until 1986.

HOW FAR SHOULD I GO?

Populations of the asexual *L. binervosum* are difficult to differentiate. To be able to identify the asexual species within this complex requires sampling several individuals, because a single unusual plant can be misleading. Unless you are on a quest to see all our endemic species (or are in the Channel Islands), there is a strong case for only identifying the sexual species of *Limonium* and lumping the rest as *L. binervosum* agg. For those wanting to push on, geographic location is a good starting point, and Stace 2019, and Sell and Murrell 2018 will provide more help.

CHAPTER 5

Whitebeams, rowans and service trees

The small trees and shrubs in the genus *Sorbus* include the common and distinctive Rowan, which is also known as the Mountain Ash. Its alternative common name relates to the fact that its leaves are divided into several leaflets which superficially resemble those of the Ash tree. However, the two species are not closely related. There are three other native British species of *Sorbus* that reproduce sexually, these are the Whitebeam, and both the Service Tree and the Wild Service Tree. In addition to these, there are several introduced species of *Sorbus* which are widely grown as ornamentals, and many rare asexual clones and hybrids which are usually only of interest to a dedicated group of enthusiasts. Many of these asexual clones have very restricted global distributions and because of this some have gathered a local following. If you are interested in the unusual and unique, *Sorbus* is a good place to look, because there are around 16 species that only occur in the British Isles. In fact, the genus *Sorbus* contains more British endemics than any other. If you are wanting to practice your ID skills, you may not need to trek to the Avon Gorge as you could well encounter other unusual *Sorbus* species as amenity trees on your high street.

WHY IS THIS GROUP OF PLANTS COMPLEX?

As mentioned above, within the genus *Sorbus* there are four native species, *S. aucuparia* (Rowan), *S. aria* (Common Whitebeam), *S. torminalis* (Wild Service Tree) and *S. domestica* (Service Tree) that all contain two sets of 17 chromosomes, and which reproduce by the production of sexual seeds. After that, things start to get more complicated.

Sorbus species are challenging to identify because many of them are derived from different combinations of crosses between these native sexual species. This complexity is confounded by frequent increasing of chromosome numbers. As with many plant groups in which hybridisation is common, the *Sorbus* have escaped hybrid sterility by producing asexual seeds. These asexually reproducing lines can be regarded as new species because they cannot cross with other species, and they are morphologically distinctive. The occurrence of asexual seed production within the *Sorbus* varies. Some are thought to always produce asexual seeds, while other species only sometimes produce asexual seeds, and others only rarely produce asexual seeds.

Rather curiously, in order to be able to successfully produce asexual seeds, many species of *Sorbus* require to be pollinated first. However, since they are unable to self-pollinate and technically all members of the same species are in fact genetically the same individual, this means that they must be pollinated by another species before they are able to produce asexual seeds. It is therefore perhaps not surprising that new hybrids and indeed new species are probably still being produced.

Into this complicated network of hybrids and chromosome duplication, further genetic complexity has been brought by humans. Horticulturalists have introduced several other species of *Sorbus*. Their attractive flowers, fruits, leaves and fairly small stature make attractive species to plant in gardens and public spaces. *S. intermedia*, the Swedish Whitebeam, and *S. hybrida*, the Swedish Service Tree, are both good examples of this.

HOW CAN I TELL THEM APART?

This group of plants is one of the more challenging in our flora. Once you can identify the more common sexual species, it may be helpful to find yourself a tame *Sorbus* expert to help you gain confidence or alternatively look out for a training course. The important features to look for are the leaf shape (are the leaves entire, lobed or made of separate leaflets?), is the undersurface of the leaf hairy and white or green? At the species level, you will probably need to count the number of veins on the leaves. The colour, size and shape of the fruit are all important characters to record too, as is the presence or absence of spotty markings called lenticels on the fruit's surface. Within the flower, if the female parts are fused or not can aid in identifying the species, but it may be less helpful in assigning your specimen to its subgenus (because crosses between subgenera are common).

Let's focus on the subgenus level

Even at this level of resolution many of the following descriptions are unhelpfully variable, because the species within the subgenus can vary in many of the key characteristics.

1. *Sorbus* (rowans)

The leaves of these species have about 5 to 12 pairs of leaflets which are slightly hairy and green on their undersurface. **The central part of the flower comprises 3 to 5 female styles that are not fused or only fused at the base.** The petals and sepals arise midway along the ovary. The sepals are covered in glands. The fruit are white, orange or red and may have a few small spots.

Characteristics of our six *Sorbus* subgenera

Subgenus	Leaves				Fruit				
	leaflets	shallow lobes	deep lobes	white undersurface	orange- red	green- brown	with spots	2–4 cm	sepals with glands
Sorbus (rowans)	■				■		■		■
Aria (whitebeams)				■	■				
Tormaria (broad-leaved whitebeams)		■		■	■	■	■		
Cormus (service trees)	■					■	■	■	
Torminaria (wild service trees)			■			■	■		■
Soraria (false rowans)	■	■	■	■	■		■		+/–
Your specimen									

S. aucuparia (Rowan)

This native species grows in woods, moorland and rocky places across the British Isles, where it can reach 18 m tall. **Its leaves are typically divided into between 5 and 9 pairs of leaflets.** Flowers contain 3 to 4 styles. Its round fruit occur in clusters of 8 to 12, with each individual fruit measuring about 1 cm long. When ripe the fruit are a distinctive orange-red colour and may have a few small spots.

2. *Aria* (whitebeams)

The leaves of these species are typically oval in shape with a toothed margin and are only rarely lobed. **The lower surfaces of the leaves are hairy and a distinctive white or grey colour.** Within the flowers there are 2 to 3 female styles that are not fused. The petals and sepals arise from above the ovary. The sepals have no glands on their surface. Fruit are red in colour and usually free from spots.

S. aria (Common Whitebeam)

Grows in woods and scrubland mostly on calcareous soils; probably only native in southern England, but now widely planted across the British Isles. Can grow up to 20 m tall. Its leaves are oval, typically twice as long as they are wide. **The undersurface of the leaves is covered in dense hairs, giving a distinctive white appearance.** The leaves typically have between 9 and 14 pairs of veins. The leaf margin is toothed (sometimes with teeth on the teeth) and usually unlobed. The bright red fruit measure 1–1.5 cm long and occur in clusters of between 10 to 20. The surface of the fruit may be covered in a few to many small, evenly dispersed spots.

3. *Tormaria* (broad-leaved whitebeams)

These trees have simple oval leaves that are divided into shallow lobes. The lower surfaces of the leaves are hairy and whitish in colour. Flowers contain 2 or 3 female styles, which may be fused. The petals and sepals arise midway along the ovary; sepals have no glands. **The fruit may be yellowish to reddish brown or orange-brown in colour; they are covered in many conspicuous spots.**

S. bristoliensis **(Bristol Whitebeam)**
Endemic to the Avon Gorge, where it may grow to 15 m tall. Its oval leaves are 1.2–1.8 times longer than they are wide and have 8 to 10 pairs of veins. The undersurface of the leaves are hairy and white in appearance. **The leaf margins are dissected by short lobes which point towards the leaf apex; this lobing only starts around a third of the way up the leaf.** The flowers have five creamy white petals, measuring 1–1.3 cm across; they contain two styles and about 10 pink stamens. **The round fruit measure 9–12 mm long/wide, are bright orange in colour and may be covered in spots towards their base.**

4. *Cormus* (service trees)

These trees have leaves with paired leaflets rather like those in subgenus *Sorbus*. The undersurface of the leaves are green and slightly hairy. **The central part of the flower comprises 5 fused styles.** Petals and sepals arise midway along the ovary. The sepals are without glands. **The fruit are greenish-brown in colour, sometimes flushed red, and have many obvious spots on their surface.** There is a single species in this subgenus in Britain.

S. domestica **(True Service Tree)**
An extremely rare tree that is probably only native in very few sites, but is sometimes planted elsewhere; it can grow to 20 m tall. Its leaves are 15–25 cm long and have 6 to 8 pairs of leaflets. The white flowers are 1.3–1.8 cm across, with 5 fused styles and about 20 cream-coloured stamens. **Fruit are large and pear-shaped, measuring 2–4 cm long. The fruit are green-brown in colour with a red blush and are covered in many large spots.**

5. *Torminaria* (wild service trees)

The leaves of these trees are divided into deep and distinctively pointed lobes. The undersurface of the leaves are green and slightly hairy. Within the flower there are two female styles which may be fused. Petals and sepals arise above the ovary. The sepals are glandular. **Fruit are brown and covered in many obvious spots.**

S. torminalis **(Wild Service Tree)**
This rare native tree occurs in woodlands and hedgerows across much of England and Wales, where it may grow to a height of 25 m. Its leaves are oval and slightly longer than they are wide. The leaves are divided into pointed lobes, which dissect the leaf to almost halfway. The undersurface is green and slightly hairy. **The leaves**

Examples of leaves from each of the six *Sorbus* subgenera

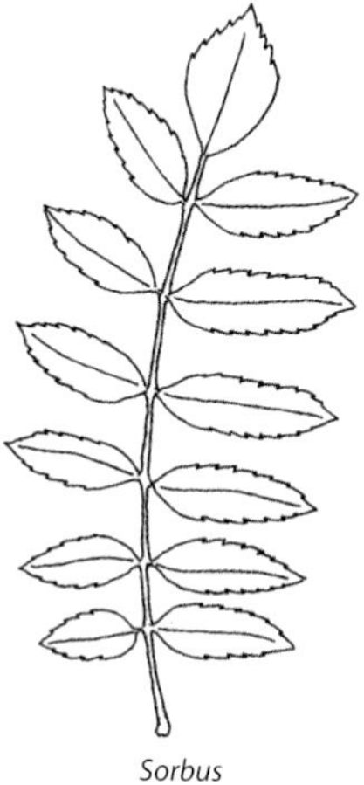

Sorbus
S. aucuparia (Rowan)

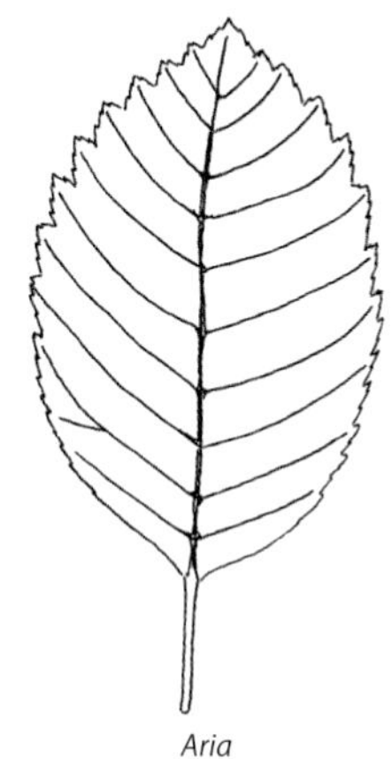

Aria
S. aria (Common Whitebeam)

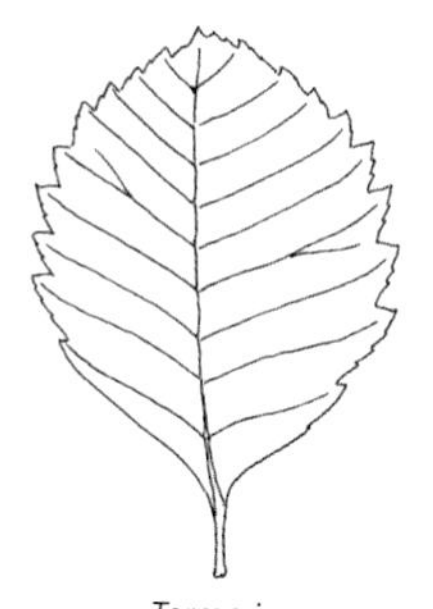

Tormaria
S. bristoliensis (Bristol Whitebeam)

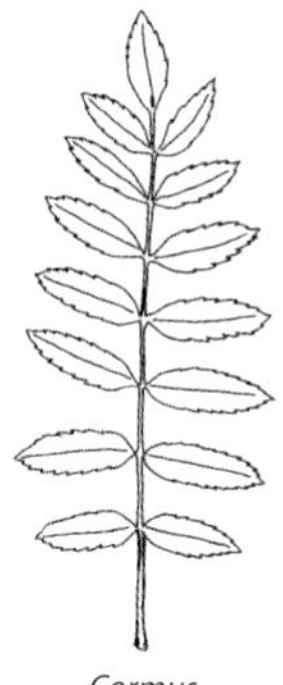

Cormus
S. domestica (Service Tree)

Torminaria
S. torminalis (Wild Service Tree)

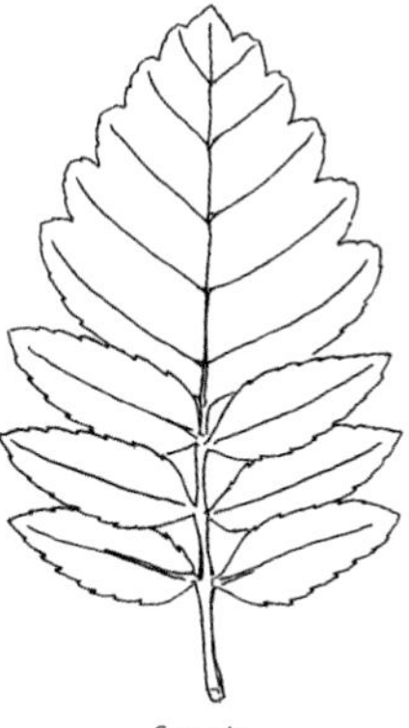

Soraria
S. hybrida (Swedish Service Tree)

look somewhat like a maple. Flowers have 5 white petals, measuring 1–1.5 cm across; they contain two styles and 20 creamy-white stamens. **The russet brown fruits are 1–1.5 cm long, occurring in clusters of about 10 and covered in many large spots.**

6. *Soraria* (false rowans)

These trees have leaves with paired leaflets, or have leaflets at their base and are fused near their apex (other Sorbus species have simple, lobed leaves). The leaves are hairy and white in appearance on their lower surface. The central part of the flower comprises 2 to 4 female styles that are usually not fused. The petals and sepals arise midway along the ovary. The sepals are sometimes covered in glands. **Fruit may be orangey-red to crimson with a few small spots.**

S. hybrida (Swedish Service Tree)
This introduced small tree may be found in parks and gardens and as an escape scattered throughout western Britain north to Aberdeen; it may grow to 12 m tall. **Its leaves have free leaflets closest to the base (usually 2 pairs), but these appear fused to become lobes and finally just teeth close to the apex. The leaves have 7 to 12 pairs of veins.** The undersurface of the leaves is white and hairy. The central part of the flower comprises 2 to 4 female styles that are usually not fused. The oval fruit measure 1–1.5 cm long and are scarlet in colour. The skin of the fruit has a few small spots.

S. intermedia (Swedish Whitebeam)
This introduced tree is widely planted and sometimes self-sown across most of England; grows to a height of 10 m. **Its leaves are roughly oval in shape with shallow lobes which are toothed. The lower third of the leaf margin is unlobed and untoothed. The leaves have 6 to 9 pairs of veins.** The upper surface of the leaves is dark green and the undersurface white and hairy. The central part of the flower comprises three female styles that are not fused. The oblong fruit measure 1.2–1.5 cm long and are scarlet with a few small spots.

HAVE OTHERS RECOGNISED THIS LEVEL OF VARIATION?

Victorian botanists included these trees in the same genus as apples and pears, with only three species being recognised. Since then, the genus *Sorbus* has attracted the interest of many modern taxonomists, particularly those at the Cambridge Botanic Garden. Perhaps because of this it has experienced an unusual amount of taxonomic change. Only since around the turn of the millennium has DNA evidence truly revealed the origins of many of our endemic species.

The Rowan and the Whitebeam have been well known since ancient times and have a long history of use, appearing widely in folk traditions. In contrast, the Service Tree was first recorded in Britain as a single tree in the Wyre Forest, Worcestershire, in 1678. It is still only known from a handful of sites, with new records still being reported into this century. The populations of many of our endemic asexual species of *Sorbus* typically comprise very few individuals. Thus, even without their taxonomic complexity, it not surprising that many of them were not recognised as new species until recently.

HOW FAR SHOULD I GO?

It has been argued by specialists of some other apomictic genera that the species-level classification has more readily been awarded within the *Sorbus* than in other groups. This may be jealousy, because they have not been able to have their own favourite variant canonised, or it may genuinely explain why there are more British endemic *Sorbus* than any other group of plant. Even *Sorbus* experts have recognised that endemic asexual clones are an evolutionary dead-end, and in terms of conservation we should perhaps focus on the evolutionary process by which new asexual lines are created, rather than worrying too much about preserving every endemic species indefinitely. Does this mean that it is not worth being able to identify these species? Not necessarily: some of these endemic *Sorbus* species have attracted significant local interest, and efforts are made to monitor, protect and propagate them. Even if you accept the argument that the evolutionary process rather than the particular species is most important in the long term, you still need to be able to recognise the different species and any new ones that may arise in future.

If you are interested in the high-level resolution of *Sorbus* species it is worth consulting BSBI Handbook No. 14, *Whitebeams, Rowans and Service Trees of Britain and Ireland: A Monograph of British and Irish Sorbus*, by Tim Rich, Libby Houston, Ashley Robertson and Michael Proctor (2010).

CHAPTER 6

Yellow composites – things that look a bit like a dandelion

HAWK'S-BEARDS, SOW-THISTLES, CAT'S-EARS AND OXTONGUES

Most botanists are reasonably confident that they can identify a dandelion – even if they don't appreciate just how many species of dandelions there are. However, for many of us this confidence can be shattered when we look in a wildflower guide and discover that there are lots of other yellow-flowered things that look a bit like a dandelion. These lookalikes can be challenging to identify. They include members of several closely related groups, many of them common, some of them less so. This is an informally defined group of species, but here you will find all the yellow composites with flowerheads similar in form and size to dandelions.

WHY IS THIS GROUP OF PLANTS COMPLEX?

The good news is that not all these yellow-flowered Dandelion-like things are complex or difficult to identify. So, with a few tips you are likely to make rapid progress.

Perhaps what makes these plants daunting is the fact that they are all very similar shades of yellow. Plants with pink, purple and mauve flowers all seem to be different shades. In contrast, to human eyes, yellow-flowered species can all appear remarkably similar. This may be a function of the stability of the biochemistry of yellow pigments, or of our eyes' inability to detect colour variation in this part of the spectrum.

Part of the frustration in identifying these species is simply the number of possible alternatives you need to compare, and their unhelpful common names which seem obsessed with hawks. All these dandelion lookalikes are members of the daisy family, the Asteraceae, which includes about 24,000 species – more than any other plant family on earth. This problem of expansiveness is often overlooked by taxonomists, but it is not trivial.

Finally, one of the groups of Dandelion-like plants, the hawkweeds (*Hieracium* spp.), are genuinely complex. Like the dandelions themselves, the hawkweeds produce seeds asexually. This form of reproduction is always associated with the taxonomic complexity that delights some botanists and confounds others. Many of our endemic species produce asexual seeds – a notable example of this is *Hieracium attenboroughianum* (Attenborough's Hawkweed), a plant found only in the Brecon Beacons and named in honour of Sir David Attenborough.

HOW CAN I TELL THEM APART?

The first challenge is to determine which group of plants your specimen is a member of. To do this, we need to look at the leaves, including examining any hairs they may have (so a hand lens is useful), check if flower stalks are branched and look at the fluffy pappus attached to the seeds (again, a lens will help).

Once you have managed to identify the specimen to genus level, we will attempt to identify it to species, see below.

Is it a dandelion (*Taraxacum*)?

Check the flower stem: **dandelion stems are unbranched, hollow tubes that ooze white latex when picked**; they have a single flowerhead at the top. Dandelion leaves are mostly hairless and are only ever found in a basal rosette. Dandelions are covered in detail in Chapter 2.

A guide to differentiating groups of yellow composites

						Stems		
Group	**Latex**	**Leaves untoothed**	**Flowers before leaves**	**Rosettes**	**Taller than 50 cm**	**hollow**	**branched**	**with leaves**
Dandelion (*Taraxacum*)	■			■		■		
Sow-thistle (*Sonchus*)	■				■	■	■	■
Goatsbeard (*Tragopogon*)		■			■			
Viper's-grass (*Scorzonera humilis*)		■					░	■
Coltsfoot (*Tussilago farfara*)			■					
Oxtongue (*Picris* or *Helminthotheca*)				■	■		■	
Cat's-ear (*Hypochaeris*)				■			■	
Hawkbit (*Leontodon* or *Scorzoneroides*)				■				
Hawk's-beard (*Crepis*)					■		■	■
Hawkweed (*Hieracium* or *Pilosella*)	■			░	■		■	■
Your specimen								

If it is tall and a bit like a thistle it might be a sow-thistle (*Sonchus*)?

As their name suggests, sow-thistles the most thistle-like of the dandelion lookalikes, so they are pretty easy to identify. **Their flower stems are hollow and branched** (so not like a dandelion). The stems produce white latex when picked, which turns orange with time. The fleshy stems are blue-green or bright green in colour and their prickly leaves have lobes where they join the stalk. They may grow to well over a metre in height.

Is it a Goatsbeard (*Tragopogon*)?

Goatsbeard plants produce **large dandelion-like clocks** (seedheads). They are easy to identify, because **the green bracts around the flowerheads are frequently longer than the yellow petals.** Flowers are solitary on stalks that have few branches. Each seed has a distinctive flattened, stalked parachute.

It is unlikely to be Viper's-grass (*Scorzonera humilis*)

This is a rare species, now restricted to very few sites. However, if you are in a damp meadow and have a tallish erect plant up to 100 cm with few branches, then check the **bracts surrounding the flowerhead. If there are lots of these in multiple rows, looking a bit like a pineapple,** it could be Viper's-grass.

If it is in flower in February or March it could be Coltsfoot (*Tussilago farfara*)

Although somewhat dandelion-like, the **flowers of Coltsfoot appear before its leaves in early spring.** The leaves are a rounded, heart shape and downy-white on the undersurface.

If it is bristly it might be an oxtongue (*Picris* or *Helminthotheca*)

If your plant is rather tall for a dandelion, perhaps growing to almost a metre but usually less, and is **covered in translucent bristles** you might be looking at an oxtongue. The branched stems are also covered in bristles. **Leaves have distinctive wavy edges** and may have toothed margins. The seeds have two rows of hairs in the pappus.

What about cat's-ears (*Hypochaeris*)?

If the **solid flower stem has few branches, is leafless and the basal rosette leaves** are usually covered in unforked hairs, you may be looking at a cat's-ear. Next check the flowerheads, these are surrounded by many bracts of different lengths. Looking **inside the flowerhead you should find bract scales at the base of individual florets.** The seed has a fluffy pappus, made of both simple and feathery scales.

Is it a hawkbit (*Leontodon* or *Scorzoneroides*)?

The *Leontodon* Hawkbits have **solid flower stems which are unbranched and leafless, with very hairy leaves** only in a basal rosette and the **leaf hairs are forked.**

In contrast, Autumn Hawkbit (*Scorzoneroides autumnalis*) has solid flower stems, some of which are branched and usually free from hairs. **The rosette leaves are distinctly and deeply lobed, the ends of the leaves are pointed. The leaf surface is usually hairless or may have a few unbranched hairs.**

Could it be a hawk's-beard (*Crepis*)?

Hawk's-beards have solid, branched flower stems with hairy or hairless leaves. Look at the **flowerhead: they are surrounded by just two rows of bracts, one shorter row and one longer.** The white fluffy material on the seeds is simple (not feathery). Double check it is not an oxtongue.

Is it a hawkweed (*Hieracium* or *Pilosella*)?

Hawkweeds are the really complex and variable ones. They have branched, leafy flower stems which produce white latex when broken. They are usually hairy, often with multicellular hairs. Their dandelion-like flowerheads are surrounded by many bracts of different lengths in overlapping rows.

Now let's try to identify the species

Sonchus (sow-thistles)

Three of these species are most commonly encountered growing in disturbed ground. Those are the three species you are likely to find.

S. oleraceus (Smooth Sow-thistle)
An annual plant of wasteland. **The lobes on stem leaves are pointed where they join the stem.** The plant lacks hairs except on the flowerheads and just below them. Its white latex slowly turns orange.

S. asper (Prickly Sow-thistle)
An annual plant of wasteland and rough ground. **The lobes on stem leaves are rounded where they join the stem. The leaves are glossy and often undivided.** The plant lacks hairs except on the flowerheads and just below them. Its white latex quickly turns orange.

S. arvensis (Perennial Sow-thistle)
A perennial plant of bare ground, marshes and fens, where it **produces rhizomes and spreads forming patches.** The lobes on the stem leaves are rounded. The stems are sometimes covered in glandular hairs.

S. palustris (Marsh Sow-thistle)
A scarce perennial plant of marshes and fens. Stems grow from a thick root-system. Plants are hairy at the top but lack hairs at the base. **The leaves have arrow-shaped lobes where they join the stem.**

Sonchus (sow-thistle) basal and stem leaves

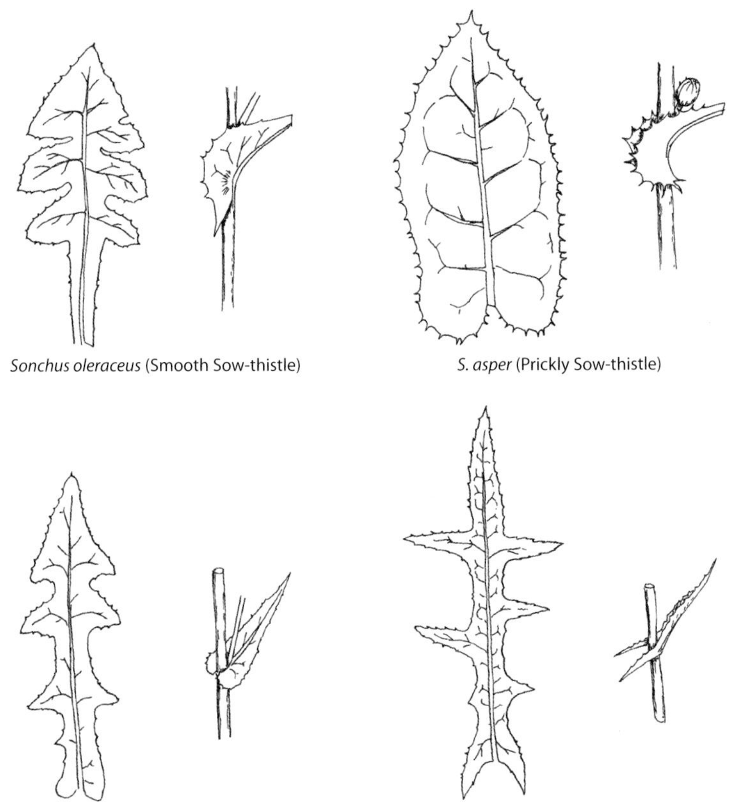

Sonchus oleraceus (Smooth Sow-thistle) *S. asper* (Prickly Sow-thistle)

S. arvensis (Perennial Sow-thistle) *S. palustris* (Marsh Sow-thistle)

Tragopogon

***Tragopogon pratensis* (Goatsbeard)**

There is only a single species of *Tragopogon* that is native to the British Isles. Plants grow to 75 cm tall in grassy habitats. Flowerheads open in the mornings but close by noon. The leaves are linear and grass-like.

***Tragopogon porrifolius* (Salsify)**

This is an introduced species that is grown as a root vegetable and very occasionally escapes into the wild. It is similar to Goatsbeard, except its flower is purple.

Scorzonera

***Scorzonera humilis* (Viper's-grass)**

There is only a single species of *Scorzonera* that is native to the British Isles and it is a rare plant of a few marshy grasslands. Plants grow to 100 cm tall. Its leaves are not like a dandelion, being untoothed and lanceolate. Flowerheads are dandelion-like, occurring singly on tall stems. The flowerheads are surrounded by multiple rows of bracts.

Scorzonera hispanica (Black Salsify)
This is an introduced species that is grown as a root vegetable and occasionally escapes into the wild. It is similar to Viper's-grass but grows to 1.5 m.

Tussilago

Tussilago farfara (Coltsfoot)
These plants flower early in the year (Feb–April) before the other dandelion lookalikes. The flower stalks are unbranched, covered in purple scales and support a single flowerhead. After flowering, the flower stalk bends over like a swan's neck, then straightens again when the seeds are ready to disperse.

Picris and *Helminthotheca* (oxtongues)

Until recently taxonomists placed all the oxtongues in the genus *Picris*. They have now been split into two, which can be distinguished as follows. Plants in the genus *Picris* have **lanceolate bracts around their flowerheads, and only their inner row of pappus hairs are feathery**. In contrast plants in the genus *Helminthotheca* have oval-shaped bracts and both rows of pappus hairs are feathery.

Picris hieracioides (Hawkweed Oxtongue)
Hawkweed Oxtongue plants grow to 100 cm tall. **The leaves are lanceolate, with small teeth along their wavy margins**. There are several rows lanceolate bracts around the flowerheads. The seeds have two rows of pappus hairs but **only the inner row is feathery.** Plants are commonly found in open grasslands south of a line between Cardiff and Newcastle upon Tyne.

Helminthotheca echioides (Bristly Oxtongue)
Bristly Oxtongue plants grow to 80 cm tall. **The narrow leaves are covered in many bristles.** There are several rows of oval bracts around the flowerheads. The seeds have **two rows of pappus hairs, both of which are feathery.** Plants are found in disturbed ground. They occur most frequently south of a line between Blackpool and Newcastle upon Tyne, including most of Wales and eastern Ireland.

Hypochaeris (cat's-ears)

Cat's-ears are commonly found growing in old grasslands. They grow to about 50–60 cm. There are just three species in the British Isles, all of which are perennials.

H. radicata (Common Cat's-ear)
The leaf margins of this plant have shallow lobes. The end of the leaf is blunt, and the leaf surface covered in rough hairs. The flower stems are without hairs except

near the base and have leaves that are reduced to small scales. Flowerheads open every day. A common plant of old grasslands.

H. maculata (Spotted Cat's-ear)
The leaf margins of this plant have very shallow lobes to being almost unlobed. The **leaves are covered in distinctive purple blotches** and the leaf surface is covered with rough hairs. The flower stems are free from leaves and are usually hairy. Mostly a plant of calcareous or sandy soils.

H. glabra (Smooth Cat's-ear)
The leaf margins of this **plant have lobes producing an almost saw-like appearance.** The end of the leaf is blunt, and the **leaf surface is free from hairs.** The flower stems are also free of hairs and may have small scale-like leaves. Flowerheads only open in bright light. This species is mostly found in the West of England growing in mature grasslands and open ground.

Hypochaeris (cat's-ears) rosette leaves

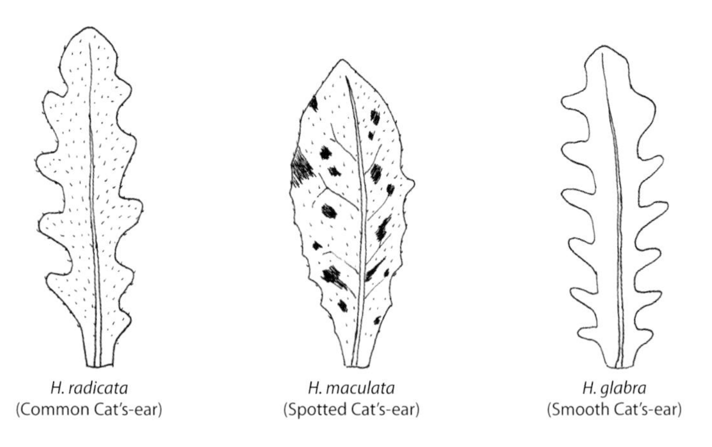

Leontodon and *Scorzoneroides* (hawkbits)

There are three species of hawkbits found in the British Isles. Previously, they were all placed in the genus *Leontodon*. Recent taxonomic revision has separated the Autumn Hawkbit from the other two. The three remain united by the common name Hawkbit, which adds to the confusion of muddling them with Hawk's-beards and Hawkweeds.

Scorzoneroides autumnalis (Autumn Hawkbit)
Until recently known as *Leontodon autumnalis*. It is a common plant of unimproved dry grasslands. **The rosette leaves are distinctively deeply lobed and usually free of hairs – when hairs occur, they are unbranched.** The flower stalks measure 5–60 cm and are often branched, they may be hollow in the upper section. **There**

are no stem leaves. The capitula measure 1.2–3.5 cm across and its ray florets are often reddish on the outer surface.

The leaves of other hawkbits are hairier, and the hairs tend to be forked.

Leontodon hispidus (Rough Hawkbit)
The flower stems are unbranched and grow to about 60 cm. **These stems are usually notably hairy, including some hairs which are divided into three, the hairs may be up to 5 mm in length.** The rosette leaves have sharp tooth-like lobes. **The bracts surrounding the flowerhead are very hairy.** Grows in base-rich sites, often calcareous grasslands.

L. saxatilis (Lesser Hawkbit)
The stems of Lesser Hawkbit are unbranched and grow to about 40 cm. These stems are covered by a few or no hairs. The rosette leaves have tooth-like lobes. The leaves are variable, sometimes being hairy and sometimes free of hairs. Occasionally these hairs split into three, but most are bifurcated and 1–3 mm long. **There are distinctive red spots at the base of hairs on the midrib of each leaf. The bracts surrounding the flowerhead are variable, usually being hairless but sometimes hairy.** Grows in unimproved acid grasslands.

Scorzoneroides and *Leontodon* (hawkbits) rosette leaves

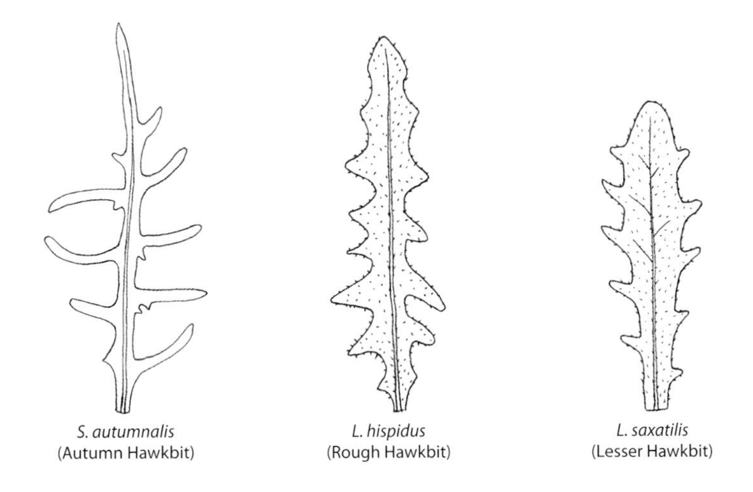

Crepis (hawk's-beards)

There are several native and introduced species of hawk's-beards, some of which are common and widespread, while others are rarer. They tend to be short-lived annuals or biennials, although some are longer-lived. Thus, you will often encounter these plants as non-flowering rosettes, which can be tricky to identify. When they

Characteristics of *Crepis* species

Species	Plant height cm		Life history		Stem leaves			Capitulum bracts	
	30–60	>60	Short-lived	perennial	absent	lobed	unlobed	hairless	hairy
C. vesicaria (Beaked Hawk's-beard)									
C. biennis (Rough Hawk's-beard)		>100							
C. capillaris (Smooth Hawk's-beard)									
C. paludosa (Marsh Hawk's-beard)									
C. setosa (Bristly Hawk's-beard)									
C. foetida (Stinking Hawk's-beard)					rare				
C. mollis (Northern Hawk's-beard)									
C. praemorsa (Leafless Hawk's-beard)									slightly
Your specimen									

flower, they can be distinguished by having branched stems with leaves. They have **two rows of bracts around their flowerheads** and the fluffy pappus attached to their seeds is made from several rows of pure white, unbranched hairs.

Let's start with the common examples

C. vesicaria (Beaked Hawk's-beard)

Biennial plants growing up to 80 cm tall. The stem leaves are deeply lobed and clasp the stem. The bracts around the flowerheads are covered in glandular hairs. Plants grow commonly on disturbed and rough ground, road verges and walls.

C. biennis (Rough Hawk's-beard)

Biennial plants growing to 1.2 m tall. Its stem leaves are irregularly and sharply lobed, sometimes clasping the stem. The bracts around the flowerheads are covered in glandular hairs. Occurs in rough grasslands, including road verges.

C. capillaris (Smooth Hawk's-beard)

Annual or biennial, to 75 cm tall but generally smaller. Usually lacks hairs or only slightly hairy. Stem leaves are variable: they sometimes have alternate long and short teeth, while others are unlobed. They do not clasp the stem. The bracts around the flowerheads may lack hairs, although forms with glandular hairs are known. Common in grasslands and on waste ground.

Crepis (hawk's-beard) stem leaves

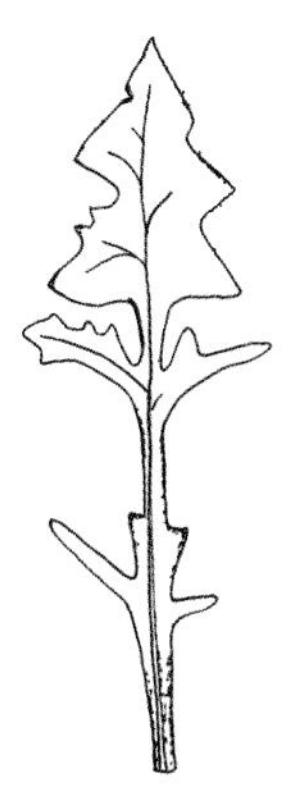

C. vesicaria
(Beaked Hawk's-beard)

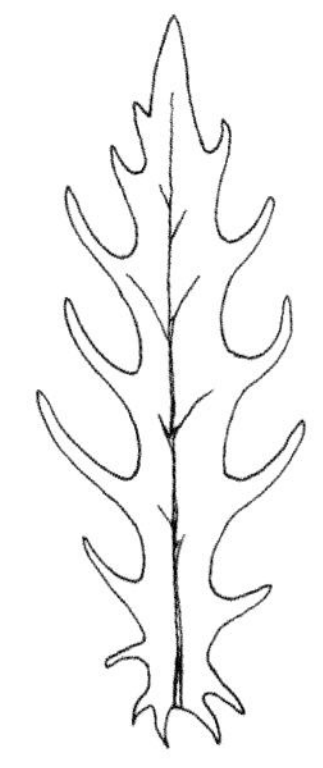

C. biennis
(Rough Hawk's-beard)

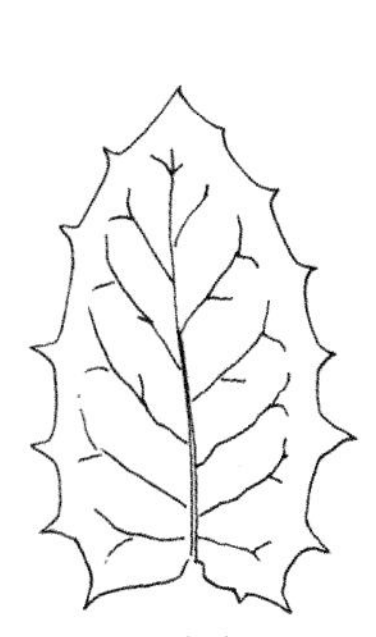

C. paludosa
(Marsh Hawk's-beard)

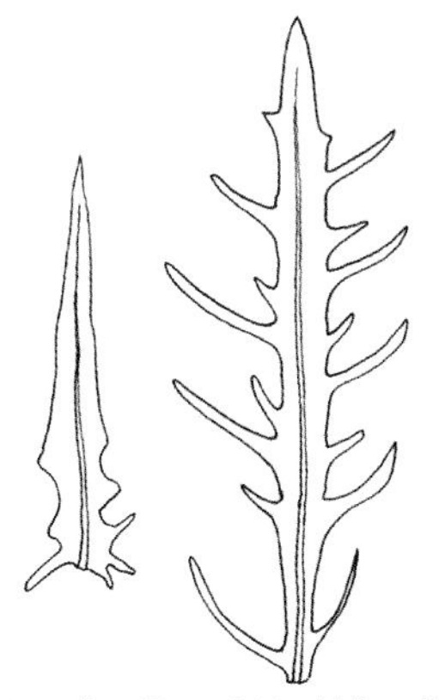

C. capillaris (Smooth Hawk's-beard)

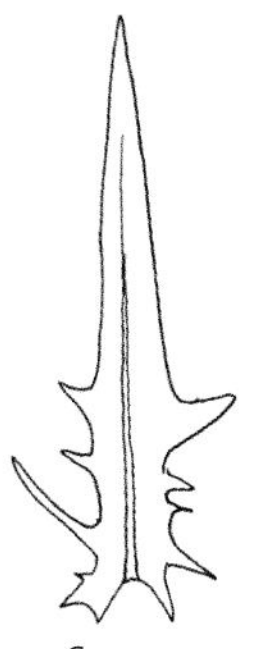

C. setosa
(Bristly Hawk's-beard)

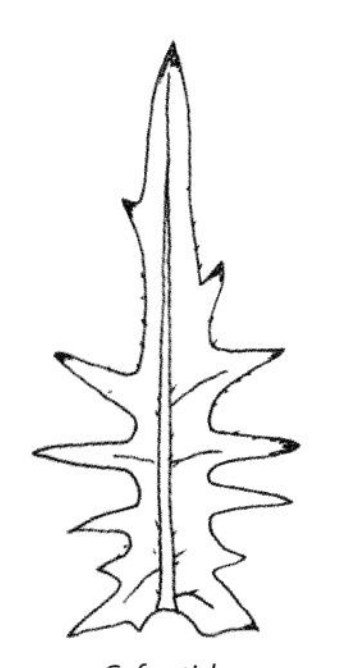

C. foetida
(Stinking Hawk's-beard)

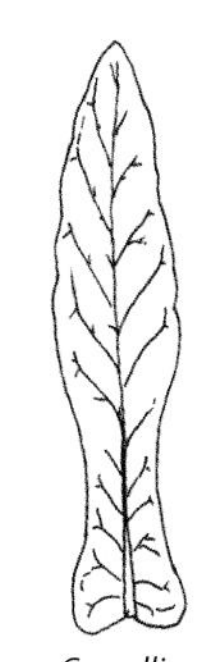

C. mollis
(Northern Hawk's-beard)

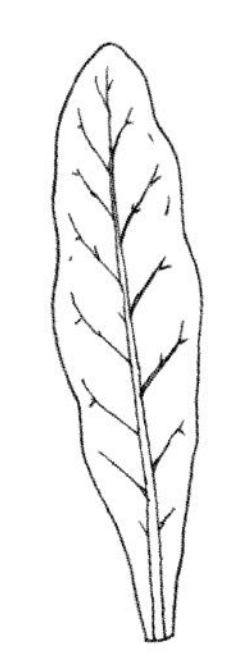

C. praemorsa
(Leafless Hawk's-beard)
*Rosette leaf

C. paludosa **(Marsh Hawk's-beard)**
Perennial to 80 cm tall. Lacks hairs or only slightly hairy. **Stem leaves are broad, with small teeth along their margin**, and clasp the stem. **The bracts around the flowerheads are covered in many long hairs.** Grows in wet places, including wet-grasslands and damp woods. North Wales, northern England, Scotland and parts of Ireland.

Now let's look at the rarer species

C. setosa **(Bristly Hawk's-beard)**
Annual or biennial to 75 cm tall. Plants are usually hairy. **Stem leaves are narrow and usually toothed near their base**, and clasp the stem. The bracts around the flowerheads are covered in hairs. May occur as a crop weed or in other disturbed ground.

C. foetida **(Stinking Hawk's-beard)**
Annual or biennial to 60 cm tall. **As its names suggests, this plant is smelly (with an odour of bitter almonds).** Plants are hairy. When they do occur, stem leaves are few; they are usually deeply lobed, sometimes clasping the stem. The bracts around the flowerheads are hairy. Grows in rough ground, at a few sites in south east England.

C. mollis **(Northern Hawk's-beard)**
Perennial to 60 cm tall. Usually lacks hairs or is sparsely hairy. **Stem leaves usually lack lobes or teeth and widen at their base.** The bracts around the flowerheads are covered in glandular hairs. A rare plant of hills, being found at only a few sites in northern Britain.

C. praemorsa **(Leafless Hawk's-beard)**
Perennial to 60 cm tall. Usually lacks hairs or only slightly hairy. **No stem leaves.** Rosette leaves usually lack teeth and lobes. The bracts around the flowerheads are usually slightly hairy. A rare plant of calcareous grasslands in north-west England.

Hawkweeds (*Hieracium* and *Pilosella*)

All hawkweeds have dandelion-like flowerheads. Until recently, taxonomists placed all of them in the genus *Hieracium*. **Members of the genus *Pilosella* produce stolons**, their stems usually lack leaves, and the basal leaves are usually oval with smooth edges. In contrast, *Hieracium*s do not produce stolons, their stems usually have leaves and the leaves are generally lobed or toothed.

Pilosella (mouse-ear-hawkweeds)

There are several native and introduced mouse-ear-hawkweed species, two of which are common and widespread. They are all perennials which spread via stolons, often forming patches. Some *Pilosella* species always produce seeds sexually, while others sometimes produce asexual seeds. In the sexually reproducing species, hybridisation is frequent. These habits make their identification more challenging.

Let's start with the common ones

P. officinarum (Mouse-ear-hawkweed)

A common and widespread species of short grasslands. It reproduces via both sexual and asexual seeds, probably linked to its highly variable number of chromosomes. As a result, it is morphologically highly variable too. **Unbranched flowering stems support a single flowerhead.** Flowers are a lemon-yellow colour rather than the brighter, richer yellow associated with most dandelion-like plants. The flowerheads are very hairy. The leaves are a simple oval shape, usually with a few long hairs on the green upper surface. **The lower surface of the leaf is paler in colour, being covered in lots of short hairs. From above this gives the impression of a white line around the edge of the leaf.**

P. aurantiaca (Fox-and-Cubs)

This is an easy species to identify because its **flowers are a highly distinctive orange foxy-brown colour.** There are many flowerheads crowded together at the top of a single stem. A garden escape which is now common on rough ground across most of the British Isles, although is less common in north-east Scotland and Ireland.

Now let's look at the more unusual species

P. peleteriana (Shaggy Mouse-ear-hawkweed)

A rare native plant of calcareous grasslands and sea cliffs. It looks like a more solid and larger version of the common Mouse-ear-hawkweed with a single flowerhead per stalk. The flowerheads are covered in dense hairs, usually some of which are star shaped. **Stolons are short and stout and covered in crowded full-sized leaves.**

Characteristics of *Pilosella* species

	Flowerheads / stalk			Flower colour		Stolons		
Species	1	2–4	many	yellow	orange	short	long	above & below ground
P. officinarum (Mouse-ear-hawkweed)	■			■			■	
P. aurantiaca (Fox-and-Cubs)			■		■			■
P. peleteriana (Shaggy Mouse-ear-hawkweed)	■			■		■		
P. flagellaris (Spreading Mouse-ear-hawkweed)		■		■			■	
P. caespitosa (Yellow Fox-and-Cubs)			■	■				■
Your specimen								

Occurs in a handful of locations on the south coast, in the Peak District and on the Welsh border. Frequent on the Channel Islands.

P. flagellaris **(Spreading Mouse-ear-hawkweed)**
An uncommon plant of short grassy banks. **It generally has 2 to 4 flowerheads per stalk.** The flowerheads are covered in glandular hairs, non-glandular hairs and others which are star shaped (stellate). Scattered in England, but relatively frequent in an area north of Nottingham and around Glasgow and Edinburgh.

P. caespitosa **(Yellow Fox-and-Cubs)**
An introduced plant with a scattered distribution. Grows on grass banks and verges. Its stolons can often be seen above ground. **There are many flowerheads crowded together at the top of a single stem.**

Hieracium (hawkweeds)

All hawkweeds are perennials which only reproduce by producing asexual seeds, in the same way as dandelions. This results in the production of many different clonal lines. In the British Isles these have been divided into 16 sections, including more than 400 microspecies. They are not a group for the novice botanist, or for tackling out of season. But the following will help you get started. Ideally, you should look at a few specimens to avoid being baffled by atypical plants.

Some botanists have divided the hawkweeds into groups, which can be a helpful place to start:

- **Many-leaved hawkweeds** are tallish plants (usually >30 cm), with many leaves all the way up their stems which are covered in hairs. **Minimum stem leaf number = 6, and no basal leaves when flowering.** They produce clusters of many flowerheads (2–3 cm in diameter). Grow in grasslands and heaths.
- **Few-leaved hawkweeds** are medium-height plants (generally 20–100 cm tall). They produce leaves in basal rosettes and a few up their stems, which are covered in hairs. **Maximum stem leaf number = 6.** They produce clusters of a few flowerheads (2–3 cm in diameter). Occur in grasslands, in rocky ground and on walls.
- **Alpine hawkweeds** are short hairy plants. Their leaves are mostly in a basal rosette, with very few on their stems. The flowerheads (2.5–3.5 cm in diameter) are usually solitary. The grow in rocky mountainous places.

Many-leaved hawkweeds sections

- *Sabauda* common plants of roadsides and rough grasslands. Plants may grow to 1 m tall. **Lower stem leaves have stalks, upper ones do not.**
- *Hieracioides* common plants of coastal areas, sandy heaths and rocky ground. May grow to 75 cm tall.
- *Foliosa* locally common plants of grasslands and rocky places. May grow to a metre tall, **their stems are often reddish. Leaves have broad bases which clasp the stem.**

Hieracium (hawkweeds) informal groupings

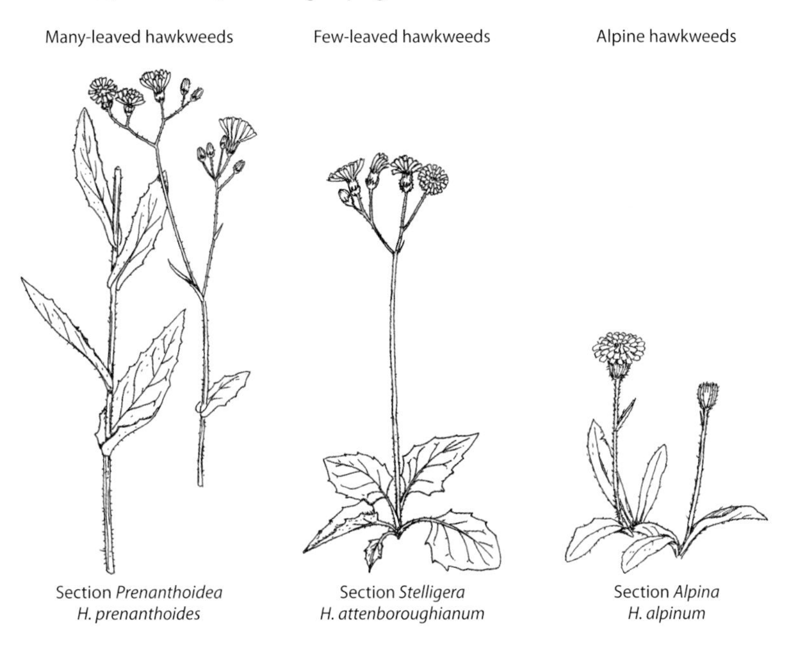

Section *Prenanthoidea*
H. prenanthoides

Section *Stelligera*
H. attenboroughianum

Section *Alpina*
H. alpinum

- *Tridentata* locally common plants of grasslands and rocky places. May grow to a metre tall, their stems are often reddish. **Leaves are narrow and do not clasp the stem.**
- *Prenanthoidea* includes some locally common species of rough grasslands and woodlands, often on limestone. **Leaves narrow towards the base before broadening and clasping the stem.**
- *Alpestria* **this is an unusual section that has characteristics of all three groups.** They have relatively few stem leaves but may grow to 1 m tall, while producing few flowerheads. In rocky places and hilly areas.

Characteristics of many-leaved *Hieracium* sections

	Stem leaves					Capitula	
Section	**range**	**hairy**	**toothed**	**clasping stem**	**petioles**	**number**	**bracts curved out**
Sabauda	15–50				lower	many	rarely
Hieracioides	>15					varied	
Foliosa	10–30				lower	varied	
Tridentata	6–30				lower	varied	
Prenanthoidea	6–30					many	
Alpestria	2–15				lower	few	
Your specimen							

Few-leaved hawkweeds sections

- *Hieracium* some of these are common plants of roadsides and rough areas. May grow to 20 cm. **Look out for long petioles and purple spots on the undersides of leaves.**
- *Vulgata* many of these are common plants of roadsides and rough grasslands. May grow to 20 cm. They have few basal leaves for a few-leaved hawkweed. Stem leaves have petioles and **many have dark purple patches on their upper leaf surfaces.**
- *Stelligera* are plants of roadsides and rocky areas, grassy banks and cliffs, often on limestone. May grow to 60 cm. **Look out for lots of blue-green basal leaves with long petioles.**
- *Oreadea* are plants of roadsides and rocky areas, grassy banks and cliffs, usually on limestone. May grow to 60 cm. **Look out for bristly basal leaves which are blue-green in colour, often with purple spots with long petioles.**
- *Amplexicaulia* are uncommon plants of rough ground. May grow to 60 cm. **Look out for basal leaves with winged petioles and stem leaves that clasp the stem.**
- *Cerinthoidea* are plants of roadsides, rocky areas, cliffs and upland grasslands. May grow to 50 cm. **Look out for up to 7 blue-green basal leaves and stem leaves that clasp the stem.**
- *Andryaloidea* are uncommon plants with few records. May grow to 30 cm. **Look out for up to 5 whitish, hairy basal leaves with winged petioles.**
- *Villosa* are uncommon plants with a few records from dunes and walls. May grow to 50 cm. **Look out for plants that are covered in lots of long white hairs and with few flowerheads.**

Characteristics of few-leaved *Hieracium* sections

	Basal leaves			Stem leaves					Capitula	
Section	<5	hairy	petioles	range	hairy	toothed	clasping stem	petioles	number	bracts curved out
Hieracium		■	■	0–2	■			■	2–20	
Vulgata	■	■		2–8	■	■		■	<5	
Stelligera			■	0–1					2–15	
Oreadea	■	■	■	2–10					<5	
Amplexicaulia		■	■	2–6	■		■		2–15	
Cerinthoidea		■	■	1–7	■	■	■		2–8	■
Andryaloidea	■	■	■	2–5	■			■	2–8	
Villosa		very		4–8	very		■		2–4	
Your specimen										

Alpine hawkweeds sections

- *Alpina* are small hairy plants of mountainous areas. **Flowerheads are solitary.**
- *Subalpina* are like those in section *Alpina*, except they have fewer rosette leaves and more stem leaves. **They may have branched stalks with more than one flowerhead.** These are also plants of hilly areas.

Characteristics of alpine *Hieracium* sections

	Basal leaves			Stem leaves					Capitula	
Section	**number**	**hairy**	**petioles**	**range**	**hairy**	**toothed**	**clasping stem**	**petioles**	**number**	**bracts curved out**
Alpina	<10			0–4					1	
Subalpina	<5			1–4					2–5	
Your specimen										

HAVE OTHERS RECOGNISED THIS LEVEL OF VARIATION?

Once you get over the initial shock of the diversity within this group, you realise that most of these plants are easy to identify. It is therefore no surprise to learn that early botanists also recognised most of these species. For the most part, historical common names mirror our modern taxonomy, with the exception of three genera that have been divided in modern times.

This relatively simple story unravels when we look at the hawkweeds. Modern British botanists have been fond of describing asexual *Hieracium* clones as species in their own right. This practice has been much less favoured elsewhere in Europe and North America.

Culpeper's herbal of 1658 mentions eight common species of hawkweed, some of which clearly correspond to species we now recognise. He comments that they have similar herbal properties (virtues), thus he has no real motivation to investigate them further.

The hawkweeds' habit of producing asexual seed was not appreciated until relatively recently. In fact, this lack of understanding contributed to the Austrian biologist Gregor Mendel's (1822–1884) early genetics work on peas being neglected for about 50 years. Unfortunately, Mendel chose to replicate his pea experiments using hawkweeds, not appreciating that their seeds are not produced through normal sexual processes. The learned monk's frustration is palpable in his writings. He attributed his failure to replicate his earlier results to the small delicate nature of hawkweed flowers and his inability to perform crosses without self-pollination occurring.

HOW FAR SHOULD I GO?

Most non-botanists would probably feel more than satisfied by being able to recognise these plants as a yellow thing that looks a bit like a dandelion but which is not actually a dandelion. Hopefully, the above has given you confidence to realise that most of these species can readily be identified with a few tips and a little practice.

You will, however, need to decide what level of resolution you are content with when looking at the asexual hawkweeds. You may wish to stop at the level of the three broad groupings used here. For some, though, hawkweeds become an addiction. Why this compulsion is so prevalent in Britain is unclear. It predates Brexit and so you are unlikely to be tarred with that brush if you go down this path.

SECTION II

Hybrids

(species that have sex with other species)

There are a few ways in which plant species can evolve. Sometimes it occurs slowly over long periods of geographic isolation; sometimes it can occur rapidly by a process of chromosome duplication; and on other occasions, species can divide within a region, when subpopulations are selected for different characteristics. The mode of speciation can greatly influence an individual plant's subsequent ability to hybridise with individuals from other populations and consequently how easy it is to classify it as one species or another.

The plant families found in this section share the ability to readily cross with other species. The definition of a species generally includes the inability to interbreed with other species; or at least when interspecies hybrids do occur, they are sterile. The commonest reason for this condition being violated occurs when species that have been ecologically or geographically separated from one another over a long period of time come back into contact before a breeding barrier has fully evolved. The usual explanations for such family reunions occurring include climate change following an ice age, or as a result of human activity transporting species around or blurring habitat boundaries.

The occurrence of hybrid individuals makes the classification of plants into neatly defined species much more difficult. In some cases, hybrids are easily identified and recorded because they have a high degree of sterility, which results from having obtained unbalanced chromosomes from their different parent species. However, in more complex cases hybrids are fully fertile and can successfully back-cross with one or both parent species. In such cases we could be said to be watching an evolutionary process in action, in which species boundaries merge and once-truly-different species reunite. Alternatively, the exact opposite may be happening, and we may be observing the process of species diverging across an ecological gradient. In these examples, populations at extreme ends of such environmental gradients may be morphologically distinct and unable to hybridise. In isolation, these extreme populations behave exactly like more conventional genetically isolated species. Yet they may be able to exchange genes indirectly via a chain of interconnected intermediate populations. When species are either gradually dividing or reuniting through hybridisation, we probably need to define species boundaries on a case-by-case basis.

CHAPTER 7

Docks and sorrels

Most people are familiar with these short-lived herbaceous perennial plants because they are commonly found growing close to human habitation, in gardens and fields. Docks are usually considered weeds, except when being applied as an ineffective folk remedy (plantains are far more effective) to ease nettle stings. In contrast, sorrels are occasionally cultivated for culinary purposes. However, there is no real botanical difference between the two: sorrels are in fact just small docks. Together there are around 20 species found growing in Britain, which for the most part are not too challenging to identify. They are included here because of their fondness for hybridisation and because there are many introduced species that are probably under-recorded.

WHY IS THIS GROUP OF PLANTS COMPLEX?

Docks and sorrels are sometimes overlooked because they lack petals. This absence does not help in their identification, because petal shape and colour are often critical features in the process. **Instead of petals, the outer parts of their flowers are small, lack colourful pigments and are called tepals.** The most important differences used to characterise members of this family involve their fruits.

The species are generally quite distinct, but these differences don't instantly jump out at the observer. You will need to use a hand lens, take a little time and know what you are looking for.

Earlier botanists inferred the taxonomic relationships between these species from the structure of their seed capsules. More recent DNA analysis suggests that these groupings were wrong, and that the species are more closely related than was previously thought. The earliest members of this family appear to have had hermaphrodite flowers and to have chromosomes in sets of ten. Over time, one, two or three of these chromosomes have been lost, and this has been followed by periods of evolution linked to the duplication of entire sets of chromosomes. During this process, some species have developed XY sex chromosomes (rather like those in humans) and individual plants have entirely male or entirely female flowers. When chromosome duplication occurs, only the X chromosomes appear to increase in number, so that XXY and XXXY plants are formed. In these cases, a reduction in the production of male flowers seems to be linked to the ratio of X and Y chromosomes. In other species of *Rumex*, there are two different Y chromosomes and male plants have XY_1Y_2. In this lineage, the Y chromosomes appear to have been duplicated in the same ratio as the others, and the expression of sex is more straightforward. Given this level of evolutionary complexity, it is not surprising that earlier taxonomists made a few mistakes.

The ecology of *Rumex* ensures that they thrive in habitats that are highly modified by humans. For this reason, these species have tended to move around the globe in our footsteps. Thus, in addition to our native docks there are several introduced species that occasionally occur – apparently at random. So, it's always worth keeping an eye out for exotic aliens.

A final level of complexity is added by the fondness of these species for hybridising. There are more than twice as many different hybrid combinations that have been found as there are pure species. Fortunately, the hybrids are sterile and rarely occur except as isolated plants, usually with one or both parent species.

HOW CAN I TELL THEM APART?

As docks lack dramatic flowers, then we need to use other features to differentiate these species. The sorrels can be readily identified by their smaller, distinctive lobed leaves. Similarly, the Curled and Fiddle Docks have distinctive leaves. However, for the most part, these species are separated on the morphology of their seed capsules. These rather beautiful and intricate structures measure just a few millimetres across, so you will probably need a hand lens. You have to look for bumps at the base of the seed capsules. These are called tubercles. It is important to note if one or all three of them is swollen.

Those not familiar with this group sometime misidentify the unrelated Horse-radish (*Armoracia rusticana*) as a dock. While Horse-radish leaves are superficially similar, they are more robust, have a rather thick mid-vein and arise from a creeping rhizome, rather than from a rosette. They also smell faintly of Horse-radish if crushed.

Let's start with the commoner more widespread species

R. acetosa (Common Sorrel)

This is a common plant that occurs in old grasslands across the British Isles. It is a highly variable species, with several subspecies being described. Plants are either male or female. Flowering stems have only a few or no branches and grow to a height of 50 cm (occasionally taller in more fertile, tall grasslands). Flowers have tepals that measure 2.5–4 mm. The seed capsules are a roundish heart shape, with small swollen bumps (tubercles) near their base. **The fleshy leaves grow up to 10 cm long and are dock-like but with two backward-pointing lobes at their base that often overlap each other.**

Characteristics of native docks and sorrels

Species	Curled leaf margin	Leaf bases			Number of tubercles			Seed capsules toothed
		lobed	round	tapered	0	1	3	
R. acetosa (Common Sorrel)		■			■			
R. acetosella (Sheep's Sorrel)		■			■			
R. obtusifolius (Broad-leaved Dock)			■			■		■
R. crispa (Curled Dock)	■		■			■	■	
R. conglomeratus (Clustered Dock)			■				■	
R. sanguineus (Wood Dock)			■			■		
R. palustris (Marsh Dock)				■			■	■
R. maritimus (Golden Dock)				■	■		■	■
R. hydrolapathum (Water Dock)				■				
R. longifolius (Northern Dock)	■			■				
R. pulcher (Fiddle Dock)			■				■	■
R. alpinus (Monk's-rhubarb)			■		■			
R. rupestris (Shore Dock)			■				■	
R. aquaticus (Scottish Dock)			■		■			
Your specimen								

R. acetosella (Sheep's Sorrel)

A common species that occurs in short acid grasslands across the British Isles. It is highly variable, with two subspecies in Britain. Plants are either male or female. Flowering stems are branched and grow to a height of 30 cm. The flowers are small, red-orange and can be highly variable, with tepals that measure up to 2 mm. The seed capsules are small, round and easily dislodged from the stem. **The leaves are shaped like arrowheads, with two sideways-pointing lobes at their base.**

R. obtusifolius (Broad-leaved Dock)

This very common species is found just about everywhere, although it is absent from a few areas of Ireland. Grows in enriched soils, in both disturbed ground and rough grasslands, where it can reach up to 1.5 m tall, but it is frequently shorter when cut. An erect perennial with branched stems. The leaves are broad and shiny with an undulating margin. The base of the leaves is heart shaped; the lower leaves may be up to 25 cm long. On the undersurface, the leaf veins are often hairy. Its flowers are hermaphrodite, with tepals that are 3–6 mm long. **The seed capsules are variable; the native form is roughly an elongated triangle with several short teeth at the base, with a single swollen tubercle (bump) and two non-swollen tubercles.**

R. crispa (Curled Dock)

This very common species is found just about everywhere, although it is absent from a few areas of Ireland and the highlands of Scotland. It often grows in

Details of seeds of Docks and Sorrels

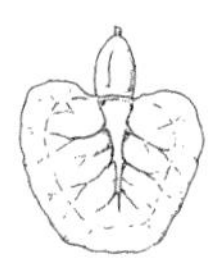

R. acetosa
(Common Sorrel)

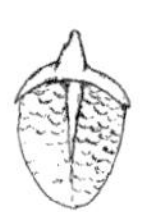

R. acetosella
(Sheep's Sorrel)

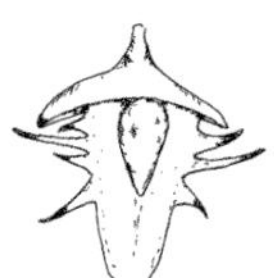

R. obtusifolius
(Broad-leaved Dock)

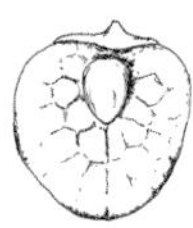

R. crispa
(Curled Dock)

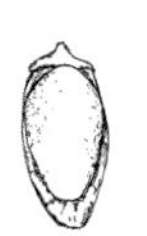

R. conglomeratus
(Clustered Dock)

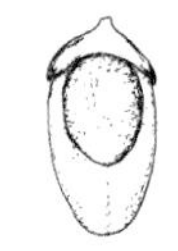

R. sanguineus
(Wood Dock)

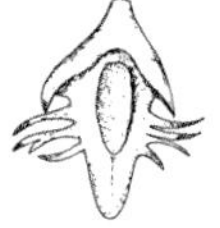

R. palustris
(Marsh Dock)

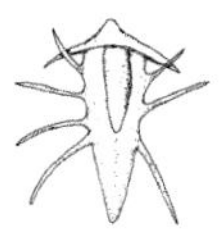

R. maritimus
(Golden Dock)

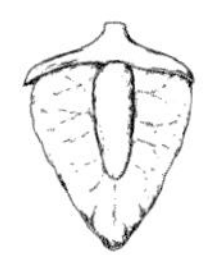

R. hydrolapathum
(Water Dock)

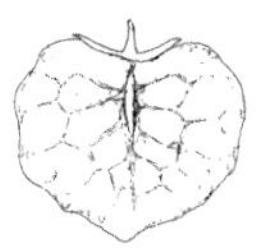

R. longifolius
(Northern Dock)

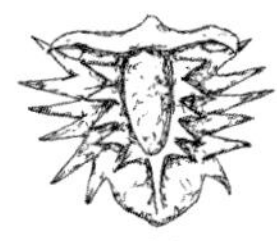

R. pulcher
(Fiddle Dock)

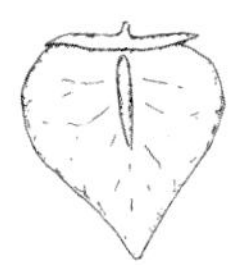

R. alpinus
(Monk's-rhubarb)

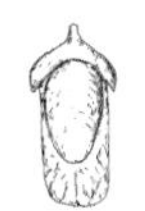

R. rupestris
(Shore Dock)

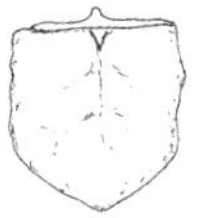

R. aquaticus
(Scottish Dock)

R. scutatus
(French Sorrel)

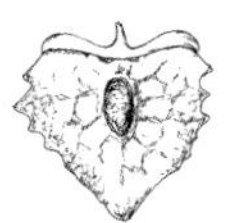

R. cristatus
(Greek Dock)

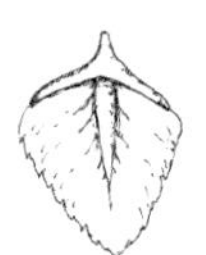

R. salicifolius
(Willow-leaved Dock)

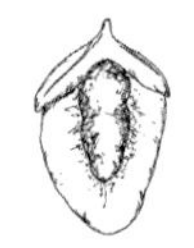

R. cuneifolius
(Argentine Dock)

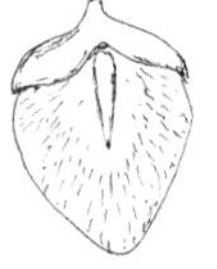

R. confertus
(Russian Dock)

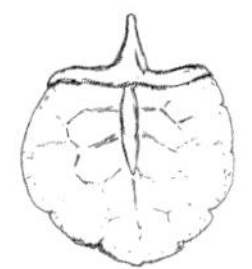

R. patientia
(Patience Dock)

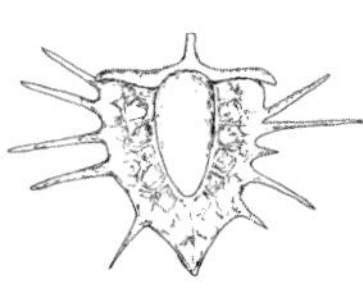

R. dentatus
(Aegean Dock)

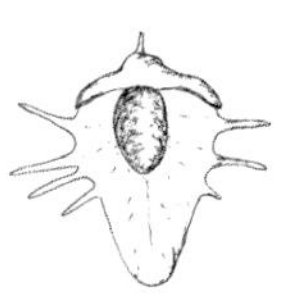

R. obovatus
(Obovate-leaved Dock)

R. brownii
(Hooked Dock)

enriched soils, in both disturbed ground and around farms, where it can reach up to 1 m tall. An erect perennial with branched stems. **Its leaves are distinctively long and rather narrow, with a curled or wavy margin.** The leaf bases are round; the lower leaves may be up to 30 cm long, stem leaves are much shorter.

Flowers are hermaphrodite, with tepals 3–6 mm long. **The seed capsules are roughly heart shaped, with a smooth margin. At the base there is typically a single swollen bump, or tubercle, and two non-swollen tubercles. However, there is a maritime variant that has three swollen tubercles and narrow, rather fleshy leaves.**

R. conglomeratus (Clustered Dock)
A widespread species, although much less frequent in Scotland, a few areas of Ireland and central Wales. Typically grows in waterlogged soils adjacent to water. Can reach up to 60 cm tall, occasionally taller. It is a short-lived perennial; **a less robust plant than the Broad-leaved and Curled Docks.** Its leaves are long and narrow, with a slightly undulating margin. The base of the leaves is round; the lower leaves may be up to 20 cm long. Flowering stems have much smaller leaves. **The flowering stems are distinctly wavy, with many branches at an angle of 30–90°.** The flowers are hermaphrodite, with tepals 2–3 mm long. **The seed capsules are small, narrow and untoothed, each with three well-developed swollen bumps (tubercles).**

R. sanguineus (Wood Dock)
Another widespread species, although it is much less frequent in Scotland and some areas of Ireland. A plant of damp shady areas, often in woods and hedgerows, where it can grow up to 1 m tall. This is a rather delicate, erect perennial that **looks somewhat like the Clustered Dock, except its branches arise at an angle of less than 30° and there are no leaves between the whorls of flowers.** Its leaves are long and fairly narrow, with a slightly undulating margin. The base of the leaves is round; the lower leaves may be up to 40 cm long. The leaf veins are usually green, but they may be red or even purple. Flowers are hermaphrodite, with tepals 3–5 mm long. **The seed capsules are small, narrow and untoothed. At the base is a single well-developed swollen bump, or tubercle, while the other tubercles are small or absent.**

R. palustris (Marsh Dock)
Almost entirely a plant of the east of England, with outliers on the Somerset Levels and Gwent Levels. It is very rare in Wales and Scotland and absent from Ireland. Typically grows in waterlogged soils adjacent to water. A short-lived perennial that can grow up to 60 cm tall, but occasionally up to 1 m. Its leaves are narrow and willow-like, with a very slightly undulating margin. The base of the leaves is tapered; the lower leaves may be up to 30 cm long. Flowering-stem leaves are smaller and narrow. The flowering stems are much branched; **these branches are widely spreading, before curving upwards.** Its flowers are hermaphrodite, with tepals 3–4 mm long. **The seed capsules are long and narrow with a few long teeth at their base; each capsule has three relatively large swollen bumps (tubercles).** Sometimes confused with Golden Dock, but **the dry flower spikes of this species are rusty brown rather than golden brown.**

***R. maritimus* (Golden Dock)**

Occurs scattered across much of England, but is much less common in Wales, Scotland and Ireland. It typically grows in waterlogged soils adjacent to water. An annual or short-lived perennial that can grow up to 60 cm tall, usually less. **The plant is a distinctive yellow-green colour with golden-yellow flowers.** Its willow-like leaves are long and narrow. The base of the leaves is tapered; the lower leaves may be up to 20 cm long. **Flowering stems are much branched and support many small narrow leaves and flowers.** Its flowers are hermaphrodite, with tepals 2.5–3 mm long. **The seed capsules are long, narrow with long teeth, each with an elongated swollen bump (tubercle).**

***R. hydrolapathum* (Water Dock)**

A widespread species in much of England, but far less common in Wales, Scotland and Ireland. As the name suggests, it is a species of water margins and wetlands, but does not persist in closed vegetation. **An impressive erect perennial plant that may grow to as much as 2 m tall.** Its huge leaves are long and broad, with a slightly undulating margin. The base of the leaves is somewhat tapering; the lower leaves may be 1 m long or more. Its flowers are hermaphrodite, with tepals 5–8 mm long. **The seed capsules are found in dense clusters, are triangular in shape and untoothed. At the base there are three elongated swollen smooth bumps (tubercles).**

***R. longifolius* (Northern Dock)**

As its name implies, this species is widespread in northern England and Scotland, but absent from southern England, Wales and Ireland. Typically grows in wet open ground, often adjacent to water. A robust erect perennial that can reach up to 120 cm in height. **Its leaves are long and narrow, with a wavy (curled) margin rather like the Curled Dock.** The bases of the leaves are somewhat tapering; the lower leaves may be up to 80 cm long. **Flowering stems have short erect branches which form a dense mass of flowers.** Its flowers are hermaphrodite, with tepals 4.5–5 mm long. **The seed capsules are round and untoothed, and lack swollen bumps (tubercles).**

***R. pulcher* (Fiddle Dock)**

Widespread south of a line between the Severn Estuary and the Wash, but rare elsewhere. A perennial that grows in old dry grasslands where it can reach 50 cm tall, but much wider because it is **much branched with the branches at an angle of 90°. Its leaves are highly distinctive in having a 'waistline' so they somewhat resemble the shape of a violin (hence the common name).** The lower leaves may be up to 20 cm long. Its flowers are hermaphrodite, with tepals 4–5.5 mm long. **The oval seed capsules are warty and have several usually long teeth. Each capsule has a textured swollen bump (tubercle).**

R. alpinus **(Monk's-rhubarb)**
A scarce species primarily found in eastern Scotland and northern England, absent from Wales and Ireland. **An erect rhizomatous perennial, unusual for a larger dock in that it forms patches.** Grows in grassy places near buildings and roads, where it can reach a height of up to 80 cm, occasionally taller. Its leaves are long and broad, with a wavy margin. The base of the leaves is round; the lower leaves may be up to 80 cm long. **The flowering stems have only few branches.** Its flowers are hermaphrodite, with tepals 5–6 mm long. **The seed capsules are round, untoothed and lack swollen bumps (tubercles).**

There are a couple of rarer native docks to consider

R. rupestris **(Shore Dock)**
A rare coastal plant of Devon, Cornwall, South Wales and Anglesey. An erect, branched perennial plant that can grow up to 50 cm, occasionally taller. Its leaves are long and narrow, with a wavy margin. The base of the leaves is round; the lower leaves may be up to 20 cm long. **The flowering stems branch at an angle of 25–50°.** Its flowers are hermaphrodite, with tepals 3–4 mm long. **The seed capsules are small, narrow and untoothed, each with three large well-developed swollen bumps (tubercles).**

R. aquaticus **(Scottish Dock)**
A very rare species only found around Loch Lomond. **A tall erect perennial that can reach as much as 2 m in height.** Its leaves are long and broad, with an undulating surface and margin. Leaves are broadest at the base, which is round and may reach a length of up to 1 m. The flowering stems have much smaller leaves and **change angle at every branch.** Its flowers are hermaphrodite, with tepals 5–8 mm long. **The seed capsules are long, triangular, untoothed and lack swollen bumps (tubercles).**

Finally, you may have one of the rarer and more unusual introduced species

In addition to our native docks, there are many introduced species that occasionally turn up more or less randomly across the British Isles. These are probably under-recorded. If you find a specimen that does not neatly fill one of the above descriptions, you may have stumbled upon something more exotic.

R. scutatus **(French Sorrel)**
This introduced species is occasionally found in rough ground, most frequently in north-west England. It has never been recorded in Ireland. Plants are either male or female. Flowering stems are highly branched and grow to a height of 50 cm. Flowers have tepals that measure 5–8 mm. The flowers have a red-pink margin and green centre. **The seed capsules are heart shaped, and lack small**

swollen bumps (tubercles) near their base. The leaves grow up to 10 cm long; they are dock-like, but with two very broad lobes at their base which point outwards.

R. cristatus **(Greek Dock)**
This is one of our more abundant introduced species; it is widespread in south-east England but with very few records elsewhere. Grows to a height of 2 m. Its leaves are long and wavy. **The seed capsules are large and shield shaped, with a markedly toothed margin and a single large round tubercle.**

R. salicifolius **(Willow-leaved Dock)**
Seeds of this species seems to have been introduced with grain. There are very few records, largely around London and a few other cities, and mostly not recent. Grows to 50 cm tall, and as its name suggests, has willow-like leaves. **Its seed capsules are shield shaped, lack teeth and have swollen warty bumps on all three sides.**

R. cuneifolius **(Argentine Dock)**
This South American species is restricted to a few sand dunes in south-west England and South Wales. It grows to a height of 30 cm. Its leaves are oval and tough. **Its seed capsules are narrow, oval and lack teeth; each capsule has a large swollen bump.**

R. confertus **(Russian Dock)**
There are just a couple of records of this species from south-east England. It grows to 1.2 m tall. Its leaves are long and broad. The flowers are relatively large, with tepals measuring 6–9 mm. **The seed capsules are also large, smooth, lack teeth and have only a single small tubercle.**

R. patientia **(Patience Dock)**
Mostly found in urban areas, particularly London and northern England. It grows to a height of 2 m. Its leaves are long with a tapered base. **The seed capsules are large and round with an intact margin and a single small, smooth tubercle.**

R. dentatus **(Aegean Dock)**
There are a few widely dispersed records of this species, which is absent from Ireland. Its seeds are thought to have arrived as a contaminant of wool. Grows to 70 cm tall. Its leaves are long and wavy. **The seed capsules are triangular, with several long teeth and one or sometimes three smooth tubercles.**

R. obovatus **(Obovate-leaved Dock)**
There are very few records of this species, mostly from southern England. Its seeds are thought to have arrived as a contaminant of grain. Annual that grows to 40 cm tall. Its leaves are rounded at the end. **The seed capsules are triangular, with short teeth and three warty tubercles.**

R. brownii (Hooked Dock)
This species has not been recorded for some time. Its seeds are thought to have arrived from Australia as a contaminant of wool. Thus, it was previously found in northern woollen mill towns. **It has highly distinctive triangular seed capsules which have prominent hooked teeth along their margins and narrow tubercles.**

HAVE OTHERS RECOGNISED THIS LEVEL OF VARIATION?

Culpeper's *Herbal* of 1798 simply says about dock: 'Many kinds of these are so well known, that I shall not trouble you with a description of them.' From a herbalist's perspective, he considered them all to have the same virtues and thus only bothered to include a description of the Common (or Broad-leaved) Dock. Basically, if you are looking for a dock leaf to rub on a nettle sting, it really does not matter which species you choose, as their effects are psychosomatic. However, Culpeper does include both Common Sorrel and Sheep's Sorrel elsewhere in the text – you don't want to make sorrel sauce from a common dock!

Most of our native docks were well known to and described by Linnaeus. Bentham and Hooker's *British Flora* of 1858 includes ten of the 14 native species covered here. Interestingly, they included *Rumex aquaticus* (calling it Smooth-fruited Dock, now usually called Scottish Dock), which according to some reports was not discovered until 1935.

Although most of our native docks have been well known for a long time, their taxonomy has changed radically – with previous subgroupings being merged back into the genus *Rumex* as DNA evidence has replaced morphological data from the seed capsules.

HOW FAR SHOULD I GO?

This group is common and widespread. Despite or perhaps because of this, these plants have been somewhat overlooked. This situation has probably been exacerbated by their lack of showy flowers. For these reasons, many of the introduced species are probably under-recorded – particularly because they often turn up in unattractive urban habitats.

There are a few reasons to reverse the neglect that these species have experienced. Firstly, they are potentially agricultural weeds, and it is important to know if newly introduced species are actively spreading. Secondly, recording the occurrence of such species tells us more about the social history of our islands. For example, species whose seed contaminated imported wool are still associated with northern former mill towns. And finally, there is an elegant beauty to their seed cases that will reward anyone who spends the time to admire them.

When looking for unusual alien docks, it is worth being aware that hybrids are not uncommon. Their hybrid nature is easy to establish because they are sterile and produce no seeds. They nearly always occur as single hybrid plants growing with one or both their parents. Hybrid populations that back-cross and merge with the parent species are unknown.

CHAPTER 8

Pondweeds

These aquatic plants (mostly in the genus *Potamogeton*) grow in water of varying depths and flow speeds, where they may be fully submerged, floating or a combination of submerged and floating. Their floating leaves are opaque and broad, and their submerged leaves are usually long, narrow and frequently translucent. Both types of leaves have distinctive parallel veins. These plants have characteristic stipules, which are sheaths of tissue found at the base of their leaves that act to protect the young developing foliage. Plants are long-lived, some overwintering as rhizomes, some with resting buds technically known as turions. Although you may not have encountered some of the truly aquatic species, others are commonly found in garden ponds and in shallow pools in boggy areas. They are considered one of the more difficult groups of plants to identify; however, their reputation is rather worse than the reality.

WHY IS THIS GROUP COMPLEX?

As with most complex groups of plants, there are multiple reasons that pondweeds can be challenging. Flowering can be sporadic in these species, so they are primarily identified using leaf characters. Unfortunately, leaf shape and size are highly variable and greatly influenced by water depth and flow speed. Secondly, different populations of the same species can look rather different from each other, because they are generally founded by the chance movement of a fragment of plant material that subsequently dominates a waterbody, with a single genotype persisting and accumulating mutations over a long period of time. Thirdly, the aquatic nature of these plants might be a deterrent to some botanists. Sampling in deep water or in fast-flowing rivers can be tricky. Finally, this group is prone to hybridisation. As long-lived perennials, sterile hybrids can survive for many years. Some plants are thought to be hybrids between native species and formerly native species that have not lived here for many centuries. In addition, their reluctance to flower can hide their sterility. Together, all this makes it rather difficult when you are trying to determine their hybrid origins.

HOW CAN I TELL THEM APART?

As mentioned, pondweeds are primarily identified based on vegetative characters. To tell them apart you will need to determine if they have both floating leaves and submerged leaves, or just submerged leaves. It is important that you look closely at the leaf shape. Take note of the leaf base, the leaf tip and the length of the leaf stalk. It can be important to measure length of the stipule (the tube of protective material where the leaf joins the stem), so a small ruler is handy. With many species you will need to look very closely at the leaf veins, so a hand lens will be useful, and occasionally even a microscope.

Photographing these plants for later identification is not always practical, so you may wish to collect and preserve samples. This is also not straightforward. With fully aquatic species, you will need to float your specimen in a tray of water, arranging it onto a piece of paper (it would be difficult/impossible to arrange the specimen so you can see the important features once it is extracted from the water). Once removed from the tray you will need to use a sheet of baking parchment to prevent your sample from sticking to the upper sheet of paper. The sample can then be pressed as with any terrestrial species, but the intermediate absorbent sheets will need more regular changing.

When attempting to identify pondweeds, the first thing to do is determine which of the three groups below your specimen belongs to. **However, care is needed: in deep water, floating leaves may be absent and, conversely, in dry periods, submerged leaves can dry out and rot away.**

- **Floating-leaved pondweeds:** These plants have opaque floating or aerial leaves that are generally broad and wider than 1 cm and invariably widest in the middle of the leaf. They usually also have submerged leaves that are narrower than their floating leaves.
- **Broad-leaved submerged pondweeds:** The leaves of these plants are always submerged and often translucent. The leaves are usually wider than 1 cm and have more than three longitudinal leaf veins each side of the midrib.
- **Narrow-leaved pondweeds:** The leaves of these plants are usually narrow and thread-like with parallel sides. The leaves are usually less than 5 mm across and have one or two longitudinal veins each side of the midrib. When identifying these species, you will need to look carefully at the longitudinal veins as they approach the leaf tip.

Let's start with the floating-leaved pondweeds

Potamogeton natans **(Broad-leaved Pondweed)**
One of our commonest pondweeds, found throughout Britain and Ireland in ponds, rivers and drainage ditches. It has both opaque floating leaves and very long narrow, opaque submerged leaves. The stalks of the floating leaves measure 15–60 cm and the leaf bases are rounded. **At the point where the floating leaf**

Characteristics of floating-leaved pondweeds

Species	Floating leaf			Submerged leaf tip pointed
	leaf stalk longer than blade	leaf base rounded	leaf tip pointed	
P. natans (Broad-leaved Pondweed)				
P. polygonifolius (Bog Pondweed)				
P. alpinus (Red Pondweed)				
P. gramineus (Various-leaved Pondweed)				
P. coloratus (Fen Pondweed)				
P. nodosus Loddon Pondweed)				
P. epihydrus (American Pondweed)				
Your specimen				

attaches to the stalk, there is a distinctive slightly thickened collar that allows the leaf to articulate when the water level changes. The floating leaves are elliptical and measure 2.5–12 cm long by up to 5 cm wide. There are about 20 longitudinal veins but the cross-veins are indistinct. **The submerged leaves are almost reduced to a midrib and look like a leaf stalk.** They measure 15–30 cm long but less than 3 mm wide. **Occasionally there are no floating leaves, and these plants can be classed as being narrow leaved and submerged.** The stipules are usually separate from the leaf and 5–12 cm long. The plant's main stems measure up to 100 cm.

P. polygonifolius (Bog Pondweed)

Another very common pondweed that is found across Britain and Ireland, although it is less frequent in eastern England. It occurs in acid conditions in bogs, shallow ponds and streams. It has both opaque floating leaves and narrow, translucent submerged leaves. The stalks of the floating leaves measure about 10 cm. The base of the floating leaves is tapered. **The floating leaves are variably elliptical and measure 2–6 cm long by up to 4 cm wide; they have about 20 longitudinal parallel veins and visible cross-veins.** Submerged leaves are translucent and narrowly elliptical with a rounded apex. They measure 8–20 cm long and 1–3 cm wide. The stipules are blunt and 2–4 cm long. The leafy stems measure 20–50 cm.

P. alpinus (Red Pondweed)

Occurs across Britain and Ireland, although it is less frequent in the south and west. It is usually found in still waters associated with peaty, acid conditions. **Sometimes fully submerged so may be classed in the next group.** It may have both opaque floating leaves and narrow, translucent submerged leaves. **The floating and submerged leaves can be rather similar, and both types of leaf have short or no stalks and a reddish tinge. The base of the leaf is tapered and typically has prominent, transparent air cells.** The floating leaves have an elongated elliptical

Floating and submerged leaves of floating pondweeds

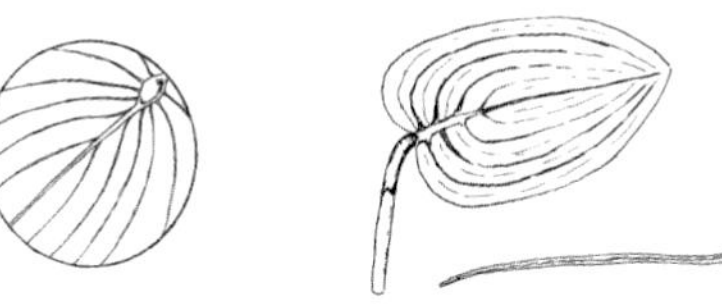

Potamogeton natans (Broad-leaved Pondweed)

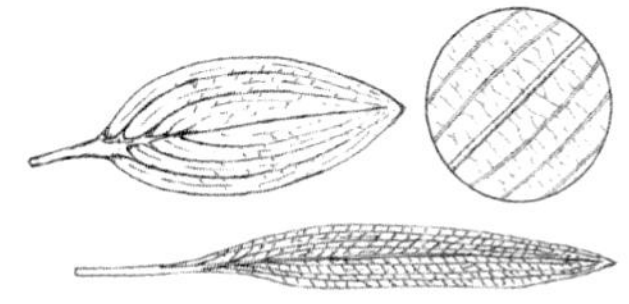

P. polygonifolius (Bog Pondweed)

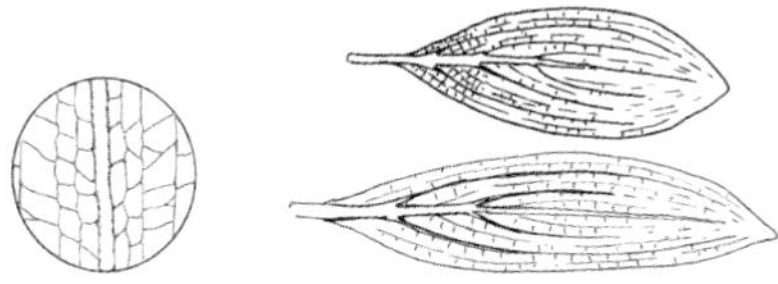

P. alpinus (Red Pondweed)

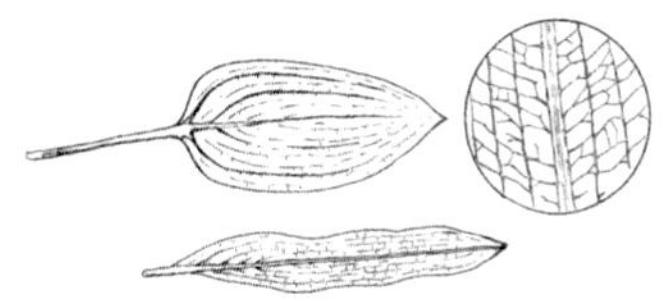

P. gramineus (Various-leaved Pondweed)

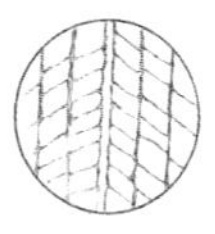

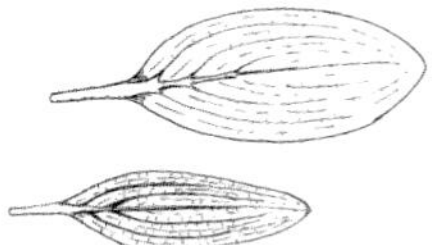

P. coloratus (Fen Pondweed)

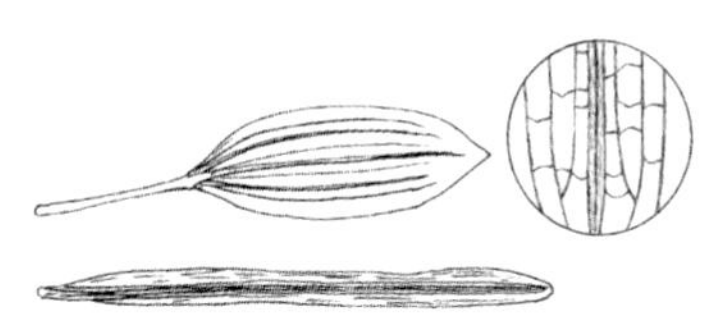

P. epihydrus (American Pondweed)

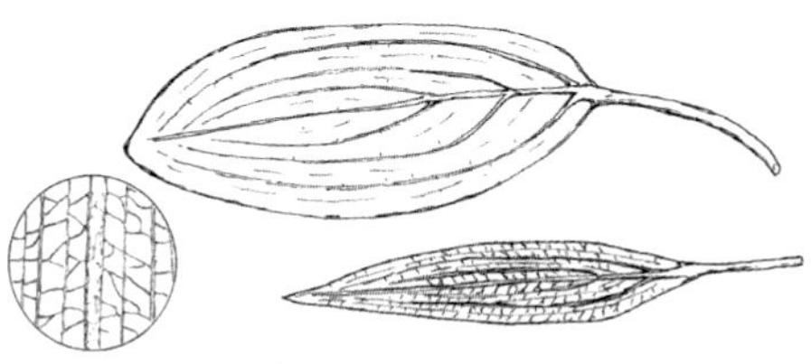

P. nodosus (Loddon Pondweed)

shape and measure 3–8 cm long by up to 3.5 cm wide. The submerged leaves are translucent and narrowly elliptical with a rounded apex. They measure 6–20 cm long and 1–3 cm wide, and have a midrib and 6 to 10 longitudinal parallel veins and clearly visible cross-veins. The stipules are robust, blunt and 2–6 cm long. The leafy stems measure up to 2 m.

P. gramineus **(Various-leaved Pondweed)**
Found across Britain and Ireland, although less frequent in the south, it usually occurs in still waters associated with acid conditions. **Sometimes fully submerged and lacking floating leaves and may thus be classed in the next group.** In shallower water, has both opaque floating leaves and narrow, translucent submerged leaves. The stalks of the floating leaves are usually longer than the leaves themselves. The base of the floating leaves is rounded. The floating leaves are oval and measure 3–9 cm long by up to 4 cm wide. They have about 19 longitudinal parallel veins and clearly visible cross-veins. The submerged leaves are translucent and narrowly elliptical with a rounded apex, they measure 2.5–8 cm long and 0.5–3 cm wide. They have a midrib and 4 to 10 longitudinal parallel veins and have clearly visible cross-veins. **Each side of the midrib are one or two rows of empty cells, giving the appearance of a translucent band.** The stipules are lanceolate with rounded tips, and 2–5 cm in length. The leafy stems are highly branched and measure up to 1 m.

P. coloratus **(Fen Pondweed)**
Scattered across Britain and Ireland, although most frequent in Ireland and eastern England. Usually found in fenland ponds and ditches. It has both opaque floating leaves and narrow, translucent submerged leaves. **Floating leaves may be just below the surface and look similar to the submerged leaves.** The stalks of the floating leaves are shorter than the leaf length and submerged leaves are almost stalkless. The base of the floating leaves is rounded. The floating leaves are oval and measure 2–8 cm long by up to 5 cm wide. They have up to 20 prominent longitudinal parallel veins and a **distinctive network of highly visible cross-veins.** The submerged leaves are translucent, long and narrow. They measure 6–15 cm long and 1–3 cm wide. The stipules are blunt and 2–4 cm long. The leafy stems measure 20–50 cm. Has a particularly 'sepia' hue.

P. nodosus **(Loddon Pondweed)**
Only found in a few rivers in the south of England. It has both opaque floating leaves and narrow, translucent submerged leaves. **The leaf stalks are often much longer than the leaf blade.** The base of the floating leaves is tapered. These floating leaves are long and oval, measuring 6–15 cm long by up to 6 cm wide; they have prominent longitudinal parallel veins and visible cross-veins. The submerged leaves are translucent, long and relatively wide, with a tapering base. They measure 10–20 cm long and 1.5–4 cm wide. The stipules are long, lanceolate and 7–10 cm in length. The unbranched leafy stems measure up to 30 cm.

P. epihydrus (American Pondweed)

This introduced pondweed is only found in a few sites in the Outer Hebrides, Skye and a few canals in Lancashire and Yorkshire. It has both opaque floating leaves and narrow, translucent submerged leaves. The floating leaves have short stalks that are similar in length to the leaf blade. The floating leaves have tapered bases; they are narrowly elliptical and measure 3–8 cm long by up to 3 cm wide. They have a midrib and about six longitudinal parallel veins without visible cross-veins. The submerged leaves are translucent and ribbon-like with a rounded apex; they measure 8–24 cm long and 1–2 cm wide. **The submerged leaves have a band of thicker tissue along their midrib, which appears paler.** The stipules are flattened, generally rounded and 3.5 cm long. The main stems are flattened, usually unbranched, and measure up to 1 m.

Characteristics of broad-leaved submerged pondweeds

	Submerged leaves			
Species	**opposite**	**cross-veins visible**	**leaf base round**	**leaf tip pointed**
P. crispus (Curled Pondweed)				
P. lucens (Shining Pondweed)				
P. praelongus (Long-stalked Pondweed)				
P. perfoliatus (Perfoliate Pondweed)				
G. densa (Opposite-leaved Pondweed)				
P. alpinus (Red Pondweed)*				
P. gramineus (Various-leaved Pondweed)*				
Your specimen				

* Red Pondweed and Various-leaved Pondweed may have both floating and submerged leaves or just submerged leaves.

Next the broad-leaved submerged pondweeds

Potamogeton crispus (Curled Pondweed)

This fully submerged species is common across most of Britain and Ireland except the north of Scotland and Mid Wales. It is found in both still and flowing water. **It only has submerged leaves, which are highly distinctive, having a wavy margin and a serrated edge and lacking a stalk.** They are an elongated oblong shape, 3–10 cm long and up to 1.5 cm wide. **Given that the leaves are relatively narrow, it might be considered a narrow-leaved pondweed.** The leaves are shiny or translucent, with a midrib and 2 or 4 longitudinal veins. The stipules are 1–2 cm long, but they rapidly break down and are lost. The stems can grow up to 120 cm long; they are flattened, often with angled sides, with the broader side furrowed when mature.

P. lucens (Shining Pondweed)

This submerged species is common across most of England and Ireland but is uncommon in Scotland and Mid Wales. Found in both still and slow-flowing rivers

Leaves of broad-leaved submerged pondweeds

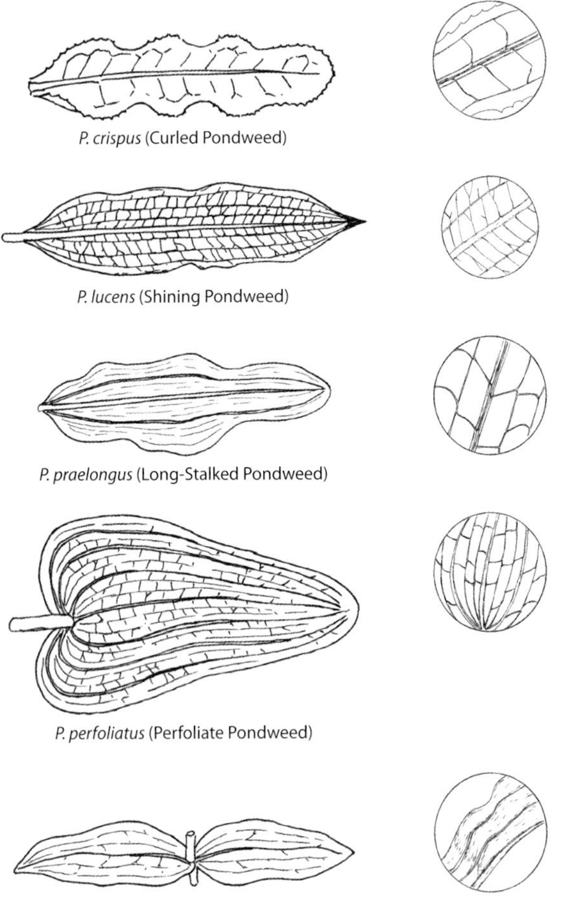

in base-rich areas. **Only has submerged leaves, which are distinctively translucent and glossy.** These are lanceolate, oblong shaped and measure 10–20 cm long and up to 6 cm wide. **The leaves have a short stalk and about three clear longitudinal veins and a similar number of faint veins on each side of the midrib.** There is a clear network of cross-veins visible. Stipules are 3–8 cm long, with a blunt keel-shaped tip. The stems are tough and can be very long, growing up to more than 2 m.

***P. praelongus* (Long-stalked Pondweed)**

This submerged species occurs across most of Britain and Ireland, most common in the north and absent from south-west England. Found in both still and flowing waters. It only has submerged leaves, which are translucent but not shiny. They are an elongated, oblong shape and measure 6–18 cm long and up to 5 cm wide, usually with a distinctly wavy margin. **The leaves lack stalks and usually have one clear longitudinal vein on each side of the midrib plus 5 to 7 less distinct veins.** The stipules are 0.5–6 cm long, often extending longer than the next leaf, with a blunt tip. The stems are branched and very long, growing up to more than 2 m. **The common name is related to the flower stalk, which can be up to 40 cm long – extending well above the water surface.**

P. perfoliatus **(Perfoliate Pondweed)**
This submerged species occurs across most of Britain and Ireland. Found in both still and flowing waters. It only has submerged leaves, which are thin and translucent. They are oval in shape and measure 2–6 cm long by up to 4 cm wide. **The leaves lack stalks, but instead the base of the leaves clasp around the stem.** The leaves usually have three or four clear longitudinal veins on each side of the midrib (which may not be obvious), plus a few less distinct veins. The stipules are about 1 cm long, are very delicate and soon disappear. The stems are branched and very long, growing up to more than 2 m.

Groenlandia densa **(Opposite-leaved Pondweed)**
This highly distinctive submerged species is no longer considered to be a *Potamogeton*. Locally common in England, but rare in Scotland, Wales and Ireland. Grows in ponds, ditches and streams, often in fast-flowing water. **It only has submerged leaves, which are found in opposite pairs, clasping the stem.** They are translucent, lanceolate in shape with a wavy margin and have small teeth near their tip. The leaves measure 1.5–2.5 cm long by up to 1.5 cm wide. They have a midrib and 2 to 4 longitudinal veins, with few cross-veins. There are no stipules. The stems bifurcate and are between 10 and 30 cm long.

Finally, the narrow-leaved pondweeds

Potamogeton pusillus **(Lesser Pondweed)**
This fine-leaved submerged species occurs across most of Britain and Ireland. It is found in both still and flowing base-rich waters. It only has submerged leaves, which are thin and grass-like. The leaves lack stalks. They are linear with pointed tips, and measure 4–7 cm long and up to 2 mm wide. **The leaves typically have a midrib and two longitudinal veins that fuse with the midrib at a narrow angle below the tip of the leaf.** The midrib is NOT usually bordered by bands of clear empty cells towards its base, but when this occurs the bands are only a single cell deep. The stipules are pale brown in colour, 0.5–1.7 cm long; **they are tubular for more than half their length.** The stems are highly branched near their base, slender, slightly flattened and up to 1 m long. This species is sometimes confused with Small Pondweed.

P. berchtoldii **(Small Pondweed)**
Our commonest fine-leaved pondweed, occurring across most of Britain and Ireland. Found in both still and flowing waters. It only has submerged leaves,

Characteristics of narrow-leaved pondweeds

Species	Leaf tip pointed	Longitudinal veins		Longitudinal veins merge before leaf tip
		2	4	
P. pusillus (Lesser Pondweed)		■		■
P. berchtoldii (Small Pondweed)	■	■		■
P. trichoides (Hairlike Pondweed)	■	■		
P. obtusifolius (Blunt-leaved Pondweed)		■		
P. friesii (Flat-stalked Pondweed)	■	■	■	■
P. compressus (Grass-wrack Pondweed)	■	■	▒	■
P. acutifolius (Sharp-leaved Pondweed)	■	■		
P. rutilus (Shetland Pondweed)	■	■		■
S. pectinata (Fennel-leaved Pondweed)				
S. filiformis (Slender-leaved Pondweed)	■	■		
P. natans (Broad-leaved Pondweed)*				
P. crispus (Curled Pondweed)*				■
Your specimen				

* Broad-leaved Pondweed and Curled Pondweed typically have broader leaves, and their descriptions are covered in the sections above.

which are slender and grass-like and lack stalks. They are linear with pointed tips **(often with rounded ends to the pointed tips)**, measuring 2–5.5 cm long and up to 2 mm wide. **The leaves are dark green and always have a midrib and two longitudinal veins that fuse with the midrib almost at a right angle below the tip of the leaf. The midrib is bordered by bands of clear empty cells towards its base.** The stipules are twisted and 0.3–1 cm long; **they have between 6 and 8 faintly visible veins.** The stems are sometimes branched, very slender, slightly flattened and up to 1 m long.

P. trichoides (Hairlike Pondweed)

Occurs scattered across most of England, but is rarer in Wales, Ireland and Scotland. Found in both still and flowing waters. Only has submerged leaves, which are fine and hairlike. **The leaves are somewhat rigid, have a narrow tapering base, measuring 2–4 cm long and rarely more than 1 mm wide, with a long fine tip.** The leaves are translucent, dark green and always have a midrib and two longitudinal veins; the cross-veins are indistinct. **The midrib is thick and obvious, and usually bordered by bands of clear empty cells.** The stipules are twisted, between 0.7 and 1.1 cm long, narrow and semi-rigid. The stems are slightly flattened and regularly bifurcate, with their branches bearing spikes. Stems measure 20–100 cm long.

P. obtusifolius (Blunt-leaved Pondweed)

Occurs across Britain and Ireland but is less common in Wales and Scotland. Found in both still and flowing water. Only has submerged leaves, which are dark

Leaves of narrow-leaved pondweeds

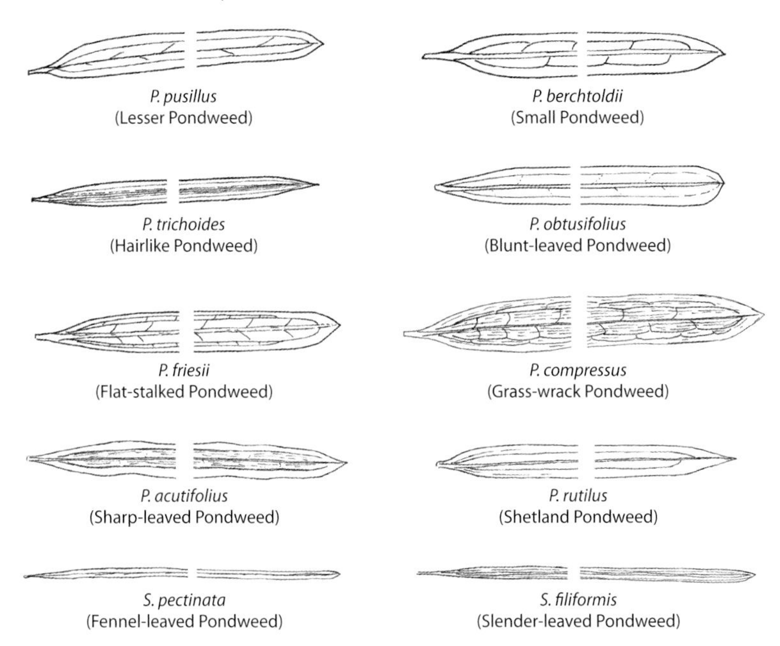

green or tinged reddish, thin and highly translucent. **The leaves lack stalks, and are strap-like with rounded ends**, they measure up to 10 cm long and up to 4 mm wide. The leaves typically have a broad, pale midrib and two main longitudinal veins that fuse with the midrib near the tip of the leaf, but this is not always easy to see. **The faint cross-veins join the midrib at an angle of about 45°.** The stipules are broad and blunt. They measure 1.3–2 cm long, and have many faint veins. The slender stems have numerous short branches. They are slightly flattened and grow up to 1 m long.

P. friesii (Flat-stalked Pondweed)

Found in scattered locations across Britain and Ireland in lakes, ponds and canals. Only has submerged pale green leaves, which are thin and translucent. The leaves lack stalks, and are ribbon-like with rounded ends and pointed tips. The leaves are 4–6.5 cm long and up to 3.5 mm wide; they typically have a midrib and four main longitudinal veins. **The outer veins are closer to the inner veins than the inner veins are to the midrib. The inner veins fuse with the midrib close to the leaf tip. The outer veins fuse with the inner veins further back along the leaf.** The stipules are white, and measure 0.7–1.5 cm in length; they are tubular but split open as they age. The stems are slender, very flattened, and up to 1 m long. They have many short leafy branches.

P. compressus **(Grass-wrack Pondweed)**
Locally abundant across central England and Wales but rare in Scotland and absent from Ireland. Found in both still and flowing water. Only has submerged leaves, which are thin and translucent. The leaves lack stalks, are strap-like with pointed tips, and are 10–20 cm long and up to 4 mm wide. The leaves typically have a midrib and two main longitudinal veins (sometimes four) that fuse with the midrib near the leaf tip. In addition to the main veins, there are many faint longitudinal strands and **distinct wavy cross-veins. The stipules measure 2.5–3.5 cm long; they are twisted and round ended.** The stems are branched, very flattened, sometimes winged and up to 2 m long.

P. acutifolius **(Sharp-leaved Pondweed)**
Rare and now only occurs in south-east England. Found in both still and flowing water, usually associated with calcareous soils. Only has submerged leaves, which are thin and translucent. The leaves lack stalks and are strap-like, with rounded ends which terminate in a point. They measure 5–13 cm long and up to 4 mm wide. **The leaves typically have a midrib and two main longitudinal veins that do not fuse with the midrib but have free ends near the leaf tip. In addition to the main veins there are many faint longitudinal strands.** The stipules measure 1.5–2.5 cm long; they have many veins and terminate in a point. The stems are highly branched, very flattened and up to 2 m long.

P. rutilus **(Shetland Pondweed)**
Only occurs in the north of Scotland including Shetland, where it grows in lakes. It has bright green submerged leaves, which are thin and translucent. **The leaves lack stalks, and are very narrow and linear, terminating in a point. They measure 3–6 cm long and only about 1 mm wide. The leaves typically have a midrib and two main longitudinal veins that fuse with the midrib well before the leaf tip, or just run out.** The stipules measure 1–2 cm long; they are tubular at their base and fibrous. The stems are highly branched, very slender and short – being only 3–60 cm in length.

Stuckenia pectinata **(Fennel-leaved Pondweed)**
Until recently known as *Potamogeton pectinatus*. Widespread across Britain and Ireland, but is less frequent in Scotland and Wales. Grows in both still and moving waters. **Stem leaves are always wider than branch leaves, and lower leaves are wider than upper leaves.** It has translucent green submerged leaves that are thin and hair-like and form a tangled mass. Leaves are 5–20 cm long and up to 2 mm wide. The leaves typically have a pale midrib and two or four longitudinal veins and indistinct cross-veins. The leaves' features are not easy to see as they are so reduced. The stipules measure 1–7 cm long **and have a distinctive long sheath of tissue which clasps the stem, which measure 2–5 cm long.** The slender stems have many branches. These are round in cross-section, and up to 2 m long. This species is sometimes confused with the Slender-leaved Pondweed.

Stuckenia filiformis **(Slender-leaved Pondweed)**
Until recently considered to be a *Potamogeton*. Almost confined to Scotland and Ireland, where it grows in both still and moving waters including brackish conditions. Has yellow-green translucent submerged leaves, which are thin and thread-like. The leaves are 5–20 cm long and up to 1 mm wide; they typically have a pale midrib and two main longitudinal veins. They also have just two faint cross-veins. But the features of the leaf are not easy to see as they are so reduced. **The stipules measure 0.5–3 cm long; they are tubular when young and have a distinctive sheath of tissue which clasps the stem, measuring 0.5–1.5 cm long.** The stems have a few branches near their base. They are round in cross-section, and 15–30 cm long.

HAVE OTHERS RECOGNISED THIS LEVEL OF VARIATION?

Of the 22 species covered here, only half of them were included in Bentham and Hooker's Victorian *Flora of Britain*. Earlier herbals appear to have almost entirely overlooked the group; perhaps early botanists simply did not like getting their feet wet. Even today this group is under-recorded, with Broad-leaved and Bog Pondweeds accounting for about half of all records. This may be because these two are by far our most common species, and they are most easily accessible as they grow in wet ground as well as in waterbodies. A few pondweed species have restricted distributions and were only discovered relatively recently – for example, Shetland Pondweed was unknown before 1890.

HOW FAR SHOULD I GO?

Pondweeds can be useful ecological indicators, as some of the broad-leaved species are quickly lost following nutrient enrichment. Thus, being able to identify *Potagometon*s can be an important skill for ecological consultants. Despite their environmental plasticity, they are not an impossible group to tackle by any means. However, there are a number of hybrids you need to keep an eye out for, with some being abundant in a few waterbodies. For those wanting to know more, Chris D. Preston's BSBI Handbook No. 8. *Pondweeds of Great Britain and Ireland* is recommended.

CHAPTER 9

Roses

Members of the genus *Rosa*, within the family Rosaceae, are instantly recognisable as roses, but beyond that they are something of a law unto themselves. It is difficult to say how many species of rose can be found in Britain and Ireland. Every field guide is likely to differ in this respect. The scientific names of roses and their taxonomy are prone to change. Over time, some species have been divided and others have been combined, and a vast number of hybrid roses have been described. Even with DNA analysis this complexity has not yet been fully resolved. Perhaps more than any other group, roses will make you question the species concept as sets of neat boxes into which a group of individuals can unambiguously be placed.

WHY IS THIS GROUP OF PLANTS COMPLEX?

Among animals the addition or loss of a single chromosome can result in dramatic consequences. In contrast, plants frequently alter their chromosome number to their apparent advantage. Plants may double, triple, quadruple, and so on, the number of chromosomes that their cells contain – sometimes with no apparent effect, but sometimes instantly becoming a new species. Roses are masters of this art, but this is merely the start of their complexity.

Just one of our rose species (*Rosa arvensis*, the Field Rose) contains only a single set of chromosome pairs, one from each of its parents. A second species (*R. spinosissima*, the Burnet Rose) has twice this number. Most of our roses, however, contain five sets of chromosomes, with each of their cells holding 35 chromosomes (five sets of seven). These pentaploid roses (as they are technically known) are thought to have evolved following the hybridisation of two different species. This is the basis of the taxonomic complexity in roses.

To reproduce sexually, roses need to be able to produce pollen and ovules. This involves cells dividing equally into two. It is not possible to divide 35 chromosomes into two equal parts. Instead, roses divide their chromosomes into one set of seven and another set containing four groups of seven (28). During the cell division that produces either pollen or ovules (called meiosis), two sets of seven chromosomes pair and swap genetic material before dividing, as occurs in other organisms. The remaining 21 chromosomes are passed into one of the daughter cells with no genetic exchange. On the male side, only those cells containing seven chromosomes develop into pollen. On the female side, only the cells containing 28 chromosomes develop into ovules. The ovules therefore contain 21 chromosomes that are effectively inherited asexually, plus seven chromosomes that have undergone genetic recombination. The unwanted cells are aborted. When fertilisation occurs the seven chromosomes from the male side rejoin the 28 from the female to reconstitute viable cells with 35 chromosomes.

Roses, uniquely, do not inherit the same numbers of chromosomes from their male and female parents. If this is not complex enough, roses frequently still cross between species. The hybrid offspring of such crosses therefore vary depending on which of the parent species is male and which is female; the female parent contributes four times as much genetic material to their offspring as the male parent does. However, of the genetic material passed on to the next generation, only two-fifths has been acted upon by the sexual processes of segregation and recombination; the other three-fifths is effectively inherited asexually. Given that most species of rose have evolved through hybridisation, crosses between them are hybrids of hybrids. Their unique method of unequally dividing their sex cells ensures fertility is maintained and complexity abounds.

HOW CAN I TELL THEM APART?

Looking at roses, you are instantly drawn to the exquisite beauty of their flowers. However, the flowers are not the most helpful features when it comes to identification. The best place to start is by looking for glandular hairs on the rosehips and flower stalks (a hand lens will help). It is worth taking photographs of the hips, to preserve a record of the nature of any hairs. Other important characters include: the number and shape of their spines (again take photos), the shape and position of sepals, and whether the sepals are retained as the fruit ripens (see page 84). The serration along the edge of leaves (are the teeth single or multiple?) helps in the identification of some species, and so it is a good idea to press a few leaves.

Let's start with the easy ones

The two species of rose with lower chromosome numbers are easy to distinguish from the rest. They are relatively **common and have white flowers.**

R. arvensis (Field Rose)
(section *Synstylae*)
The female parts of the flower (the style) are fused into a pin-like structure in the middle of the flower that is retained on the fruit long after the petals have dropped. There are no glandular hairs on either the fruit or the flower stalks. The sepals are scarcely lobed and not retained on the ripe fruit. A low-growing bush usually less than 1 m tall, found in scrubland, woods and hedges.

R. spinosissima **(Burnet Rose)**
(section *Pimpinellifoliae*)
The ripe fruits of this species are a distinctive dark purple or black. There may be glandular hairs on the flower stalks and occasionally on the fruit. The sepals are not lobed and are retained on the ripe fruit. The stems are covered with many dense, straight thorns. The plants spread below ground, forming thick patches typically less than 50 cm tall. Often found in sand dunes, but also occurs inland in dry open ground.

The other native roses

These native roses, all part of section *Caninae*, nearly all contain five sets of seven chromosomes. They are divided into three subsections that can be crudely distinguished by the presence or absence or glandular hairs (stalked glands) on their fruit and flower stalks, as follows.

- **Dog-roses** (section *Caninae*, subsection *Caninae*): no glandular hairs on either fruits or flower stalks (usually).
- **Downy-roses** (section *Caninae*, subsection *Vestitae*): glandular hairs on both fruits and flower stalks.
- **Sweet-briars** (section *Caninae*, subsection *Rubiginosae*): glandular hairs on flower stalks but not fruit.

Dog-roses (subsection *Caninae*)

R. canina **(Dog-rose)**
The sepals have side lobes, they bend back away from the flower and fall before the fruit is ripe. Thorns are highly curved to hooked, with a broad base. Leaflets have single teeth along their margins and lack hairs. May grow to 3 m or more. Extremely common in hedges and woodland edges.

R. squarrosa **(Glandular Dog-rose)**
The sepals have side lobes, they bend back away from the flower and fall before the fruit is ripe. Thorns are highly curved to hooked, with a broad base. **Leaflet margins have teeth upon their teeth, which are tipped by small red glands.** May grow to 3 m or more. Scattered in hedges and woodland edges across Britain and Ireland.

R. caesia **(Northern Dog-rose)**
The sepals have side lobes, they bend away from the flower and fall before the fruit is ripe. Thorns are highly curved to hooked, with a broad base. **Leaflets usually have single teeth along their margins, lacking glands, and are hairy on their underside.** May grow to 2 m or more. Locally common in hedges and scrubland in Scotland, northern England and rarely in Wales.

Characteristics of dog-roses (subsection *Caninae*)

	Thorns			Leaf margins				
Species	**broad & hooked**	**straight**	**narrow**	**single teeth**	**teeth on teeth**	**glandular**	**hairy under leaf**	**style fused**
R. canina (Dog-rose)	■			■				
R. squarrosa (Glandular Dog-rose)	■				■	■ red		
R. caesia (Northern Dog-rose)	■			■			■	
R. vosagiaca (Glaucous Dog-rose)			■		■			
R. corymbifera (Hairy Dog-rose)	■			■			■	
R. tomentella (Round-leaved Dog-rose)		■			■		■	
R. stylosa (Short-styled Field-rose)	■			■			■	■
Your specimen								

R. vosagiaca (Glaucous Dog-rose)

The sepals have side lobes, they bend away from the flower and fall before the fruit is ripe. **Stems are often red in colour** and have highly curved, narrow thorns with a broad base. **Leaflets lack hairs and are blue-green in colour**; their margins usually have teeth upon teeth, which lack glands. These bushes have an open structure. Found in hedges, woodland edges and scrubland in northern England and parts of Wales where they grow to about 2 m tall.

R. corymbifera (Hairy Dog-rose)

The sepals have side lobes, bending back away from the flower and falling before the fruit is ripe. Thorns are highly curved to hooked, with a broad base. **The undersides of the leaflets are hairy**, their margins have single teeth. May grow to 3 m or more. Found in hedges and woodland edges across Britain and Ireland but rare in Scotland.

R. tomentella (Round-leaved Dog-rose)

The sepals have side lobes, they bend back away from the flower and fall before the fruit is ripe. Thorns are straight, with a distinct bend near the tip. **The undersides of the leaflets are hairy and often the upper surface too**; their margins have teeth upon teeth. May grow to 2 m or more. Found occasionally in hedges and scrub in England and, much more rarely, Wales and Ireland.

R. stylosa (Short-styled Field-rose)

The female parts of the flower (the style) are fused into a short pin-like structure in the middle of the flower. The sepals have side lobes, they bend back away from the flower and fall before the fruit is ripe. Thorns are hooked, with a broad base. The undersides of the leaflets are hairy, their margins have single teeth. May grow to 3 m or more. Fairly common in hedges, scrub and woodland edges in southern England, Wales and Ireland.

Downy-roses (subsection *Vestitae*)

R. tomentosa (Harsh Downy-rose)
The sepals are glandular and lost before the fruit is ripe. Thorns are straight or slightly curved. **Leaflets are densely hairy on the undersurface**, with glandular teeth upon teeth on their margins. May grow to 3 m or more. Found in hedges, scrub and open woodlands, largely in southern Britain and Ireland.

R. sherardii (Sherard's Downy-rose)
The sepals have just a few glands and are retained when the fruit is ripe. Thorns are slender and arching. Leaflets are densely hairy on the lower side and often on upper side too. The margins of leaflets are glandular, with teeth upon teeth. These shrubby plants may grow to 1.5 m or more. Frequent in hedges, scrub and open woodlands in western and northern Britain, and throughout Ireland.

R. mollis (Soft Downy-rose)
The sepals are retained when the fruit is ripe and stand erect. **Thorns are slender and straight.** Leaflets are densely hairy on the lower and upper surfaces. The margins of leaflets are glandular, **with irregular teeth**. Long, stalked glands on the **large, rounded fruit**. The plants are shrubby, spreading via suckers, and may grow to 1.5 m or more. Occasional in hedges, scrub and open woodlands in Ireland, Wales, Scotland and northern England.

Characteristics of downy-roses (subsection *Vestitae*)

Species	Thorns straight	Leaves hairy		Irregular teeth
		under	top	
R. tomentosa (Harsh Downy-rose)				
R. sherardii (Sherard's Downy-rose)				
R. mollis (Soft Downy-rose)				
Your specimen				

Sweet-briars (subsection *Rubiginosae*)

R. rubiginosa (Sweet-briar)
The sepals are retained on the ripe fruit and stand erect. **Stems are covered with two types of thorns: large, hooked ones and small straight prickles.** Leaflets have glandular hairs on the undersurface and margins with teeth upon teeth.

Flowers bright pink. May grow to 2 m or more. Frequent in calcareous scrublands throughout, also commonly planted in hedges in some parts.

R. micrantha **(Small-flowered Sweet-briar)**
The sepals curve backwards and fall before the fruit ripens. **Stems are sparsely covered with long, curved thorns. Leaflets have hairs on their undersurface but not above.** Leaflet margins have multiple teeth upon teeth. These plants have stems climbing to 3 m or more. Most commonly found in scrublands not on calcareous soils in the south and west.

R. agrestis **(Small-leaved Sweet-briar)**
The sepals curve backwards and fall before the fruit ripens. **Flower stalks only have a few glandular hairs.** Thorns are curved, with wide bases. Leaflets may have hairs or lack them; leaflet margins have teeth upon teeth which are tipped by glands. May grow to 1.5 m or more. Found in scrublands, mostly on calcareous soils, as a rare plant in England, North Wales and Ireland.

Characteristics of sweet-briars (subsection *Rubiginosae*)

Species	Sepals retained in ripe fruit	Styles hairy	Flower stalk hairs	
			many	few
R. rubiginosa (Sweet-briar)				
R. micrantha (Small-flowered Sweet-briar)				
R. agrestis (Small-leaved Sweet-briar)				
Your specimen				

Hybrids

Many of our native wild roses are often found growing together in hedges and ancient scrubby grasslands. Their unusual genetics allow them to hybridise freely, with the resulting offspring frequently being fertile or partly so. However, the presence of aborted fruit (empty rosehips containing no seeds) is a strong indicator that you have found a hybrid plant. Within a region many hybrids behave much like stable species, with recognisable niches and morphologies. Many such rose hybrids have been described, but a lot less is known about their wider distributions and ecologies.

Introduced species

In addition to our wild roses, there are numerous introduced species and horticultural cultivars that are sometimes found as garden escapes. Hybrids of our native roses further crossed with introduced species have been important in the

Fruit of native roses

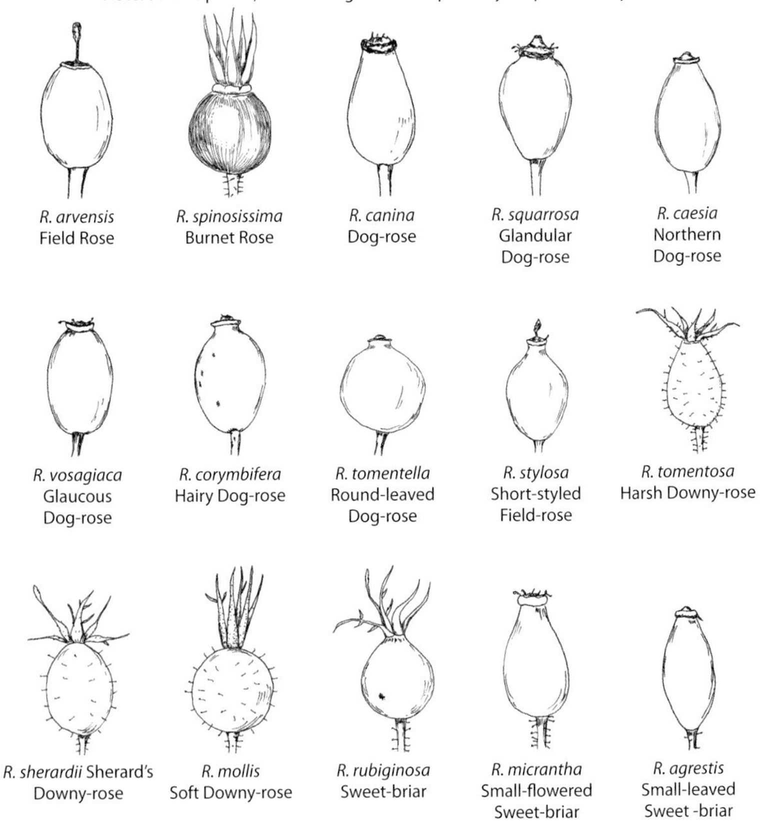

origins of cultivated roses. All of this adds complexity to their identification. As a few of these are widespread they are worth being aware of.

R. rugosa (Japanese Rose)

This is a very vigorous species that is often planted but has become invasive in some wastelands and sand dunes. It is covered in vicious spines. It produces robust pink or white single flowers, followed by distinctive large round hips that persist with their sepals intact long into the winter.

R. gallica (Red Rose of Lancaster)

Previously widely grown, this single or double rose is intensely pink to red. It is now occasionally found in hedges and scrubby areas. It suckers form dense thickets up to 1.5 m.

R. sempervirens (Evergreen Rose)
Superficially the flower of this species looks like a native dog-rose. This climbing rose is very prickly, and its evergreen leaves are highly glossy.

HAVE OTHERS RECOGNISED THIS LEVEL OF VARIATION?

Roses have been culturally significant since ancient times. However, heraldic roses seem to be based on introduced and cultivated flowers rather than being representations of native wild species. Similarly, wild roses are often mentioned in traditional folk songs, but individual species are not. It appears that historically the *Rosa* species we now recognise did not have common names in any of our endemic languages, and modern English names of roses often vary between floras.

The lack of common names for all our native roses does not mean that our ancestors were not aware of their taxonomic complexity. In *A Midsummer Night's Dream*, Shakespeare recognises two species of rose.

> I know a bank where the wild thyme blows,
> Where oxlips and the nodding violet grows,
> Quite over-canopied with luscious woodbine,
> With sweet musk-roses and with eglantine.

'Eglantine' is an archaic name for *R. rubiginosa*, derived from the old French word for prickle.

Micro-morphological variation in roses has been documented at least since the fifteenth century, in the Riddle of the Rose, a traditional folk rhyme:

> On a summer's day, in sultry weather,
> Five brothers were born together.
> Two had beards and two had none,
> And the other had but half a one.

Many variants of this riddle are known from across Europe, in many languages. The five brothers of the riddle represent the five sepals of the rose flower. Two of these have lobes, two are free of lobes and the final one has lobes on one side only. Our ancestors may not have had names for all the species we now recognise, but it is clear they were aware of the subtle differences that separate them.

HOW FAR SHOULD I GO?

Roses with hybrid characteristics are both common and widespread. Even within an individual plant their characters can be variable. Given that this is the case, there is an argument for regarding the pink-flowered pentaploid species as being just three variable species (as did Bentham and Hooker). You may only wish to identify the easier species of rose and be satisfied with naming the others to the level of subsection. This is a simple task. While doing this you may find that you encounter and recognise some of the other species described above. Roses are everywhere; they are easy to spot and, despite this, their complexity is relatively under-recorded. Thus, official distribution maps are probably not very accurate as few botanists feel confident enough to confirm the ID of sometimes even the more common species. The ease of finding new records may entice you further into this group.

Although few of our native rose species are really rare, in conservation terms it is worth being aware that the occurrence of hybrids is evidence of ongoing evolutionary processes. Conserving suitable habits with sufficient numbers of different species of rose so that hybrids are regularly generated is probably more important than conserving the individual species themselves. For those wanting more information, the BSBI Handbook No. 7 *Roses of Britain and Ireland* is out of print but worth looking for second hand, while a new and updated edition is currently in production.

CHAPTER 10

Marsh-orchids and spotted-orchids

The word orchid is often synonymous with the rare and exotic. Entranced by their beauty, some orchid enthusiasts have become obsessive in a way that is rarely seen with other plants. As a reaction to this, some other botanists actively avoid orchids in a curious act of inverse snobbery. In reality several of the marsh- or spotted-orchids are surprisingly common, and have a habit of turning up almost anywhere. Although globally there are a staggering 28,000 species of orchids, only about 60 occur in Britain. Most of these are relatively easy to identify. Of these there are just seven species of *Dactylorhiza* orchids. DNA analysis suggests that the Frog Orchid should also be classified as a *Dactylorhiza* – however, there is still a lack of agreement of exactly how many species there are. Their willingness to hybridise may make you question the species concept and perhaps even your sanity.

WHY IS THIS GROUP OF PLANTS COMPLEX?

At the heart of what makes some orchids frustratingly difficult to identify is their weird sexual habits. Each orchid produces thousands of tiny, dust-like seeds. To ensure successful pollination of so many seeds, they cannot rely on a few pollen grains sticking to bees' knees or arriving on the wind. Hence, pollination in orchids occurs via an all-or-nothing mechanism, in which a dense package of pollen (called a pollinium) is delivered, usually glued onto the head of an insect. Orchids have evolved highly specialised flowers to attract specific species of insects to carry their pollinia. Once such mutually faithful relationships have evolved, the plants become as genetically isolated from other orchids as if they had found themselves growing on a remote island. Such genetic isolation allows orchids to rapidly evolve into new species. This explains why there are so many species of orchids, and also why at a DNA level the difference between species of orchids is less than that between other groups of plants.

Although orchids have elaborate floral forms to promote high-fidelity pollination, no such system is perfect. Marsh- and spotted-orchids are among our most promiscuous species. Of our seven native species, all of them are known to hybridise with each other, except for two combinations from the possible twenty-one (hybrids between the Irish Marsh-orchid and the Narrow-leaved Marsh-orchid or the Southern Marsh-orchid have not yet been recorded in the wild). Basically, if they can hybridise, they will hybridise. Consequently, a good deal of the plants that you will encounter in the wild are intermediate in form. Many of these hybrids have been given names in their own right, but it is probably more informative to refer to them by their putative parentage.

Of our seven species of marsh- and spotted-orchids, two species (the Common Spotted-orchid and the Early Marsh-orchid) have 40 chromosomes in two sets (one from each parent). All the rest have 80 chromosomes in four sets. Hybrids between species with the same number of chromosomes are fully fertile and are able to cross with both their parent species, or with other individuals with the same number of chromosomes. The resulting populations are termed hybrid swarms, and they contain plants that have a vast range of intermediate forms. Even hybrids between species with different numbers of chromosomes have some level of fertility, so that genes can flow with relative ease between all species.

Over evolutionary time, some of these hybrids appear to have doubled the number of chromosomes they contain and thus have become species in their own right. For instance, the Southern Marsh-orchid appears to have originated in part as a cross between the Common Spotted-orchid and the Early Marsh-orchid and this doubled its chromosome number from 40 to 80. This sort of behaviour is never going to make the task of identifying these plants with surety any easier.

The task of recognising spotted-orchid hybrids is further complicated by the innate variability of pink and purple pigments in plants. The intensity of flower colour is known to vary with environmental conditions in many species. It is easy to mistake this natural variation within a species for having found a hybrid plant. For this reason, and perhaps due also to over-enthusiasm, hybrids may in fact be over-recorded.

Uniquely, a final level of complexity has been created by generations of orchid enthusiasts themselves. Their passion for discovering and cataloguing new orchid variants has led to us having a fuller and more finely resolved picture than is the case with other complex taxa. However, this has also resulted in a lack of agreement in the literature, which can be a source of frustration.

HOW CAN I TELL THEM APART?

Within a single spike, the flowers of *Dactylorhiza* orchids can vary in colour and in the extent of spottiness. Within a population, plants may vary dramatically in flower colour, plant size and in the spotting on their leaves. If you wish to identify these orchids, you will therefore need to look at several flowers from several plants, and you are likely to conclude that some individuals are truer to type than are others. Particularly large and robust plants are probably hybrids, and these are more likely to have a mixture of characteristics.

Characteristics of marsh/spotted-orchids

Species	Leaves		Lower lip			
	>2cm wide	spotty	cut >1/3	narrow	tooth projecting	central spotting only
D. fuchsii (Common Spotted-orchid)						
D. maculata (Heath Spotted-orchid)						
D. incarnata (Early Marsh-orchid)						
D. purpurella (Northern Marsh-orchid)					almost	varied
D. praetermissa (Southern Marsh-orchid)						
D. traunsteinerioides (Narrow-leaved Marsh-orchid)						
D. kerryensis (Irish Marsh-orchid)						
Your specimen						

You need to record if there is spotting on the leaves, although this is rarely a definitive characteristic. It is also helpful to measure the width of the widest leaf. However, it is not really possible to tell these species apart when they are not in flower. You need to look closely at the lower lip in the flower: use a hand lens to examine the relative size of the central tooth and the extent of spotting on the lower lip. You need to decide if this is restricted to the central area or not. Unfortunately, this can be ambiguous.

Let's look at the more widespread and distinctive species

Dactylorhiza fuchsii (Common Spotted-orchid)

The commonest species of orchid in Britain and Ireland, occurring everywhere except the Highlands of Scotland. It grows in meadows, marshes and damp woods, especially on base-rich soils, where it may reach a height of up to 50 cm. **Its leaves, which can be up to 4 cm wide, are nearly always covered in elongated, dark-purple spots.** The flowers are pale pink with purple spots. **The lower lip is cut about halfway**, with the central tooth projecting below. The spots on the lower lip are usually restricted to its inner section.

D. maculata (Heath Spotted-orchid)

Widespread in Britain and Ireland, occurring everywhere except the intensive agricultural areas of central England; the commonest orchid in Wales and Scotland. Grows in a variety of habitats including grasslands, marshes and moors, especially on waterlogged acid soils, where it may reach a height of up to 40 cm. Its leaves, which can be up to 2 cm wide, are nearly always covered in round, dark-purple spots. **The flowers are very pale pink to white with pink-purple spots. The lower lip is rather round and cut to about a quarter of its length,** with the central tooth not projecting below the two side lobes. The spots on the lower lip are not usually restricted to the inner section.

D. incarnata (Early Marsh-orchid)
This widespread but less common species occurs across most of Britain and Ireland. Highly variable, with the most distinctive forms being recognised as subspecies. Accounts of this species often differ in what should be considered as a colour form, a subspecies or a true species in its own right. Grows in wet meadows, marshes and dune slacks, mostly on base-rich soils, where it may reach a height of up to 40 cm. Its leaves can reach a width of 2 cm. The leaves of many of its subspecies lack spots. **The flowers are distinctive as from the front they appear narrower than in the other *Dactlyorhiza* species because they fold back on themselves at each side. Their side petals (technically lateral sepals) arch upwards, like hovering wings.** Flower colour is variable and includes white, pink, red and purple. The lower lip is slightly cut, with the central tooth projecting somewhat below. The spots on the lower lip are usually restricted to the inner section.

There are five subspecies of Early Marsh-orchid that appear to have different ecological niches – although these are sometimes considered as just colour forms because they are primarily recognised on the basis of their flower colour.

D. incarnata ssp. *incarnata*
Plants of this form grow to a height of 20–40 cm. The leaves are usually not spotted or may have a few small spots on their upper surface. **The flowers are a pale, salmon-pink colour.** Wet meadows, base-rich fens and marshes.

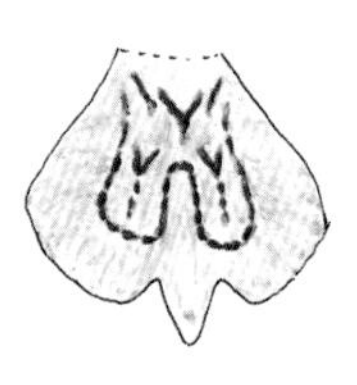

D. incarnata ssp. *coccinea*
Plants of this form grow to a height of 20 cm. The leaves are not spotted. **Flowers are a crimson red colour.** Coastal areas, usually in dune slacks.

D. incarnata ssp. *pulchella*
Plants of this form grow to a height of 20–40 cm. The leaves are usually not spotted or may have a few small spots on their upper surface. **Flowers are a pink-purple colour.** Found in acid bogs and more neutral marshes.

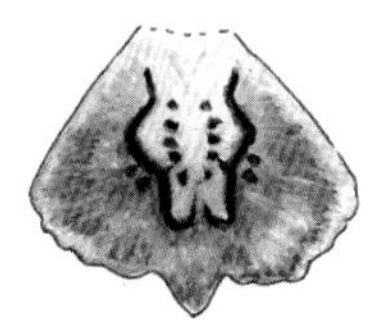

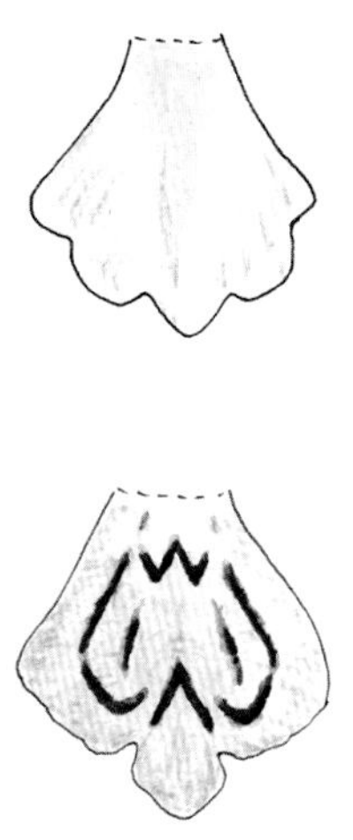

D. incarnatra ssp. *ochroleuca*
Plants of this form grow to a height of 20–50 cm. The leaves are not spotted. **Flowers are white or pale cream in colour, with a wavy lower lip.** Very rare, at just two sites in East Anglia.

D. incarnata ssp. *cruenta*
Plants of this form grow to a height of 15–40 cm. **The leaves usually have dark spots on their upper and lower surfaces. Flowers are pale pink, with distinctive darker markings.** Found in marshes in the Burren in Ireland and in the north of Scotland. This subspecies has sometimes been regarded as a separate species known as the Flecked Marsh-orchid.

D. purpurella (Northern Marsh-orchid)
Occurs in Ireland, mostly to the north, and in England, Wales and Scotland north of a line between Swansea and Hull. Grows in wet meadows, marshes and dune slacks, typically on base-rich soils, where it may reach a height of up to 25 cm. Its leaves may be up to 2.5 cm wide; they usually lack spots or have a few small spots. Flowers are a rich, reddish-purple colour with darker spots. **The lower lip is diamond shaped, only slightly cut, with the central tooth not projecting or only slightly projecting below.** The spots on the lower lip are not usually restricted to the inner section.

D. praetermissa (Southern Marsh-orchid)
Absent from Ireland and Scotland. In England and Wales occurs below a line between Blackpool and Middlesbrough, although it is spreading northwards. Therefore, **there is an increasingly wide band in northern England and Wales where both the Southern and Northern Marsh-orchids may occur together.** Grows in wet meadows, marshes, dune slacks and even waste ground on a wide range of soils, where it may reach a height of up to 50 cm. The leaves may be up to 2.5 cm wide; they usually lack spots although rarely they may have rings. Flowers are a rich, reddish-purple colour with darker spots. **The lower lip is rather round and only slightly cut, with the central tooth projecting below.** The spots on the lower lip are usually restricted to the inner section.

There are also a couple of less common species

D. traunsteinerioides (Narrow-leaved Marsh-orchid)
Occurs at only a few sites widely spread across Britain and Ireland. One of our more difficult species to identify, because of its rarity and variability. Only grows

in very wet, base-rich sites, where it may reach a height of up to 30 cm. The leaves can be up to 1.8 cm wide, are usually unspotted, but brown spots are found in Scottish populations. **Flowers are similar in form to those of Early Marsh-orchid, in that they appear longer and narrower than the other species and their side petals arch upwards, like hovering wings.** The flowers vary in colour from pale pink to rich purple. The lower lip is cut between a quarter and a half of its length, with the central tooth projecting well below. They also vary in the amount and depth of colour of spotting, but generally the spots on the lower lip are not restricted to the inner section.

D. kerryensis **(Irish Marsh-orchid)**
Only found in Ireland but its distribution there is somewhat unclear. Grows in short grasslands, often near loughs or by the sea, on neutral or base-rich soils, where it may reach a height of up to 30 cm. Its leaves which can be up to 2.8 cm wide, are usually covered in round, dark-purple spots. Flowers are rose pink to purple, with darker pink-purple spots. The lower lip is rather round and cut about a quarter of its length, with the central tooth projecting below. **The centre of the lower lip tends to be pale pink to white, with spots that are not usually restricted to the inner section.**

HAVE OTHERS RECOGNISED THIS LEVEL OF VARIATION?

The simple answer to this question is 'not until recently', and even now there is still some controversy about the taxonomy of these orchids. Modern floras may differ in the scientific names they use and perhaps even in the number of species they include. This is perhaps not surprising, since the definition of a species usually involves populations being morphologically distinct as a result of limited geneflow between them. Neither of these conditions are particularly true of *Dactlyorhiza* orchids because of their habit of frequent hybridisation.

Somewhat unexpectedly, Bentham and Hooker's Victorian tome *Handbook of the British Flora* included just two of these species, which were assigned the common names Spotted-orchid and Marsh-orchid, both in the genus *Orchis*. The great North–South divide that led to the first descriptions of the Northern and Southern Marsh-orchids as separate species occurred in 1920. We have to suspect that this story still has a way to travel before its complexity is fully resolved and accepted.

HOW FAR SHOULD I GO?

There are possibly as many answers to this question as there are orchid enthusiasts. Even modern DNA techniques do not always help resolve the origins of some complex hybrid populations. However, after spending a little time looking at these species, you will start to feel comfortable recognising a fairly typical-looking example of one species or another. Equally quickly, you will start to see individuals

Front and side views of Marsh/Spotted-orchid flowers, with examples of variation in lower lips

D. fuchsii (Common Spotted-orchid)

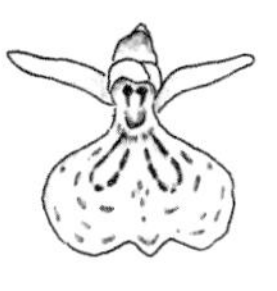  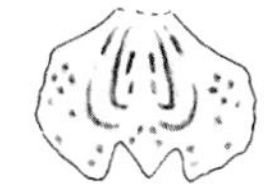 

D. maculata (Heath Spotted-orchid)

  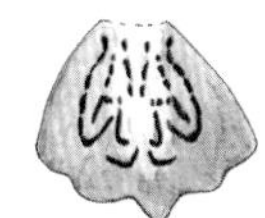 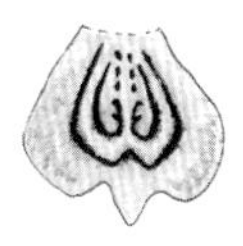

D. incarnata (Early Marsh-orchid)

  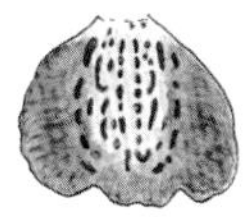 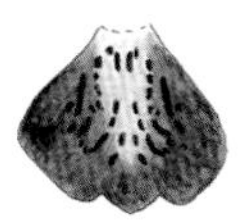

D. purpurella (Northern Marsh-orchid)

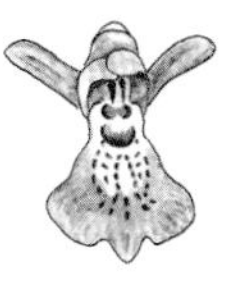  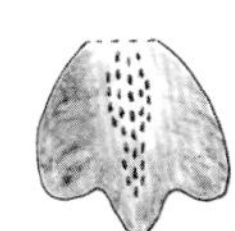 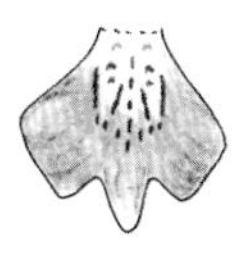

D. praetermissa (Southern Marsh-orchid)

  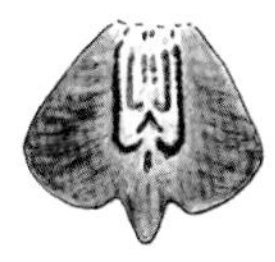

D. traunsteinerioides (Narrow-leaved Marsh-orchid)

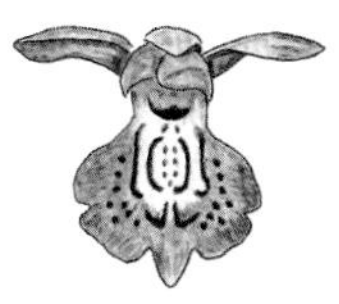  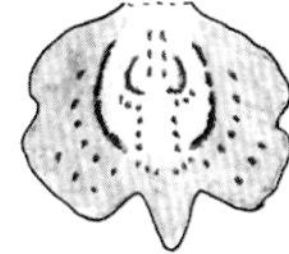 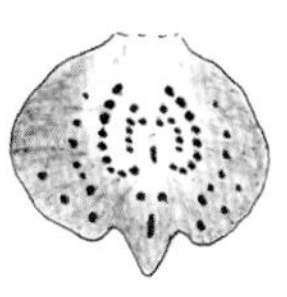

D. kerryensis (Irish Marsh-orchid)

to which it is not really possible to assign a name. If your specimen has a mix of different characters and is taller and more robust than other plants, then this vigour may indicate that you have found a hybrid. If the site contains both its likely parents, you may have some confidence in attributing parentage. Beyond that, identifying plants whose heredity seems to be determined by randomly dipping their net into the gene pool can be something of a dark art.

Even if we do not attach unique name-tags to every minor variant, it is important to appreciate the nature of this complexity and how it has arisen if we are going to be able to conserve the ongoing evolutionary process it represents. To that end, it is worth taking a few minutes to appreciate just how beautiful these flowers are. Then it really is no great burden spending the time trying to get to grips with the variability found within this group of plants. For those wanting to go further, it is worth looking at *Britain's Orchids: A Field Guide to the Orchids of Great Britain and Ireland* (2020) by Sean Cole and Mike Waller.

CHAPTER 11

Water-crowfoots

These white-flowered aquatic buttercups tend to be overlooked by both botanists and non-botanists, but for different reasons. For non-botanists, it requires an excessive effort simply to look at these plants because they grow in open water, or sometimes in unappealing drainage ditches. For those of us more botanically motivated, water-crowfoots can be discouragingly difficult species to tackle because several of them possess a rather unhelpful habit of producing radically different-shaped leaves, depending on whether they are floating or submerged. Floating leaves are broad, flat and lobed, sitting on the water surface like miniature waterlilies. Submerged leaves are threadlike and highly branched. This remarkable piece of biology means that many people will be more familiar with water-crowfoots from textbooks than real life.

WHY IS THIS GROUP OF PLANTS COMPLEX?

As one of our more challenging groups of plants, water-crowfoots have multiple ways of obscuring their identify. As already mentioned, they have an amazing botanical superpower: the ability to radically alter their leaf morphology depending on environmental conditions. Their appearance also changes significantly between winter and summer.

In addition to this shapeshifting nature, water-crowfoot identification is complicated because many of these species are known to hybridise, giving rise to multiple intermediate forms. While some individuals are busy hybridising, others regularly reproduce vegetatively without sex. It is not uncommon for aquatic plants to fragment and be carried by the water to colonise new areas. Thus, single genetic individuals can dominate entire waterbodies and, if they are distinct enough, may eventually be recognised as a new subspecies.

If this is not enough, there is significant variation in the number of chromosomes known within and between species of water-crowfoot. Variation in chromosome number between parent species tends to result in sterility in their hybrid offspring. The observation of empty, flattened fruits can therefore be helpful in confirming the hybrid origin of intermediate individuals.

HOW CAN I TELL THEM APART?

Since one of the most unusual features of this group is their ability to produce different-shaped leaves under different environmental conditions, there should be no surprise that it is impossible to identify them based on vegetative features alone. Even so, occurrence (or not) of two different leaf types can be a helpful identification hint. When both floating and submerged leaves are present, you will need to look at both.

As the common names suggest, habitat can be a good starting point. A hand lens will be helpful when looking at the shape of the nectar pit found at the base of the petals and checking for hairs on the developing fruit. The flowers are fragile, so be warned.

Types of nectar pits on water-crowfoot petals

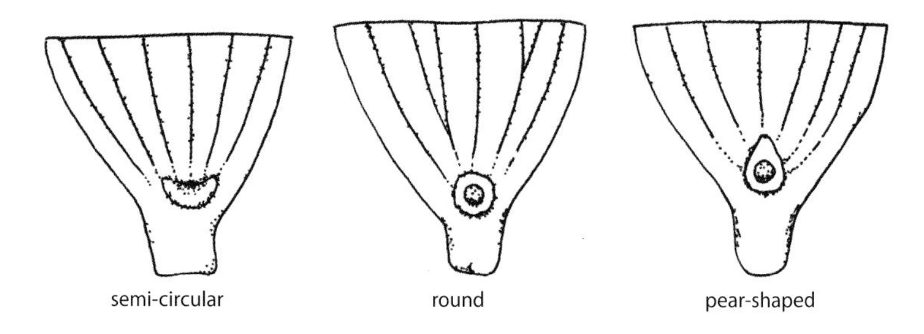

Let's start with the species that only have floating leaves

Ranunculus hederaceus **(Ivy-leaved Crowfoot)**
An annual or perennial that grows horizontally in shallow water or on mud. Found across the British Isles, more frequently in the west, except in northern Scotland where it is more common in the east. **The leaves are lobed and, as its name suggests, somewhat resemble ivy.** Its petals are small, measuring only 2.5–3.5 mm, and do not overlap. At the base of the petal is a semi-circular nectar pit. The sepals are not bent backwards, and the fruits lack hairs.

R. omiophyllus **(Round-leaved Crowfoot)**
An annual or perennial that grows horizontally in shallow ponds or on mud. Most frequently found across Wales, western parts of England and eastern Ireland. **The leaves are usually divided into three lobes (sometimes five), being cut to about halfway.** Petals measure 5–6 mm and do not overlap. At the base of the petal is a semi-circular nectar pit. The sepals are bent backwards, and the fruits lack hairs.

Characteristics of water-crowfoot species

Species	Petal size mm			petals overlap	Nectar pit			sepals turned back	fruit hairy
	<5	5–10	>10		semi-circular	circular	pear-shaped		
Ranunculus hederaceus (Ivy-leaved Crowfoot)	■				■				
R. omiophyllus (Round-leaved Crowfoot)		■			■			■	
R. fluitans (River Water-crowfoot)		■	■	■			■		or not
R. trichophyllus (Thread-leaved Water-crowfoot)	■	■		■	■				■
R. circinatus (Fan-leaved Water-crowfoot)	■	■			■				■
R. aquatilis (Common Water-crowfoot)		■		■		■			■
R. tripartitus (Three-lobed Crowfoot)	■				■			■	■
R. peltatus (Pond Water-crowfoot)			■	■			■		■
R. penicillatus (Stream Water-crowfoot)			■	■			■		■
R. baudotii (Brackish Water-crowfoot)		■		■	■			■	
Your specimen									

Species that only have floating leaves

Species that only have submerged leaves

Species that have both floating and submerged leaves

Next, the species that only have submerged filamentous leaves

R. fluitans (River Water-crowfoot)

A perennial that grows in rivers with moderate rates of flow. Scattered across England and rare in Wales, Scotland and the north of Ireland. **Its filamentous leaves are long, often measuring more than 8 cm. The leaves are longer than the stem internodes.** Its broad petals are large, measuring 7–13 mm, and overlapping. At the base of the petal is a pear-shaped nectar pit. The sepals are not bent backwards, and the developing fruit may be downy to hairless.

R. trichophyllus (Thread-leaved Water-crowfoot)

An annual or perennial that grows in slow rivers, canals and ponds across the British Isles, more commonly in the east. **Its filamentous leaves are shorter than the stem internodes, their branching is open and rather tree-like.** Its broad petals measure 3.6–6 mm and overlap. At the base of the petal is a semi-circular nectar pit. The sepals are not bent backwards, and the fruits are hairy.

***R. circinatus* (Fan-leaved Water-crowfoot)**
A perennial species that grows in slow rivers, ditches and deeper waterbodies across the British Isles, more commonly in the east. **Its filamentous leaves are rigid and shorter than the stem internodes. The leaves branch in only one plane, giving a fan-like appearance.** Its petals measure 4–10 mm and touch but do not overlap. At the base of the petal is a semi-circular nectar pit. The sepals are not bent backwards, and the fruits are hairy.

Finally, the species that can have floating *and* submerged filamentous leaves

***R. aquatilis* (Common Water-crowfoot)**
An annual or perennial species that grows in ponds, ditches and slow rivers across the British Isles, less commonly in the north and west. **Its floating leaves are divided into three or five lobes, cut to about halfway, with the lower lobes touching or nearly touching. Its filamentous leaves are ridged and their branching is open and rather tree-like.** Its broad petals measure 5–10 mm and overlap. **At the base of the petal is a circular nectar pit.** The sepals are not bent backwards, and the fruits are downy.

***R. tripartitus* (Three-lobed Crowfoot)**
This rare species is an annual or perennial of shallow ponds and ditches. In the British Isles it is restricted to the far west and south. **Its floating leaves are divided into three lobes (rarely five), cut to about halfway; the widest lobes are at the base of the leaf. Sometimes lacking filamentous leaves, but when present they are extremely fine and hairlike but collapse when taken out of the water.** Its petals are small, measuring only 1–4.5 mm, and do not overlap. At the base of the petal is a semi-circular nectar pit. The sepals are bent backwards, and the fruits are hairy.

***R. peltatus* (Pond Water-crowfoot)**
An annual or perennial that grows in ponds, ditches, canals and slow rivers scattered across the British Isles; it is less common in the north of Scotland and Cornwall. **Its floating leaves are divided into about five lobes, usually cut a little more than halfway. Lower lobes do not touch.** Its filamentous leaves may be absent, but if present they may be rigid or floppy when out of water. **Its broad petals measure 12–15 mm and overlap.** At the base of the petal is a pear-shaped nectar pit. The sepals are not bent backwards, and the fruits are hairy.

***R. penicillatus* (Stream Water-crowfoot)**
A perennial species that grows in swift-flowing streams and rivers across the British Isles, but is rare or absent from Ireland and northern Scotland. May lack floating leaves, but when they occur these are divided into about five lobes,

Leaves of water-crowfoot species

R. hederaceus
(Ivy-leaved Crowfoot)

R. omiophyllus
(Round-leaved Crowfoot)

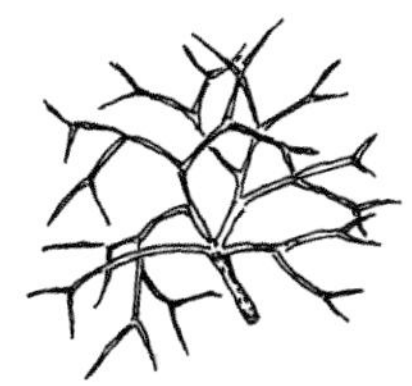

R. trichophyllus
(Thread-leaved Water-crowfoot)

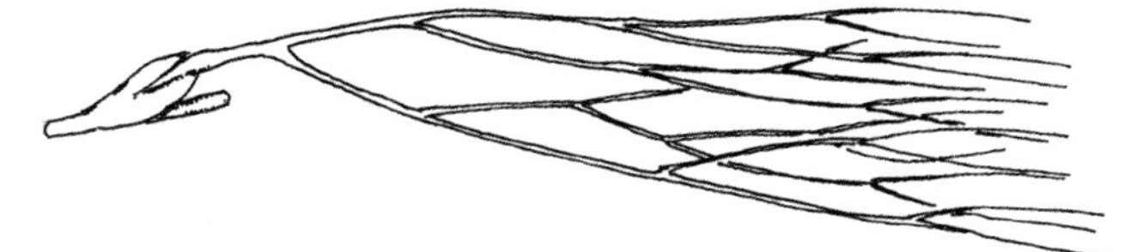

R. fluitans (River Water-crowfoot)

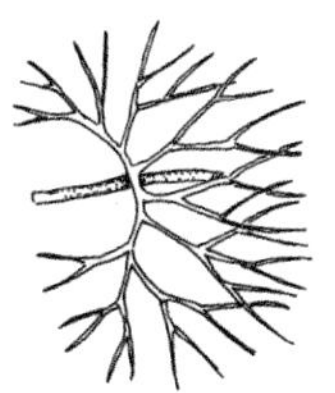

R. circinatus
(Fan-leaved Water-crowfoot)

R. aquatilis (Common Water-crowfoot)

R. tripartitus
(Three-lobed Crowfoot)

R. peltatus
(Pond Water-crowfoot)

R. baudotii
(Brackish Water-crowfoot)

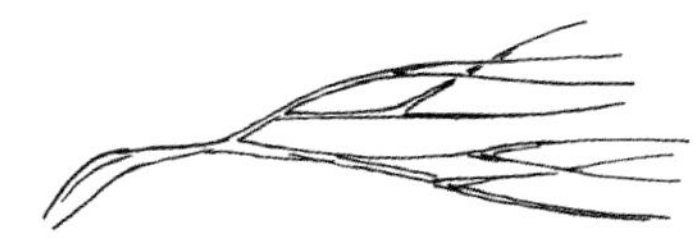

R. penicillatus (Stream Water-crowfoot)

usually cut to around halfway. **Its filamentous leaves are floppy when out of water.** Its broad petals measure 10–15 mm and overlap. At the base of the petal is a pear-shaped nectar pit. The sepals are not bent backwards, and the fruits are hairy.

R. baudotii **(Brackish Water-crowfoot)**
An annual or perennial that **grows in brackish ponds and ditches; not surprisingly, it has a scattered coastal distribution.** Floating leaves may be absent, but when present are divided more than halfway, usually into three lobes. **Its filamentous leaves are rigid when out of water, with open rather tree-like branching.** Its broad petals measure 5–10 mm and overlap. At the base of the petal is a semi-circular nectar pit. The sepals are bent backwards down the stem, and the fruits are hairless.

HAVE OTHERS RECOGNISED THIS LEVEL OF VARIATION?

The diversity of this group was not recognised until relatively recently. The medieval herbals tend to omit all species of water-crowfoot, although the term crowfoot was sometimes used for more familiar terrestrial yellow-flowered buttercups. Carl Linnaeus only recognised two species of water-crowfoot: Common Water-crowfoot and the terrestrial form of Ivy-leaved Crowfoot. So it remained, with just these two species being included in the standard Victorian British flora (Bentham and Hooker). Finally, in 1871 a botanist called James Robinson doubled the number of species to four. The remaining species were only described during the twentieth century.

HOW FAR SHOULD I GO?

If you want to go further than Linnaeus's two species of water-crowfoot, there is no escaping the fact that you are probably going to get wet. You will need to sample their flowers and any submerged leaves. However, if you don't mind a little watersport and the plants are flowering, you should be able to identify the ten species covered above. Unfortunately, since all these species are known to hybridise you may encounter intermediate individuals. It should be possible to confirm their hybrid nature by the occurrence of intermediate leaf forms and by their inability to produce viable seed.

CHAPTER 12

Willows

This is a group of trees and shrubs familiar to most people. They are widely planted, and some species may be considered semi-domesticated. Willows are frequently found associated with rivers and ponds. Several species are very common and even non-botanists recognise 'Pussy Willow' and 'Weeping Willow', although these common names are only weakly associated with actual species. Furthermore, the familiar species are just the tip of the iceberg. Many willows have limited distributions and there are a vast number of hybrids – making this group potentially one of the most challenging in the British flora. **Willows are probably the most promiscuous and thus taxonomically tricky of all British plant families.** When the father of taxonomy Carl Linnaeus described the Crack Willow, he appears to have been looking at a different species in his herbarium. To this day, taxonomists actively debate the status of this common species of willow. If experts disagree over the nomenclature, then it is perhaps not surprising that willows can frustrate less experienced botanists and even cause them to question the validity of the process.

WHY IS THIS GROUP OF PLANTS COMPLEX?

Willow inflorescences are known as catkins. They lack petals, and are pollinated by wind, insects and perhaps even birds. The random nature of pollination within willows facilitates the production of many hybrids. And the frustrating complexity of trying to identify these plants primarily arises from the ease with which they form hybrids, blurring the boundaries between species.

The situation is highly complex because almost all the different species of willow seem to have the ability to hybridise, including introduced species and possibly even some that are now extinct in the wild. Two-way hybrids are also known to hybridise again, so that some plants may contain genes from three parents. More than 70 two-way hybrids and more than 20 three-way hybrids have been described in the British Isles, and there could well be others out there waiting to be confirmed. Most willow hybrids are found growing close to their parents – but this is not always the case, and it can be difficult to definitively determine parentage in the field. For this reason, no attempt is made to cover the hybrids here, but hopefully the use of tabular keys will help you identify potentially hybrid individuals.

Unlike most plants, willows have separate sexes: individuals have either all-male or all-female inflorescences. In many species, the catkins open before the leaves. Thus, if you need both flowers and leaves to identify your specimen, it may require that you collect and preserve the catkins and revisit the plant once its leaves have opened. You should also be aware that **the first leaves to open are atypical in form.** For identification purposes, you will need to look at mature leaves growing part-way along a twig. Trying to identify willows therefore demands dedication, and some guides include three separate keys – to identify male plants, female plants and non-flowering plants.

A further complication results from the ease with which willows can propagate vegetatively. Detached stems root rapidly, which is probably an adaptation to life in riparian habitats. This has enabled favoured types to be cultivated and then to proliferate, and may explain why only a single sex of some forms has ever been recorded.

Finally, willows have a long history of human management. Some species have been coppiced for weaving and others pollarded to produce posts. Both these techniques modify the form of the tree and can complicate the process of identification. Even without human interference, willows can be highly variable – growing into trees in some environments and shrubs in others.

HOW CAN I TELL THEM APART?

As willows are one of the most difficult groups of plants to correctly identify, you will need as much information as possible upon which to base your conclusion. You may therefore need to look at: leaf shape and colour, the growth form of the plant, as well as the catkins. As stated above, this might mean repeat visits to collect flowers and leaves. The fact that some species produce catkins and leaves concurrently is a helpful diagnostic trait.

Lowland long/narrow-leaved willows

Salix fragilis* (Crack Willow) Sometimes considered a hybrid *S.* x *fragilis
Widespread in Britain and Ireland, except in mountainous areas. Highly variable, many different forms and hybrids have been described. Commonly found by streams and ponds, where it may grow up to 25 m tall. **Mature leaves are glossy green and hairless on both the upper and lower sides.** The leaf margins are coarse, with irregular teeth. The leaves are typically large, measuring 8–15 cm long. The name Crack Willow relates to the fact that its twigs are fragile and break easily at the base – but other species also do this. Twigs may be yellow or brown. **The catkins appear with the leaves.** The catkins are long and pendulous, measuring 4–6 cm; they have yellow scales. Each male flower contains two anthers.

Characteristics of British and Irish willows

Species	plant height			upper leaf surface			longest leaf			catkins before leaves	anther number ≠ 2	red-purple anthers	catkin scales		
	<1 m	1–20 m	>20 m	glossy	dull	hairy	<5 cm	5–10 cm	>10 cm				yellow	dark	hairy
Salix fragilis (Crack Willow)			■	■					■				■		
S. alba (White Willow)			■		■	■		■		■			■		
S. viminalis (Osier)		■			■	■			■	■				■	
S. pentandra (Bay Willow)		■		■					■		■ 5		■		
S. daphnoides (European Violet-willow)		■		■					■	■				■	
S. purpurea (Purple Willow)		■			■	■		■		■	■ 1			■	
S. triandra (Almond Willow)		■			■				■		■ 3		■		
S. babylonica (Weeping Willow)		■		■					■				■		
S. caprea (Goat Willow)		■			■	■			■	■				■	■
S. cinerea (Grey Willow)		■			■	■		■		■				■	■
S. aurita (Eared Willow)		■			■		■			■				■	■
S. myrsinifolia (Dark-leaved Willow)		■			■			■		■				■	■
S. repens (Creeping Willow)	■			■		■	■			■ *				■	■
S. lapponum (Downy Willow)	■					■		■						■	■
S. phylicifolia (Tea-leaved Willow)	■			■				■						■	■
S. arbuscula (Mountain Willow)	■			■			■					■		■	■
S. lanata (Woolly Willow)	■					■		■		■				■	■
S. herbacea (Dwarf Willow)	■			■			■						■		
S. myrsinites (Whortle-leaved Willow)	■			■				■				■		■	
S. reticulata (Net-leaved Willow)	■				■		■					■		■	
Your specimen															

Lowland long/narrow-leaved willows

Lowland oval-leaved willows

Willows of hills and mountains

Dwarf willows (section *Chametia*)

* *S. repens* is variable, with some individuals producing their catkins before their leaves and other plants producing them at the same time.

S. alba (White Willow, Cricket-bat Willow)

This widespread species occurs across Britain and Ireland except in mountainous areas. It may grow to a height of 30 m. **Its mature leaves are dull green with silky white hairs on the upper surface and dense, long silky hairs on the lower surface.** The leaves have very fine, regular teeth along their margins and grow to a length of 5–10 cm. The young twigs are hairy but become glossy yellow or brown with age and less easily detached than in other willows. **Catkins appear before the leaves and are often curved.** Male catkins are 4–5 cm long with two anthers per flower. The female catkins measure 3–4 cm. Catkin scales are yellowish.

S. viminalis (Osier)
Widespread across Britain and Ireland except the highlands of Scotland and was historically extensively planted in osier beds. An erect shrub, growing to a height of 5 m. Its leaves are long and narrow, growing to 10–20 cm. **The upper surface of the leaves is a dull dark green, with a few hairs, in contrast the undersurface which is white, silky and covered in short hairs. The leaf margins are wavy, and often slightly rolled under.** The young twigs are hairy but become hairless and a dull yellow-brown colour; they are slender and flexible. **Catkins appear before the leaves.** The fluffy male catkins contain two yellow anthers; they are oval in shape and shorter than the female catkins. **The catkin scales have a distinctive, dark chestnut tip.**

S. pentandra (Bay Willow)
Widespread in Britain and Ireland except in mountainous areas, and less common in southern England. Grows near ponds and streams to form a broad-crowned tree of up to 10 m tall. **As the name suggests, its leaves look and smell similar to the bay tree. They are tough, glossy, dark green above, and a brighter paler green on the undersurface.** The leaves grow to a length of 5–12 cm. The leaf margins have regular, fine-pointed teeth. The twigs are red-brown and hairless, with a glossy, varnished appearance. **The catkins appear with the leaves.** The male catkins have five or more anthers, and are 2–5 cm long; the greenish female catkins are shorter, measuring 1.5–3 cm long. The catkin scales are yellowish in colour.

S. daphnoides (European Violet-willow)
Scattered across lowland Britain and Ireland. A small tree, up to 10 m tall. Its leaves are glossy dark green on their upper surface, and pale blue-green below; they grow to a length of 7–12 cm. The leaf margins have small regular teeth. **Young twigs are hairy and white or pale grey; on ageing they become glossy and purplish brown. The catkins are produced before the leaves**; they are 2–4 cm long, cylindrical in shape and crowded along the stems. Male flowers contain two anthers. Catkin scales are a rich red-brown colour.

S. purpurea (Purple Willow)
Scattered across lowland Britain and Ireland, less common in the south and west. Grows in damp places, where it becomes a shrub about 4 m tall. **Its mature leaves have a dull or silky appearance, they are dark green on the upper surface and a distinctive pale bluish-green on the undersurface. The young leaves are often tinged orange or copper.** Leaves are frequently found in pairs along the twigs, they grow 2–8 cm long. **The bark is brilliant yellow on the inside. Catkins appear before the leaves** and are often curved, narrowly cylindrical and 1.5–3 cm long. The catkins are frequently tinged with purple. **The filaments in the male are fused so that they appear to contain only a single anther per flower.** The catkin scales are bronzy.

S. triandra (Almond Willow)
Scattered across lowland Britain and Ireland, but rare outside south-east England and south-east Ireland. A shrub or small tree, growing to a height of 7 m. **Its bark has an unusual habit of peeling off, to reveal chestnut-brown patches below.** The leaves grow to a length of 4–11 cm; the upper surface is dull dark green, the undersurface greyish-green. The leaf margin has regular small teeth along its full length. **The twigs are hard to snap off and are said to taste of rosewater when chewed. Unusually, this species produces a few catkins throughout summer.** The male catkins are 2.5–5 cm long and **each flower contains three anthers.** The female catkins are about 2.5 cm long. The catkin scales are pale yellow.

S. babylonica (Weeping Willow)
Introduced from China and widely planted in gardens and parks – although garden specimens are often hybrids between *S. babylonica* and *S. alba*, or even aberrant cultivars. Included here because it is so familiar and is now found growing as an escape. **The cultivated form of this species has a distinctive 'weeping' growth form; it may reach a height of up to 12 m.** Its leaves grow to 10–12 cm long. The upper surface of the leaf is glossy and a rich bright green in colour; in contrast the lower surface is paler, duller and more blue-green. The twigs are olive-brown and hairless. The catkins appear with the leaves. Most plants are female, their pale yellow catkins are about 3 cm long and often slightly curved. Primarily female catkins will sometimes include male flowers. Male flowers have two anthers. Catkin scales are pale yellow in colour.

Lowland oval-leaved willows

S. caprea (Goat Willow)
Common across Britain and Ireland, although less so in the far north of Scotland. A tall shrub or small tree which grows to a maximum height of 10 m. **Its leaves are large and broad, growing to 5–12 cm in length; their upper surface is dull green in colour and often covered in fine hairs, the undersurface is grey-white and downy. The leaf margins are distinctly wavy.** Young twigs are brown and hairy, but the hairs are lost with age. The catkins are produced before the leaves, often crowded at the ends of twigs. The catkins are erect and short, being 1.5–2.5 cm long. The male flowers contain two anthers. The catkin scales are black and covered in hairs.

S. cinerea (Grey Willow/Sallow)
Common and widespread across Britain and Ireland. A tall shrub or small tree which grows to a maximum height of 10 m, often with many stems from the base. The leaves are highly variable, growing to a length of 2–9 cm. The upper leaf surface is dull grey-green and covered with fine hairs; the lower surface is grey-white. **Some forms have rust-coloured hairs on the lower leaf surface, especially along the veins.** Young twigs are hairy and red-brown in colour,

becoming hairless after the first year. The catkins are produced before the leaves; they are erect and short, being 2–3 cm long. The male flowers contain two anthers. The catkin scales are brown and covered in dense hairs.

Goat Willow and Grey willow are both often referred to as 'pussy willow'. These two species can be distinguished by removing the bark from their twigs. If there are raised ridges running along the twig, usually starting at a leaf scar, then you have Grey Willow; no ridges means Goat Willow.

S. aurita (Eared Willow)
Widespread across Britain and Ireland, particularly on moorland, less common in south-east England. A highly branched shrub, usually less than 2.5 m tall. **Its leaves are typically small, growing to a maximum length of 5 cm; they have highly distinctive 'ears' (leaflets) where the leaf stalk joins the stem.** The upper surface of the leaves is dull green and wrinkled, with deep-set veins. The lower leaf surface is white-grey in colour and covered in hairs. The catkins are produced before the leaves; they are erect and short, being 1–2 cm long. The male flowers contain two anthers. The catkin scales are brown and sometimes covered in hairs. Removing bark from the twigs always reveals clear raised lines on the wood.

S. myrsinifolia (Dark-leaved Willow)
Most commonly encountered in northern England, southern Scotland and the north of Ireland, often by streams and ponds but also sometimes in damp rocky places at higher altitude. A low-growing shrub or small tree, 1–3 m tall. Its leaves grow to a length of 2–6.5 cm. Their **upper surface is dull dark green; the undersurface is pale grey or blue-green. The rather papery leaves may be covered in hairs when young, but these are lost with age.** The young twigs are green or dull brown and covered in hairs which are not lost until after a year. The catkins are produced before the leaves, usually towards the end of short side shoots; they are semi-erect and 1.5–4 cm long. The male flowers contain two anthers. The catkin scales are dark brown or black and thinly covered in hairs.

S. repens (Creeping Willow)
Has a scattered distribution across Britain and Ireland, often in sand dunes; largely absent from the Midlands. Although typically a species of the lowlands, has been recorded to above 800 m. A creeping and suckering low shrub which forms thickets and rarely grows over 1 m tall. Its leaves are small, 1–3.5 cm. **The leaves are highly variable: they can be bright green on both upper and lower surfaces, or white with silky hairs on both upper and lower surfaces, or bright green above and white with silky hairs below.** The twigs are slender, hairless and variable in colour. The catkins are usually produced before the leaves, but sometimes with the leaves; they are erect and short, being 1–2.5 cm long. The male

Leaves of British willow species

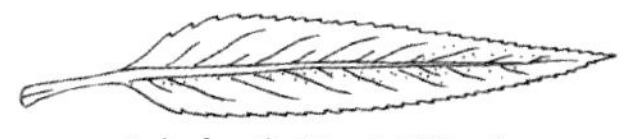

Salix fragilis (Crack Willow)

S. alba (White Willow)

S. viminalis (Osier)

S. pentandra (Bay Willow)

S. daphnoides (European Violet-willow)

S. purpurea (Purple Willow)

S. triandra (Almond Willow)

S. babylonica (Weeping Willow)

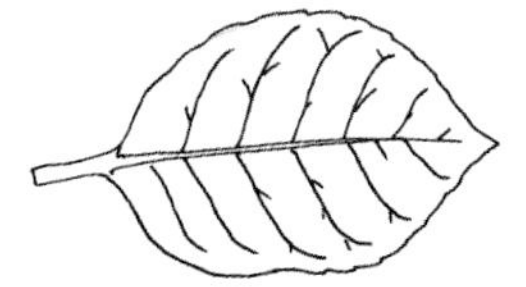

S. caprea (Goat Willow)

S. repens (Creeping Willow)

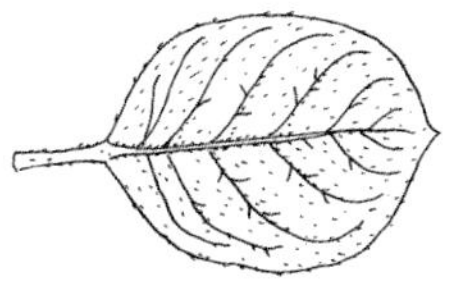

S. lanata (Woolly Willow)

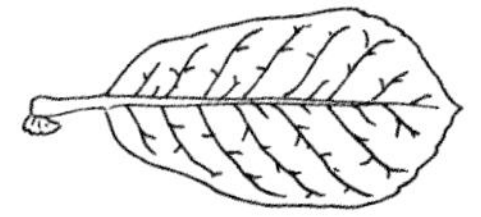

S. cinerea (Grey Willow/Sallow)

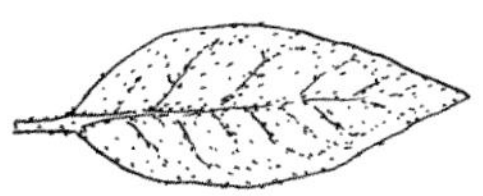

S. lapponum (Downy Willow)

S. herbacea (Dwarf Willow)

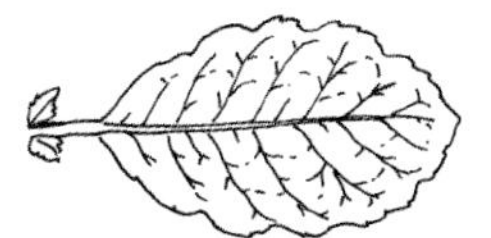

S. aurita (Eared Willow)

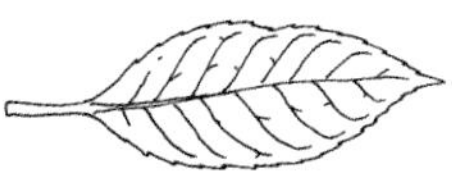

S. phylicifolia
(Tea-leaved Willow)

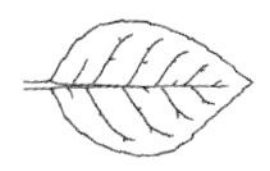

S. myrsinites
(Whortle-leaved Willow)

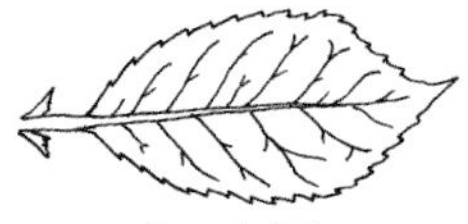

S. myrsinifolia
(Dark-leaved Willow)

S. arbuscula
(Mountain Willow)

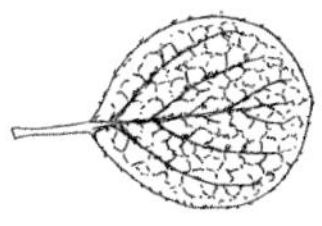

S. reticulata
(Net-leaved Willow)

flowers contain two anthers. The catkin scales are a rich brown and sometimes thinly covered in fine hairs.

Willows of hills and mountains

S. lapponum (Downy Willow)
Grows in the highlands of Scotland and rarely in the Lake District and Pennines. A low shrub, between 20 and 100 cm tall. The leaves are usually long and thin, growing to a length of 1.5–7 cm. **The upper and lower leaf surfaces are both covered in downy hairs. The upper sides are a dull, pale grey-green colour while the undersurface is more pale grey.** The twigs have obvious leaf scars, and are red-brown in colour. Young twigs are hairy, but the hairs are lost with age. The catkins appear with or just before the leaves. They are cylindrical and 2–4 cm long. Male flowers contain two anthers. Catkin scales are hairy and dark brown.

S. phylicifolia (Tea-leaved Willow)
Grows on base-rich soils in the hills of northern England and Scotland and north-western Ireland. A shrub or small tree to a height of 4 m. The young leaves may be covered in hairs, but these are lost with age, so that the **older leaves are glossy and a rich green on their upper surface. In contrast, the undersurface of the leaves are a grey-blue-green colour.** The leaves are oval and grow to a length of 2–6 cm. Twigs are generally red-brown. The catkins are produced with the leaves, and are usually found near the end of short side shoots. The catkins are cylindrical and 1.5–4 cm long. Male flowers contain two anthers. The catkin scales are dark and covered in hairs.

S. arbuscula (Mountain Willow)
Has a limited distribution, being found only in the Highlands of Scotland and very locally in Milburn Forest in the Pennines. Grows to a maximum height of just 70 cm. **Its oval leaves are 1.5–3 cm long. The upper surface of the leaves is glossy and bright green, while the undersurface is blue-green.** Younger leaves have hairs on their lower surface, but these are lost with age. The twigs are red-brown, young twigs are covered in short hairs. The catkins appear with the leaves. The catkins are small, being only 1–2 cm long. **Male flowers contain two anthers, which have a distinctive red-purple colour.** The catkin scales are very hairy and purplish-brown.

S. lanata (Woolly Willow)
This rare species grows in the mountains of Scotland. A low-growing, stunted bush, to a maximum height of less than 1 m. Its leaves are somewhat variable in shape, usually oval and growing to a length of 3.5–7 cm. **The upper surface of the leaves is a dark grey-green and covered with silky hairs. The lower surface is paler grey-green and more densely covered in hairs.** The twigs are a rich brown colour, short and stout in form, with obvious leaf scars. The catkins come before

the leaves. **The catkins appear almost round in shape and grow to a length of 2.5–3.5 cm.** Male flowers contain two yellow anthers. Catkin scales are dark brown and hairy.

Dwarf willows (section *Chametia*)

S. herbacea **(Dwarf Willow)**
Grows in the west of Ireland and mountainous regions of Wales, England and Scotland. **Our smallest species, reaching less than 8 cm tall. Its leaves are small, round and rarely more than 3 cm long.** The upper surface of the leaves is glossy and bright green, often with the leaf margin slightly red. The leaf veins appear almost transparent. The lower surface is paler and less glossy. The twigs are reddish brown and mat-forming. The catkins appear with the leaves. The catkins are small and usually found on the ends of twigs, they comprise very few flowers. Catkins are less than 1 cm long and often bright red. Male flowers contain two anthers. Catkin scales are yellowish, often tinged with red.

S. myrsinites **(Whortle-leaved Willow)**
Another species of the Scottish Highlands. **A low-growing plant, reaching less than 40 cm tall. Its leaves are oval and grow to 1.5–7 cm in length.** The young leaves are hairy, but soon become a rich glossy green on the upper and lower surfaces. The catkins appear with the leaves. There are few catkins and these are found near the ends of the twigs. Male catkins grow to a length of 1–2 cm. **The male flowers contain two purplish anthers.** Female catkins are larger and may grow to 3–5 cm long. Catkin scales are a dark reddish-purple.

S. reticulata **(Net-leaved Willow)**
Rare, only found in the Highlands of Scotland. Forms mats to a maximum height of 20 cm. Its leaves are bluntly oval and grow to 1–3 cm long. The young leaves are covered in hairs on both upper and lower surfaces. **With age, the upper surface of the leaves become hairless and dark green in colour. The lower leaf surface is pale grey-green. The leaf surface is distinctively covered in a network of prominent veins.** The twigs root easily, as they often grow down close to the soil surface or even underground. The catkins appear with the leaves. The catkins are cylindrical and measure 2–3.5 cm long; they are found at the ends of twigs. The male flowers contain two purplish anthers. Catkin scales are purplish-brown.

HAVE OTHERS RECOGNISED THIS LEVEL OF VARIATION?

Willows are rather unusual among our frustrating flora, in that their complexity appears in part to have been created by human activity. Although humans have probably only rarely purposefully hybridised willows, historically we have planted and harvested them. Growing willows from cuttings in large numbers outside their natural niche must have increased the likelihood of interspecies encounters

Reported willow hybrids with two parents. Most of these crosses have been given names in their own right, but it is more informative to refer to their parentage.

	S. fragilis	*S. alba*	*S. viminalis*	*S. pentandra*	*S. daphnoides*	*S. purpurea*	*S. triandra*	*S. babylonica*	*S. caprea*	*S. cinerea*	*S. aurita*	*S. myrsinifolia*	*S. repens*	*S. lapponum*	*S. phylicifolia*	*S. arbuscula*	*S. lanata*	*S. herbacea*	*S. myrsinites*	*S. reticulata*
S. fragilis																				
S. alba																				
S. viminalis																				
S. pentandra																				
S. daphnoides																				
S. purpurea																				
S. triandra																				
S. babylonica																				
S. caprea																				
S. cinerea																				
S. aurita																				
S. myrsinifolia																				
S. repens																				
S. lapponum																				
S. phylicifolia																				
S. arbuscula																				
S. lanata																				
S. herbacea																				
S. myrsinites																				
S. reticulata																				

and thus the rate of hybridisation. This would be particularly true when large numbers of female plants were grown away from males of their own species. Once new hybrids were identified that possessed desirable properties, these would have been preferentially propagated. In this way, hybrids would have arisen and established more rapidly than in the wild.

Although humans seem to have encouraged a diversity of forms in willows, there is little evidence that they were given unique names until botanists took up the challenge. The exception to this seems to be 'Osier', a name borrowed from Old French meaning 'basket willow'. The term 'Pussy Willow' is rather younger and appears to have been first used in the USA in the mid-nineteenth century to refer to *Salix discolor* (American Pussy Willow, or Glaucous Willow). In the British Isles, the name 'pussy willow' is most frequently attached to the Goat and Grey Willows. Given the well-known link between willow bark and aspirin, it is perhaps surprising that willows were not better documented by herbalists. Culpeper's herbal, published in 1653, mentions only a single species of willow, and comments that a decoction of the leaves or bark will take away 'dandrif'. Victorian botanists were more observant; Bentham and Hooker considered that there were 15 discrete British species but did not consider the hybrids worthy of mention.

HOW FAR SHOULD I GO?

Identifying willows is no more difficult than many other groups of plants. However, the challenge comes when dealing with the multitude of unknown hybrids. This is made trickier because repeating the same cross can produce remarkably different-looking offspring. If you want to have a degree of confidence when attributing parentage, you will need to compare catkins and leaves with known parents. This requires a level of commitment to revisit the plants when they are flowering and in leaf. You may also wish to refer to a text that covers hybrids in more detail, such as Meikle's *Willows and Poplars of Great Britain and Ireland*, BSBI Handbook No. 4.

SECTION III

Inbreeders

(species that have sex with themselves)

Although it is not as extreme as abandoning sex altogether, species that have sex with themselves tend to make the task of identification more problematic for the novice botanist. When plants routinely self-fertilise it results in there being little genetic variation within their population. However, when new plant populations are formed following long-distance seed dispersal, the colonisers will sometimes just by chance be quite different from their source population. Newly established populations are likely to continue the habitat of self-pollination, and thus chance genetic differences will be maintained. It may be counterintuitive, but inbreeding species tend to have greater levels of genetic variation between their populations that do outbreeding species. Inbreeding species are thus more likely to have been divided into subspecies than are outbreeding species. You may be very familiar with a particular inbreeding species, but the next time you encounter it you may be thrown by its being a different colour, or lacking hairs, and so on.

Inbreeding species often have small, inconspicuous flowers, which makes them less attractive to pollinators. This can also render them more challenging to identify, because floral characters are frequently important in recognising species. Ecologically, the inbreeding habit is common in casual 'weeds' of disturbed ground. The ability to produce seeds as a single parent can help establish new populations following solitary seeds being deposited in isolated patches of open soil. So, although inbreeding is often regarded as being genetically detrimental, under these conditions it can be selected for. Humans have a habit of inadvertently transporting weed seeds around the globe. These long-distance dispersal events can also make such plants difficult to identify because unusual foreign species can turn up when you are not expecting them. As well as being weeds, many inbreeding species have been domesticated by humans as crops. This adds another level of taxonomic complexity, because genes can flow between cultivated and wild species, as crops and their wild ancestors grow side-by-side in large numbers. This can blur taxonomic boundaries and make the botanist's life more 'interesting'.

Although the families included in this group are primarily inbreeders, it is seldom the case that any plant group will exclusively stick to one particular breeding system. Hence, it is not unusual for inbreeding plants to occasionally indulge in a little promiscuity. Although hybridisation is more common in outbreeding species, it is not unknown in inbreeding species. Thus, again there is overlap between the complexity in these groups and those in Section II, 'species that have sex with other species'. A similar overlap is also seen in the last two groups of plants covered here. The brassicas not only contain many inbreeding species, they could also have be placed in the last section of successful families, because they number many species.

CHAPTER 13

Eyebrights

Eyebrights are low-growing, semi-parasitic, annual plants of short-sward grasslands. Most wildflower guides only include one or perhaps two species, along with a warning that they are highly variable. At the other extreme, some enthusiasts for the group have described between twenty and twenty-five species, twenty subspecies including nine British endemics and around sixty hybrids. Unfortunately, eyebright experts frequently disagree on how to partition the observed variation into meaningful species – and the keys available can provoke even the saintly to blaspheme. However, much of the frustrating nature of eyebrights arises from the variability within them rather than that among the botanists. You will need to look at several well-grown plants and still expect to be frequently perplexed. You know when you encounter a species named *Euphrasia confusa* that this is not going to be an easy task.

WHY IS THIS GROUP OF PLANTS COMPLEX?

Eyebrights are often found growing as small, isolated populations. It is possible that, much like Darwin's finches, they have evolved to become locally adapted to their unique environmental conditions. It is also possible that populations differ because of random genetic drift – in other words, isolated populations are different because of chance colonisation events. There is DNA evidence showing that there have been multiple natural colonisation events of plants from Europe. However, we currently know little about to what extent they may have subsequently evolved to become adapted to local conditions. So, it is difficult to determine which of these two theories is correct – and maybe both are.

For instance, it has been argued that *Euphrasia micrantha* (Slender Eyebright) has evolved to its local conditions by mimicking the flowers of *Calluna vulgaris* (Heather). The idea here is that as an unusual, rare flower hidden among the heather it may otherwise be ignored by pollinators. Again, this is difficult to prove, but it raises the possibility that what have been described as true species are in fact simply locally adapted mimics.

Most plants of grassland habitats are long-lived perennials. Eyebrights manage to pull off the trick of being annual by avoiding intense below-ground competition through stealing the root systems of other plants. They invade the roots of several different species (often grasses), effectively being parasitic below ground while photosynthesising above. This curious habit contributes to the complexity of identifying eyebrights. If they are lucky enough to parasitise a vigorous host, they are likely to grow larger and produce more flowers than if they tap into a smaller, less nourishing host plant.

Another complicating factor linked to the annual lifecycle of eyebrights is that they routinely self-pollinate. It is too risky looking for the ideal partner when you have only one season to find them. Plants that self-pollinate tend to have higher levels of genetic variation between populations than within populations. This encourages taxonomists to split them.

One final factor that makes eyebrights frustratingly complex is their diminutive size. In coastal and heavily grazed sites, plants may only grow a few centimetres tall. The features you will need to compare are tiny. You may have to measure several fiddly flowers to a few millimetres, precision while on the top of a mountain in strong wind.

HOW CAN I TELL THEM APART?

You need to be aware when trying to identify eyebrights that, being semi-parasitic, they are highly plastic. Different individuals within the same species may change their appearance significantly depending on their host. For this reason, you will need to look at about six well-grown plants. Avoid anything under 5 cm, because the parts you will need to examine will just be too small. Move on.

The key features to look out for are hairs on the base of the upper leaves that occur among the flowers (floral leaves). You will also need to determine how high up the stem the plant is flowering. To do this requires that you count the number of nodes (the position on the stem where pairs of leaves attach) from the base up, ignoring the lowest node where the cotyledon leaves were. At the base of the stem, nodes are usually just scars where leaves have fallen off. This can be tricky and may take some practice. Finally, you may need to measure the length of the flower tube. This will involve pulling it off the plant and using a hand lens and ruler with millimetre increments.

Let's look at the more widespread and distinctive species

Even the most distinctive eyebright species are often tricky to differentiate from other similar species. When getting started you may wish to combine some of these and begin by comparing your specimen with the following five widespread species and pairs of species.

E. rostkoviana/anglica **(Large-flowered Sticky and English Eyebrights)**
These two have sometimes been regarded as subspecies of *E. officinalis* (which is generally just referred to as Eyebright). They grow up to 35 cm tall, with up

to five or occasionally more pairs of branches (which are sometimes themselves branched). **At the bases of the upper floral leaves there are hairs which are long and wavy with glandular tips.** The lowest flowers generally occur between nodes 6 to 10. The flowers are usually white or lilac on the lower lip and lilac on the upper lip, and measure 6.5–12.5 mm. Most commonly found in short grasslands and heaths across the British Isles.

E. arctica **(Arctic Eyebright)**

Grows to 30 cm tall, with up to five pairs of branches. **At the bases of the upper floral leaves there are glandular hairs which are short and non-wavy.** The lowest flowers usually occur between nodes 4 and 9. The flowers are generally white on the lower lip and lilac to purple on the upper lip, and measure 6–11 mm. Most commonly encountered in grasslands in the north and west. However, this species may in fact be more common in the south than distribution maps indicate, as a result of historical recording errors.

E. nemorosa/pseudokerneri **(Common and Chalk Eyebrights)**

These species grow up to 35 cm tall, with up to nine pairs of usually erect branches. There are no glandular hairs. **The lowest flowers usually occur between nodes 10 to 15.** The flowers are generally white on the lower lip and white to lilac on the upper lip, and measure 5–9 mm. Found in many habitats, including short, herb-rich grasslands and heaths across the British Isles, being commonest in the lowlands and rarely occurring in Scotland.

Characteristics of widespread eyebright species

	Glandular hairs			Flowering node		Flower size		Number of branches
Species	**long**	**short**	**absent**	**<10**	**≥10**	**<8 mm**	**≥8 mm**	
E. rostkoviana/anglica Large-flowered Sticky & English Eyebrights								<5
E. arctica Arctic Eyebright								<5
E. nemorosa/pseudokerneri Common & Chalk Eyebrights								<9
E. tetraquetra/confusa Western & Confused Eyebrights								<5
E. micrantha/scottica Slender & Scottish Eyebrights					rarely			usually unbranched
Your specimen								

Selection of five widespread eyebright species

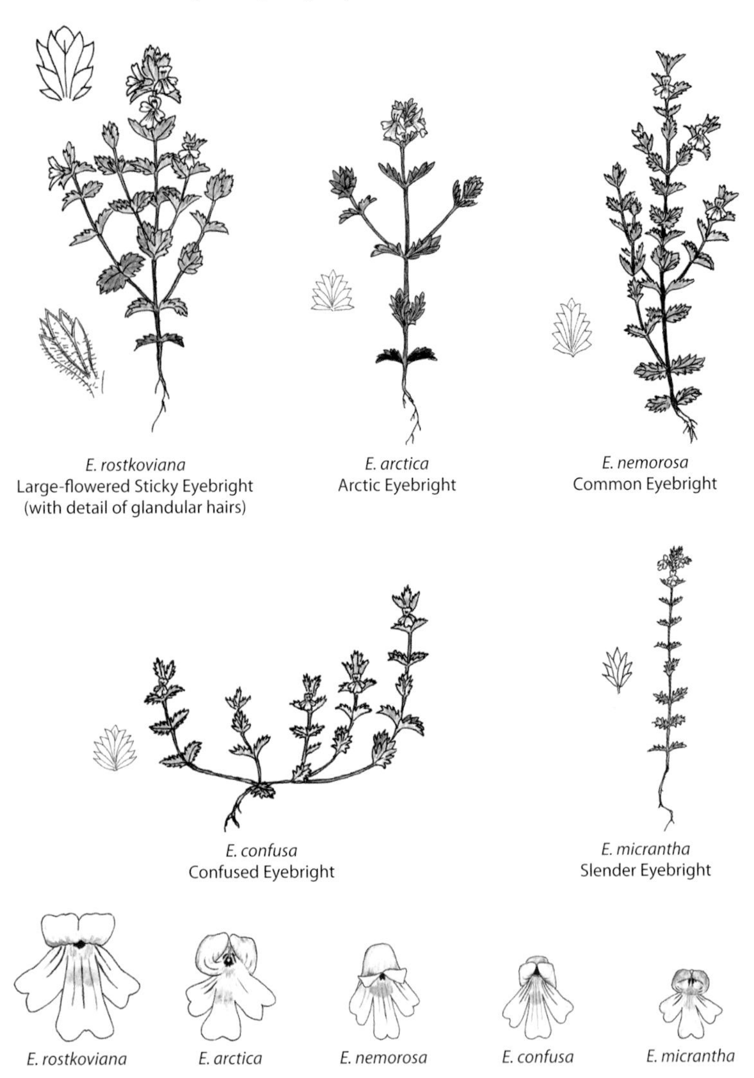

E. tetraquetra/*confusa* (Western and Confused Eyebrights)

These species grow up to 20 cm tall, with **up to five pairs of recumbent branches (often forming dense plants)**. There are no glandular hairs. The lowest flowers usually occur between nodes 5 to 10. The flowers are usually white on the lower lip and white to lilac on the upper lip, and measure 5–8 mm. Found in many habitats including short grasslands, dunes and heaths across most of the British Isles, except eastern England and northern Scotland.

E. micrantha/scottica **(Slender and Scottish Eyebrights)**
These species grow up to 25 cm tall, **they are usually unbranched** but may have up to five pairs of branches. There are no glandular hairs. The lowest flowers usually occur between nodes 3 to 7 in *scottica* and sometimes higher in *micrantha*. **The flowers are usually lilac to purple** and measure 4.5–6.5 mm. **Found in heaths growing with *Calluna*** across most of the British Isles, except central and eastern England.

HAVE OTHERS RECOGNISED THIS LEVEL OF VARIATION?

Historically, herbalists have only recognised a single species of eyebright. Bentham and Hooker's *Handbook of the British Flora* published in 1858, which was the standard work for many years, included just a single example. By 1965, with Keble Martin's *Concise British Flora*, the single eyebright had been split into twenty species. This is more than are included in Stace (2019). DNA analysis published in 2007 by G.C. French *et al.* suggests that the Cumbrian Eyebright and Cornish Eyebright (considered as stabilised hybrids of *E. officinalis*, the name often used for the species of which *E. rostkoviana/anglica* are considered subspecies), which both have two sets of chromosomes, may in fact be considered good species in their own right. However, most eyebright species have four sets of chromosomes and variation in these may be more related to their environment than to their genes.

HOW FAR SHOULD I GO?

In most regions of the British Isles, if you encounter an eyebright it is likely to be one of those covered above. Once you are comfortable with these you may wish to progress to the rarer and more unusual species. To do this you will need to consult a specialist publication, such as *Eyebrights (Euphrasia) of the UK and Ireland* (2018) by Chris Metherell and Fred Rumsey.

If you find yourself looking at a specimen that does not fit any of the above descriptions and you are in the mountains of northern Scotland, then it is probably best ignored until you have more experience and a copy of *Eyebrights of the UK and Ireland* to hand. Chris Metherell's advice is: 'ignore the rare stuff – all those hairy things in the north ... If it's a species that you can only find by going to a very special place to see, then you can ignore them if you're just beginning.'

Given French *et al.*'s conclusion that most of the variation they observed resulted from the environment rather than genetic differences, you may be content to consider that most of the variation observed in the field is simply due to plasticity – and so perhaps Bentham and Hooker were right to include just a single species. Future genetic work may clarify the situation further. However, the division between plants with two and four sets of chromosomes does appear to act as a barrier to hybridisation. So, there is a logic in recognising two separate taxa of eyebrights.

CHAPTER 14

Fumitories

As a group, members of the genus *Fumaria* are easy to identify. These are annual plants of arable fields, gardens and disturbed ground. They have delicate small flowers and highly divided leaves. A few of them are common and widespread, and perhaps because of this they are often overlooked. However, some of the ten *Fumaria* species found in the British Isles are rare, including two that occur found nowhere else on Earth. For this reason, it is worth knowing how to identify them, because it is possible that what you thought was just a common weed might be something more unusual.

WHY IS THIS GROUP OF PLANTS COMPLEX?

Many weeds have a life-history strategy that involves opportunistically growing in disturbed open ground and thus avoiding competition from other species. To be able to colonise such niches, which are often created almost at random by human activity, plants need to produce lots of small, highly mobile seeds. Once a seed has germinated and established, the seedling is likely to find itself growing on its own. Under these circumstances, isolated plants need to be able to self-pollinate to produce seeds. Plants that inbreed like this have little variation within a population, but because of chance founder effects they may vary markedly between locations. Hence, taxonomists are more likely to split inbreeding species into subspecies than they do outbreeding species. It is no surprise, therefore, to find that there are several subspecies of some *Fumaria*.

There are fewer than fifty species of plant that are found in the British Isles and nowhere else in the world. Two of them are fumitories. Endemic species such as these have either just evolved, or are just about to go extinct, or both. Newly evolved species are often poorly taxonomically differentiated and hard to distinguish from their parent species. For this reason, most of the plant species endemic to Britain and Ireland are frustratingly difficult to identify. One British endemic, Purple Ramping-fumitory, is thought to have evolved from a hybrid between Common Fumitory and Common Ramping-Fumitory. The new species arose by way of this hybrid doubling its number of chromosomes. Since both parent species are highly variable and often divided into subspecies, and hybrids between the two are still being reported, you will understand why their identification can be complex.

Finally, it is always worth remembering that species with pinky-purple flowers are prone to colour variation, because of the nature of the pigments involved. Pale and white (albino) variants of flowers are common, so flower colour is a poor guide to species identification.

HOW CAN I TELL THEM APART?

The first thing to be aware of when looking fumitories is that they can only be identified reliably during the early and middle parts of their flowering season (May–July), and you should only attempt to identify well-grown non-shaded specimens. Outside of this period their key features tend to be variable and less distinct, while plants growing in shady conditions may be etiolated and their flowers paler than usual. Your confidence in disentangling this complexity will grow, but caution is always wise. The key features to look at are the shapes of flowers and fruits. The shape of the sepals, which appear like small shields on the sides of the flowers, are particularly important. Since these features are small, a hand lens will be helpful. While flower colour is important, the most prominently coloured dark purple tips of the petals are not helpful – even White Ramping-fumitory has dark purple tips to its flowers. Further confusion may arise from the common name 'ramping', which might imply they have a vigorous growth habit; however, this does not relate to growth form but rather the tendency to climb through other vegetation. Generally speaking, fumitories have flowers that measure less than 9 mm from spur to tip, while ramping-fumitories have longer flowers, measuring more than 9 mm.

Let's start with the commoner species

Fumaria officinalis **(Common Fumitory)**
Flowers pinkish and small, measuring 6–8 mm long, usually **with about 20 or more flowers per cluster**. The sepals are toothed and white or pink in colour, measuring 1–1.5 mm wide and 1.5–3.5 mm long. The fruits are wider than they are long and have a flat top. Fruit measure about 1.5 mm across. The bract on the fruit stalk is shorter than fruit stalk itself. A common and widespread plant found in most of lowland Britain and Ireland.

F. muralis **(Common Ramping-fumitory)**
Flowers are pinkish-red. They measure 9–11 mm in length, with between 12 and 15 flowers per cluster. **The stalks of the flower clusters similar in length to the flower cluster.** The sepals are large, measuring 1.5–3 mm wide and 3–5 mm long, and are toothed mostly along the bottom. The sepals are white to pale pink. Fruit are spherical in shape with a round top, and measure 2.5 mm across. The bract on the fruit stalk is about half as long as the stalk. A common plant of arable and waste ground across Britain and Ireland, but most common in western Britain and eastern Ireland.

Characteristics of *Fumaria* species

Species	Flower size		Flowers per cluster			Stalk length to flower cluster			Sepal length		
	<9 mm	>9 mm	<10	10–20	>20	shorter	same	longer	<2 mm	2–3 mm	>3 mm
F. officinalis Common Fumitory	■				■	■			■	■	■
F. muralis Common Ramping-fumitory		■		■			■				■
F. bastardii Tall Ramping-fumitory		■		■		■				■	
F. capreolata White Ramping-fumitory		■			■			■		■	
F. reuteri Martin's Ramping-fumitory		■		■		■					■
F. densiflora Dense-flowered Fumitory	■			■	■	■				■	■
F. parviflora Fine-leaved Fumitory	■			■		■			■		
F. vaillantii Few-flowered Fumitory	■		■			■			■		
F. purpurea Purple Ramping-fumitory		■			■		■				■
F. occidentalis Western Ramping-fumitory		■		■			■				■
Your specimen											

F. bastardii (Tall Ramping-fumitory)

Flowers pale pink to pink, with dark tips to their lateral petals. They measure 9–11 mm long and have 10 to 12 flowers per cluster. The stalk supporting the flower cluster is shorter than the length of the cluster. **The sepals are small in comparison to the rest of the flower, toothed and white, measuring only 2 mm by 3 mm.** The fruit are round and measure 2.5 mm across, with no fleshy neck between the fruit and stalk. The bract on the fruit stalk is a third of the length of the stalk. A species of arable and waste ground, common in much of western Britain and Ireland, often near the coast.

F. capreolata (White Ramping-fumitory)

Flowers creamy-white with dark red tips to upper and lateral petals. They measure 10–14 mm in length, with about 20 flowers per cluster. The stalks of the flower clusters are longer than the flower cluster. **The sepals are round and large, broader than the flower** and measuring 4–6 mm wide by 2.5–3 mm long. The sepals are white in colour and toothed. The fruit are round and measure 2 mm across, they have a down-curved stalk and there is a pronounced neck between the fruit and its stalk. The bract on the fruit stalk is between a third to a half the length of the stalk. A plant of arable and waste ground, reasonably common in much of Britain and Ireland, with a tendency to be coastal.

And now the rarer, more unusual species

F. reuteri (Martin's Ramping-fumitory)

Flowers pinkish purple, measuring 11–13 mm in length. There are 15 to 20 flowers per cluster, and the stalk is shorter than the flower cluster. **The sepals are large, measuring 3 mm wide by 6 mm long. The sepals are white in colour, round to oblong and have very few teeth.** The fruits are round, with a nipple on top; they measure 2.5 mm across and have a fleshy neck. The bract on the fruit stalk is between a half to two-thirds the length of the stalk. A rare plant found on the Isle of Wight and very occasionally elsewhere in southern England and historically in parts of Cornwall.

F. densiflora (Dense-flowered Fumitory)

A small-flowered species whose flowers measure 6–7 mm in length; the petals are pinkish red and have dark tips. There are about 20 flowers per cluster, and the **flower clusters are much longer than their stalks.** The sepals are large compared to the rest of the flower, measuring 2 mm wide by 3mm long. They are white to pale pink and toothed all the way around. The fruits are round, measuring 2 mm across. The bract on the fruit stalk is longer than the stalk. Reasonably common on well-drained and chalky disturbed soils in south-east England and parts of central Scotland; rare in Ireland.

F. parviflora (Fine-leaved Fumitory)

A small-flowered species whose flowers measure 5–6 mm. Its petals are white with dark reddish tips. They have 15 to 20 flowers per cluster, and the flower clusters are much longer than their stalks. **The sepals are small, measuring 0.8 mm by 1 mm, toothed and white in colour.** The fruit are round with a nipple on top and measure 2 mm across. The bracts on the fruit stalks are as long as the stalk. Mostly found on disturbed chalky soils in the south and east of England, but also Yorkshire and very rarely in southern Scotland.

F. vaillantii (Few-flowered Fumitory)

Another small-flowered species whose **flowers measure 5–6 mm.** Its petals are pale pink with dark reddish tips. **There are usually 5 to 15 flowers per cluster,** and the flower clusters are much longer than their stalks. The sepals are small, measuring 0.5 mm wide by 1 mm long. They are toothed and pale purple in colour. Fruits are round with a somewhat flattened top and measure 2 mm across. The bracts on the fruit stalks are a half to three-quarters as long as their stalk. Mostly found on disturbed chalky soils in the south and east of England, although it is occasionally found as far north as Scotland.

Flowers (with their lengths) and seedpods of British *Fumaria* species

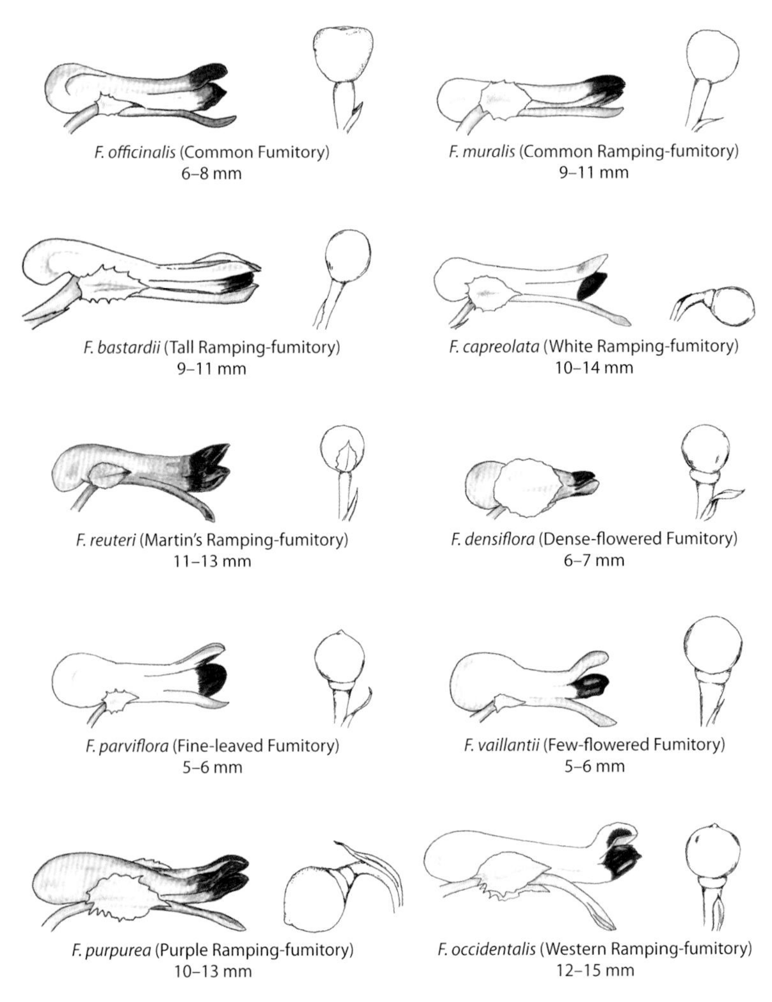

Finally, fumitories endemic to the British Isles

F. purpurea (Purple Ramping-fumitory)

Flowers are pink, becoming darker as they age; they measure 10–13 mm in length. Flowers occur in clusters of 20 to 25, and these clusters are about the same length as their stalks. **The sepals are large, measuring 3 by 5 mm**; they are white and toothed along their base. The fruits are round, measuring 2.5 mm across; **they have a distinctive down-turned stalk**. The bract on the fruit stalk is between two-thirds to the full length of the stalk. Rare and sparsely scattered, growing in arable and disturbed ground in the west of England, the east of Ireland and the east of Scotland – particularly Orkney.

F. occidentalis (Western Ramping-fumitory)
Has large white to pale pink flowers with dark red tips. The flowers are 12–15 mm long and occur in clusters of 12 to 20. The flower clusters are similar in length to their stalks. The sepals are large, measuring 3 mm by 5 mm, white in colour and toothed, mainly at the base. The fruits are round, and measure 3 mm across, with a slight nipple on top. The bract on the fruit stalk is between one-third to half the length of the stalk. A rare plant of arable and waste ground, only found in Cornwall and the Isles of Scilly.

HAVE OTHERS RECOGNISED THIS LEVEL OF VARIATION?

Fumitories have been known to herbalists since ancient times. In fact, it is possible that both the scientific and the common names (which mean 'smoke of the earth') originated with the Roman herbalist Pliny, who reported that the juice of fumitory causes the eyes to produce tears as does smoke. There are several other possible explanations, including a legend that the plants grew from vapours which arise from the ground, rather than from a seed. Fumitories are also mentioned in Shakespeare's *King Lear* and Chaucer's *Canterbury Tales*. Other references to fumitories include recipes for making yellow dyes. None of these records appear to recognise more than a single species. However, Culpeper's herbal, published in 1653, remarks that 'in the corn-fields of Cornwall it bears white flowers'.

The two British endemic species of *Fumaria* were not recognised as new species until the early years of the twentieth century. It rather looks like taxonomists and others have been slow to fully catch on to the species diversity within this genus.

HOW FAR SHOULD I GO?

Fumaria can only be identified with any confidence between May and July. This will inevitably limit opportunities to develop your field skills. However, if you are interested in finding rare and unusual species, this group is perhaps a good place to start, because these plants are both widespread, generally overlooked and there are not too many different species to choose from. Once you pick up a hand lens and begin to observe the subtle differences between these species, you will also start to appreciate their delicate beauty. For those wanting to go further, the BSBI Handbook No. 12, *Fumitories of Britain and Ireland* by Rosaline Murphy and Tim Rich is helpful.

CHAPTER 15

Violets and pansies

The Violaceae are one of the easiest groups in this book. The family includes both the violets and the pansies, which are some of our best-known and loved species. However, most non-botanists are probably unaware that there are several species of both violets and pansies which can be found growing wild, some common and widespread, others much rarer. In fact, the British and Irish flora contains a manageable nine species of violets and five species of pansies. However, there are many common colour variants, escaped garden varieties, subspecies and hybrids that potentially make this group challenging.

WHY IS THIS GROUP OF PLANTS COMPLEX?

With most puzzling plant groups, there is one reason why they are taxonomically complex. But that is not true in this case: there are actually three reasons why violets are tricky. There is a clue to the first reason in the name 'violet'. The flowers of these well-known, small herbaceous plants contain anthocyanin pigments. It is thought that because there are so many steps in the biosynthesis pathway that makes these coloured compounds this results in them being more prone to mutations than other plant pigments. Thus, pale and even white colour forms are not uncommon. Secondly, when they have the opportunity, violets readily hybridise. Flowering in early spring you cannot be too picky regarding with whom you mate. Consequently, there are almost as many hybrid violets recorded as there are pure species. Finally, violets often engage in a rather curious form of sexual reproduction. In the British Isles, spring can be a somewhat unreliable season. In some years the weather is mild: violet flowers flourish, are visited by pollinating insects and set viable seeds. However, in other years spring is cold, wet and miserable. Sex and bad weather do not mix. In these bleak years, violet flowers get nipped by the cold, rot in the wet and there are few insects around to pollinate them. Under these conditions, violets produce flowers that automatically self-pollinate without even opening. Technically this is called cleistogamy, and is the most extreme form of inbreeding. When plants routinely inbreed it results in little genetic variation within a population. Yet when new populations are established by chance, the first plants to get a hold will sometimes randomly be quite different from their founder population. Since the new population also routinely self-pollinates, any genetic differences will be maintained. For example, dispersal of seeds from a single white-flowered plant might give rise to a totally white-flowered population. In this way, new subspecies develop.

Pansies, in comparison, are relatively easy to identify as they hybridise less frequently and also do not produce cleistogamous flowers and so tend to be less prone to inbreeding and thus have less taxonomic complexity.

HOW CAN I TELL THEM APART?

One of our commonest species of violets is *Viola odorata* (Sweet Violet). As the name implies this is fragrant, and some wildflower guides suggest that possession of a scent can be used as an aid to its identification. There are two problems with this approach. The distinctive smell of violets is derived from volatile compounds designed to attract pollinators, specifically from a mixture of α-ionone and β-ionone. Unfortunately, many people have a gene that is associated with the inability to detect β-ionone, except in very high concentrations (when it smells like onions). Secondly, β-ionone temporarily blocks our sense of smell. This is less than ideal when comparing the fragrance of two different plants, but it does explain why during the Victorian period violets were a regular constituent of nosegays, designed to mask unpleasant odours.

Unfortunately, there is no single feature that will differentiate all nine species of violets; instead, we need to consider several characters. The place to start is looking at the back of the flower. A combination of the shape of the sepals (either blunt or sharp) and the colour contrast between the spur that sticks out at the back of the flower and the rest of the flower will greatly reduce the number of options possible. Next, consider the overall growth form of the plant and its leaves. Identification should now be possible, assuming you have one of the nine true species.

Viola growth forms

Leaves and flowers from base

Leaves and flowers from stem

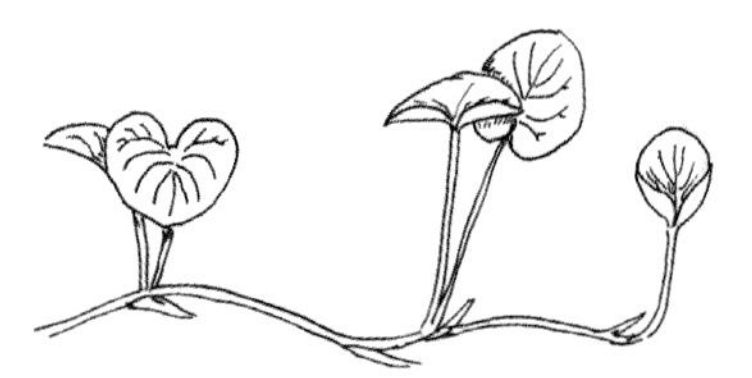

Rhizomes

Let's start with the commoner and more distinctive species of violets

V. odorata (Sweet Violet)
Flowers of this species can vary from white through to purple; the spur is usually the same colour as the rest, or darker. Its sepals are blunt. **Leaf stalks arise at the base of the plant and are covered in short, down-turned hairs.** The plants have creeping stolons and roundish heart-shaped leaves. Found in woodland, verges and scrub on base-rich soils all over Britain and Ireland, but becoming rarer in the north and west.

V. hirta (Hairy Violet)
Flowers are blue-violet; the spur is the same colour as the petals. Its sepals are blunt. Leaf stalks arise at the base of the plant in a rosette-like clump. The leaf stalks are covered in long hairs. **The hairy leaves are elongated heart shaped.** Occurs in calcareous grasslands and scrub across Britain, but more in the south; rare in Ireland.

V. riviniana (Common Dog-violet)
Flowers pale blue-violet, **with a pale-coloured whitish spur with a vertical notch at the end.** Its sepals are sharply pointed. The plant forms a rosette and has roundish heart-shaped leaves. Found in a wide range of woodlands and grasslands across much of England, becoming rarer or very rare in the far north and Scotland.

V. reichenbachiana (Early Dog-violet)
Flowers bluish-mauve; **the spur is always darker than the rest of the flower, often purple in colour.** Its sepals are sharply pointed. The plant forms a rosette, and has oval heart-shaped leaves. Common in woodlands and hedges across Britain and Ireland, but very rare in Scotland.

V. canina (Heath Dog-violet)
Flowers bluish; **the spur is pale greenish-yellow.** Its sepals are sharply pointed. Leaves and flower stalks arise from the stems. The plants have oval to round, heart-shaped through to more or less triangular leaves. On heaths, shingly lake shores and sand dunes, scattered across the British Isles.

V. palustris (Marsh Violet)
Flowers pale pinkish-purple with dark veins. The spur is a similar colour to the rest of the flower. Its sepals are blunt. The plant produces rhizomes and has round, heart-shaped leaves. In a wide range of wet habitats including bogs, fen, wet heaths and woodlands across the British Isles, especially in the west and north.

Characteristics of *Viola* species

Species	Sepals		Spur colour			Growth form				Petiole (leaf stalk) hairs		
	blunt	sharp	paler	same	darker	rhizomes	flowers from base	flowers on leaf stems	rosettes	short	long	none
V. odorata Sweet Violet	■			■	■		■			■		
V. hirta Hairy Violet	■			■			■				■	
V. riviniana Common Dog-violet		■	■						■	▒		■
V. reichenbachiana Early Dog-violet		■			■				■			■
V. canina Heath Dog-violet		■	■					■		▒		▒
V. palustris Marsh Violet	■			■		■						■
V. stagnina Fen Violet		■	■	■		■						■
V. lactea Pale Dog-violet		■		■				■				■
V. rupestris Teesdale Violet		■	■						■	■		
Your specimen												

And now the rare, more unusual species

V. stagnina (Fen Violet)

Flowers white to very pale blue. The spurs are pale and greenish. Its sepals are pointed. The plant forms a rosette and has narrowly ovate leaves. Extremely rare: restricted to fens in Oxfordshire and Cambridgeshire and a few locations western Ireland and Northern Ireland.

V. lactea (Pale Dog-violet)

The flowers of this species are pale-lilac to almost white. The spur is white. Its sepals are sharply pointed. Flower stalks arise from leaf stems. The plants have lanceolate leaves. Rare on dry heaths in the south and south-west of England and near the coast in Wales and Ireland.

V. rupestris (Teesdale Violet)

A very small species, its flowers are pale blue-violet, the spur usually paler than the rest of the flower. Its sepals are pointed. The plant forms a rosette. The leaves are usually hairy and the leaf stalks are covered in short, dense hairs. The leaves are scoop shaped. **Rare, restricted to a few limestone grasslands in the north of England.**

Flowers and leaves of our commoner violets

  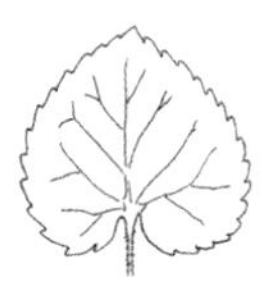

V. odorata (Sweet Violet)

V. hirta (Hairy Violet)

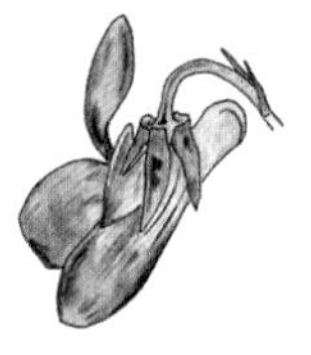  

V. riviniana (Common Dog-violet)

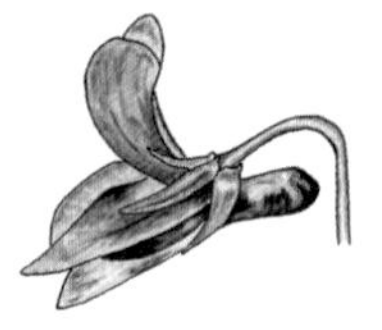  

V. reichenbachiana (Early Dog-violet)

V. canina (Heath Dog-violet)

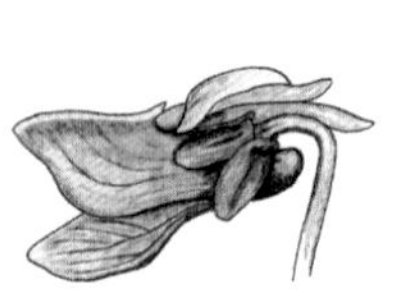  

V. palustris (Marsh violet)

Flowers and leaves of our less common Violets

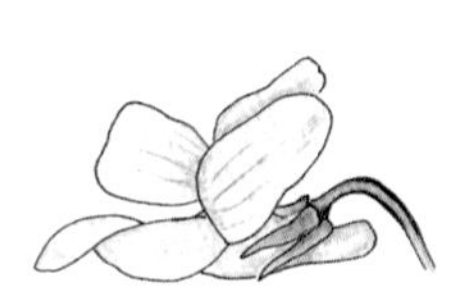

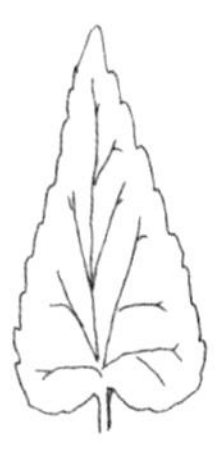

V. stagnina (Fen Violet)

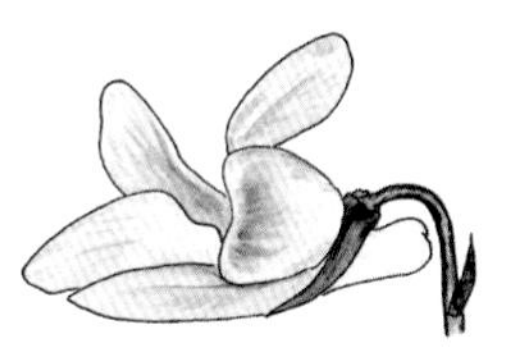
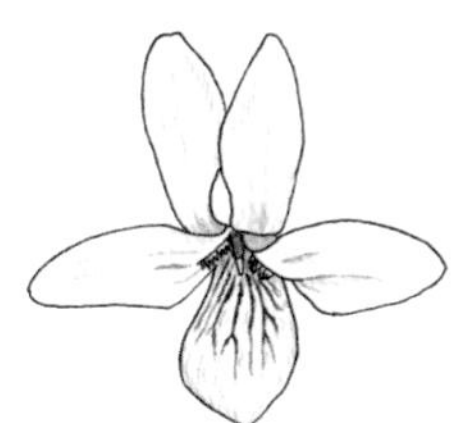
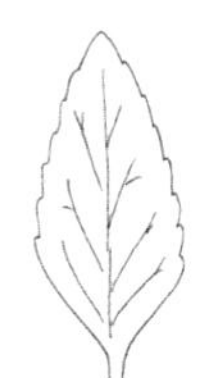

V. lactea (Pale Dog-violet)

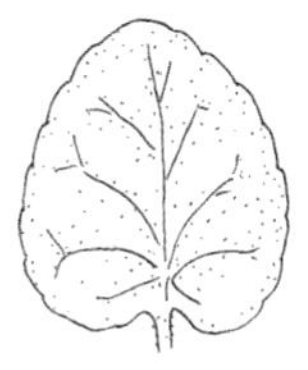

V. rupestris (Teesdale Violet)

Finally, there are a couple of violet hybrids to consider

Viola hybrids are often initially recognised by the dead, brown flowers which remain on the plant and do not form seed capsules.

V. riviniana x *V. reichenbachiana*, = *V.* x *bavarica*
These two parents commonly grow together, and occasionally this sterile hybrid is found. It is generally intermediate, and its flowers have a spur which is usually the same colour as the flower.

V. riviniana x *V. canina*, = *V.* x *intersita*
These two parents grow together in open sites such as sand dunes, stony river edges and lake shores and heathland. Here they occasionally form this sterile hybrid which is intermediate in form.

HAVE OTHERS RECOGNISED THIS LEVEL OF VARIATION?

Violets have been culturally significant since ancient times. However, early records typically do not concern themselves with the exact species involved. This changed dramatically during the Victorian period, when violets attracted the attention of both horticulturalists and highly enthusiastic botanists. The burst of interest resulted in many colour variants being identified and then cultivated. The propagation and dissemination of unusual variants must have contributed to the overall taxonomic complexity, as some have escaped back into the wild. Floras written in the first part of the twentieth century listed as many as fifty variant forms, whereas today the average member of the public is unlikely to know that there is more than one species of violet.

HOW FAR SHOULD I GO?

Several of our *Viola* species are widespread, well-known and well-loved members of the British flora; others are rare and restricted to a few sites of high conservation value. Even with the complications of their many variants, it is relatively easy to identify *Viola* to species level. Given their beauty and conservation importance, taking the time to identify violets can be rewarding. Whether you wish to go beyond this into the complexity of naming hybrids and subspecies is a matter of personal motivation. Simply knowing that hybrids may occur could prevent you spending too long pondering an oddity. However, a degree of uncertainty is always likely to be associated with attributing parentage to a *Viola* hybrid unless you have access to DNA analysis.

HOW DO I TELL VIOLETS AND PANSIES APART?

The violet family is divided into two sections: violets and pansies. These can be easily differentiated by looking at the small, leaf-like structures called stipules that are found at branching points where the leaves emerge from the stem. The stipules

Typical violet stipules with teeth rather than lobes

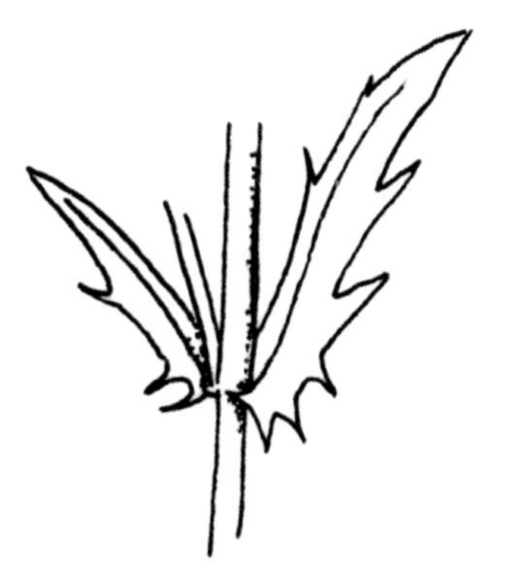

Flowers and stipules of our three commonest pansies

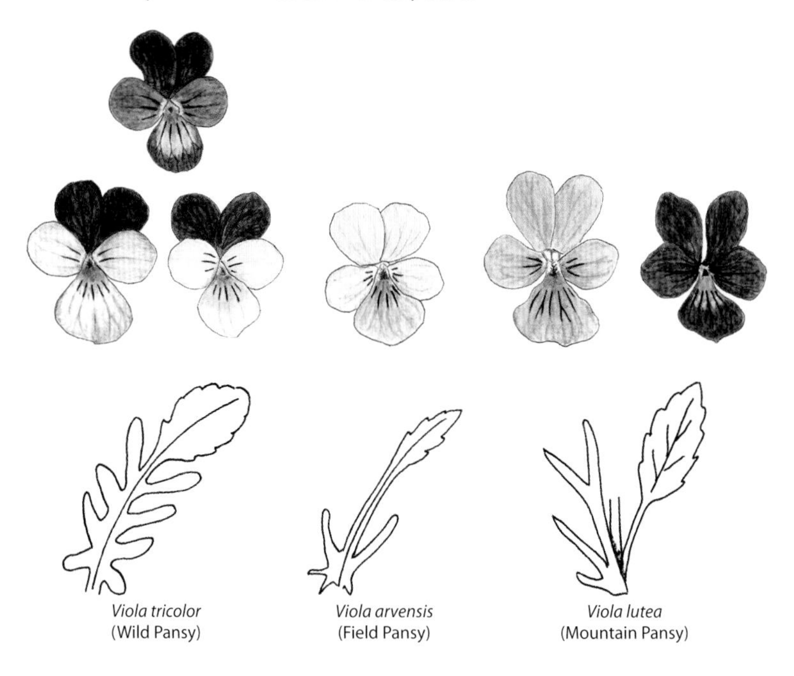

of violets are simple and unbranched but may be toothed, while in contrast the stipules of pansies are lobed and much more leaf-like.

V. tricolor **(Wild Pansy)**
An annual or short-lived perennial of grasslands, including sand dunes, arable fields and disturbed ground. Its flowers measure 10–25 mm, with petals longer than the sepals. They are highly variable in colour: the petals may be entirely yellow, or entirely purple, or any combinations of the two. Stipules lobed along midrib. Occasional throughout British Isles.

V. arvensis **(Field Pansy)**
An annual weed of cultivated ground. Flowers measure 8–15 mm, with sepals noticeably longer than petals. Usually pale cream in colour, sometimes tinged with violet. Stipules lobed from base. Common throughout Britain and Ireland, less so in far north and west.

V. lutea **(Mountain Pansy)**
A perennial plant of upland grasslands and metal mine sites, scarcely present in southern and eastern England. Its flowers are 20–35 mm long and yellow, blue, purple or a combination of the three. Stipules have few lobes along midrib.

There are another two pansies you are unlikely to find

V. kitaibeliana **(Dwarf Pansy)**
Only found in the Isles of Scilly and the Channel Islands. Somewhat like the Field Pansy, but with smaller flowers (4–8 mm) which are cream coloured with a hint of violet.

V. cornuta **(Horned Pansy)**
This plant was introduced from the Pyrenees in 1776 and is still to be found in gardens. Now also occurs occasionally across Britain, particularly in Aberdeenshire. Its lilac-mauve flowers are 20–40 mm long.

Hybrid pansies and how far should I go?

The three commoner pansy species all hybridise. In addition to being highly variable in flower colour, hybridisations means that absolute certainty in the identification of some individual plants will always be challenging. It is probably wise not to lose too much sleep over individuals that are intermediate in character.

CHAPTER 16

Short white-flowered crucifers: cresses

It is generally easy to recognise members of the cabbage family. These plants typically have four petals arranged roughly in a cross. Within the flower they usually have six stamens, typically four with long filaments and two with shorter filaments. Many members of this family are straightforward to identify, but there are two groups that are covered separately here because they include so many similar-looking species. The first of these are the low-growing white-flowered cresses, which are usually less than 50 cm tall. These often have a peppery taste. They are typically short-lived species of open ground close to sites of human activity, although some have more specialised ecological requirements. Many initially develop a rosette of leaves, before producing a flowering stem/s. The rosette and stem leaves may look very different from each other.

WHY IS THIS GROUP OF PLANTS COMPLEX?

Cresses are typically short-lived weeds of ephemeral, disturbed habitats. Each population has probably been established from a single long-distance seed dispersal event. As cresses are readily able to self-pollinate, each primary coloniser is likely to be the sole parent of all subsequent generations. Hence, any unusual characteristics that the founder plant may have had will be passed on to its offspring, and so each population of cresses may differ slightly from the others. Over time the accumulation of such random differences may be recognised as subspecies or even as new species in their own right. In some cases, such as the scurvygrasses, different populations may even have different numbers of chromosomes. This genetic variation appears to have facilitated the evolution of ecotypes/species which are adapted to particular environmental conditions.

Given that such new species have evolved as a result of the genetic isolation resulting from their habitat of inbreeding, they are still likely to be able to cross with other closely related species when the opportunity arises. It is therefore not surprising to discover that several cresses are known to hybridise, making their identification more challenging.

In addition to this genetic complexity, many of the species included in this chapter are small: their leaves are small, and their flowers are small and often inconspicuous, which means they may lack diagnostic features that would helpful to confidently confirm their identity. Certainly, a hand lens will be helpful. Finally, humans have added another level of complexity by introducing several species – for example, spilt birdseed mix can be a regular source of challenging cress species. You may need to travel no further than your back garden to be confronted by an unusual species of white crucifer.

HOW CAN I TELL THEM APART?

There are so many similar species of short white-flowered cresses that the first thing we need to do is to divide them up into smaller, more manageable subgroups. The obvious way to do this is based on the shape of their seedpods. Below, the species will be grouped into three sets: 1) species with long seedpods, 2) those with rounded seedpods, and 3) species with flattened seedpods. This division is not perfect, since a few species have only slightly flattened seedpods, or it might not be instantly clear how long is a long seedpod. However, for the vast majority of species you should have little difficulty in deciding in which group to search. The group with flattened seedpods is so large that here it is subdivided, splitting off species that also have flattened flowers. While many species of cress have rather similar small white flowers, the pepperworts and swine-cresses in the genus *Lepidium* have rather distinctive compressed flowers. In these species the flowers comprise two sets of two petals, which appear separated – as if the flower had been slightly pulled apart. These are dealt with separately in a group of their own.

Pepperwort and swine-cress flowers appear flat and compressed

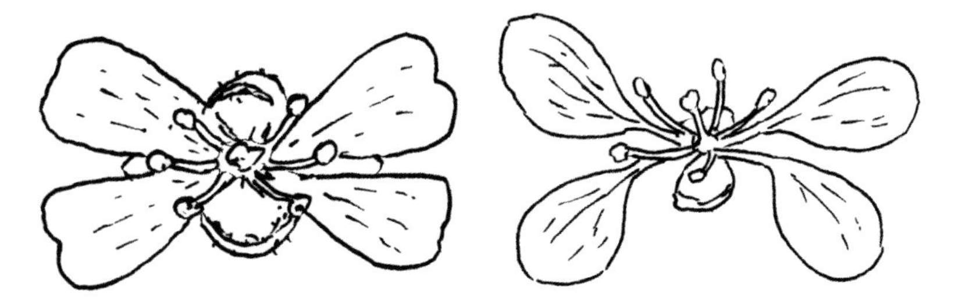

Once you have looked carefully at the shape of the seedpods, you will need to consider the shape of the basal leaves. Some cresses have leaves that are divided into separate leaflets, others have leaves partially divided into lobes, and yet others have simple undivided leaves. It is also helpful to look for the presence of any hairs on the leaves and flowering stalks.

Although the flowers of most cress species contain six anthers, a few have only four. The colour of the anthers can also be a useful feature. You will need a hand lens to inspect the anthers, and sometimes a fine-scale ruler to measure petal length will be handy.

White-flowered cresses with long seedpods

Arabidopsis thaliana (Thale Cress)
This species was the first plant to have its entire DNA genome sequenced, it has been used in many genetic studies. While going unnoticed by most people as it is a common weed of disturbed ground. Its delicate, branched stem grows to a height of 30 cm, and produces few stem leaves. **Rosette leaves are oval with small teeth and measure up to 4 cm long. The leaves are covered in hairs**, which are sometimes branched. The whole plant has a slightly greyish colour. The petals measure 2.5–4.5 mm long and are roughly twice as long as the sepals. The flowers contain 4 or 6 yellow anthers. The seedpods are long, and slightly flattened, measuring up to 1.8 cm.

A. petraea (Northern Rock-cress)
This small mat-forming perennial is an alpine plant of the mountains of Scotland and North Wales, where it grows to a maximum height of 25 cm. The stems tend to curve upwards, are unbranched and have few stem leaves. The rosette leaves usually lack hairs, **they have a rounded end and distinctive side lobes and measure up to 5 cm long.** The petals measure 4.5–9 mm long and are roughly twice as long as the sepals. The flowers contain 6 yellow anthers. **The seedpods are curved, long and slightly flattened, measuring up to 3 cm.**

Cardamine hirsuta (Hairy Bitter-cress)
This common short-lived plant occurs across the whole of Britain and Ireland in disturbed ground. Its branched stems grow to a height of 30 cm, the stem leaves are like smaller and thinner versions of the basal leaves. **Basal rosette leaves comprise 2–4 paired rounded leaflets along each side, with a larger terminal leaflet. The leaves are hairy and measure up to 10 cm long.** The petals measure 2.5–4.5 mm long and are roughly twice as long as the sepals. **The flowers contain 4 yellow anthers.** In winter-flowering plants, the flowers may not open, but are still able to self-pollinate and produce viable seeds. The seedpods are long and thin, measuring up to 2.5 cm; they tend to be held in a vertical position.

C. flexuosa (Wavy Bitter-cress)
This common short-lived plant occurs across the whole of Britain and Ireland in woods, wet ground and near waterbodies. Its hairy stems are branched and grow to a height of 50 cm; **the stem leaves are similar to the basal leaves.** Basal rosette leaves comprise 2–4 paired rounded leaflets along each side with a larger terminal

Characteristics of white-flowered cresses with long seedpods

Species	Plant height <30 cm	Leaves				Anthers		Petals >1 cm
		few stem leaves	rosette absent	hairy	with leaflets	<6	Colour	
Arabidopsis thaliana (Thale Cress)						4 or 6		
Arabidopsis petraea (Northern Rock-cress)								
Cardamine hirsuta (Hairy Bitter-cress)						4		
Cardamine flexuosa (Wavy Bitter-cress)								
Cardamine impatiens (Narrow-leaved Bitter-cress)								
Cardamine amara (Large Bitter-cress)								
Cardamine corymbosa (New Zealand Bitter-cress)								
Arabis hirsuta (Hairy Rock-cress)								
Arabis scabra (Bristol Rock-cress)								
Arabis alpina (Alpine Rock-cress)								
Arabis caucasica (Garden Arabis)								
Nasturtium officinale (Water-cress)								
Nasturtium microphyllum (One-rowed Water-cress)								
Your specimen								

leaflet. The leaves are hairy and measure up to 15 cm long. The petals measure 2.2–4.3 mm long and are roughly twice as long as the sepals. **Flowers contain 6 yellow anthers.** The seedpods are long and thin, measuring up to 2.5 cm; they tend to be carried at an angle of 45 degrees to the stem.

Hairy Bitter-cress and Wavy Bitter-cress are known to hybridise.

C. impatiens **(Narrow-leaved Bitter-cress)**
This biennial species occurs in damp woodlands and near waterbodies mostly in the West Midlands of England and into Wales. Its stems are upright, have few branches and grow to a height of 80 cm; the stem leaves are similar to the basal leaves. The rosettes don't persist at flowering. **The basal rosette leaves comprise 5–10 paired, lanceolate, deeply toothed leaflets along each side, with a similar terminal leaflet.** The leaves usually lack hairs and measure up to 25 cm

long. The petals measure 1.5–3.6 mm long and are roughly twice as long as the sepals – although petals may be absent. **Flowers contain 6 anthers, which are a greenish-yellow colour.** The seedpods are long, thin and slightly flattened; they measure up to 3 cm, and tend to be carried more horizontally as they mature.

C. amara **(Large Bitter-cress)**
This perennial species can be common in damp woodlands and near waterbodies, but it is absent from Wales, southern Ireland, the south-west of England and north-west Scotland. It grows to a maximum height of 50 cm. **There is no real basal rosette; all leaves comprise 2–5 paired, rounded leaflets along each side, with a similar larger terminal leaflet.** The lower leaves usually lack hairs and measure up to 15 cm long. The petals measure 0.5–1.2 cm long and are roughly twice as long as the sepals. The flowers are occasionally pink. **Flowers contain 6 anthers, which are a red to purple.** The seedpods are long, thin and slightly flattened; they measure up to 4 cm, and tend to be carried at an angle of 45 degrees.

C. corymbosa **(New Zealand Bitter-cress)**
This introduced annual species has recently become common in gardens, scattered across the whole of Britain and Ireland in pots and pavements. It often spreads via short creeping stems that grow to a height of 10 cm. **Stem leaves may be absent** or are similar to the basal leaves. Basal rosette leaves comprise 2–4 paired rounded leaflets along each side, with a larger terminal leaflet. **The leaves usually lack hairs and measure up to 2 cm long.** The petals measure 3–5.5 mm long and are roughly twice as long as the sepals. **Flowers contain 6 or fewer white anthers.** Later in the year, flowers may not open or lack petals. The seedpods are long and thin, measuring up to 1 cm; they tend to be carried in a fan-like cluster.

Arabis hirsuta **(Hairy Rock-cress)**
This biennial or perennial plant occurs across much of Britain and Ireland. It mostly grows in open ground within grasslands on lime-rich soils. It has upright, hairy stems with few or no branches and grows to a height of 40 cm. **There are many stem leaves, which are similar to basal leaves but are more deeply toothed.** Rosette leaves are lanceolate with small teeth or a wavy margin and measure up to 6 cm long. The leaves are covered in long forked hairs. The petals measure 4–6.2 mm long and are roughly twice as long as the sepals. The flowers contain 6 yellow anthers. The seedpods are long and slightly flattened, measuring up to 2 cm and carried vertically close to the stem.

A. scabra **(Bristol Rock-cress)**
This rare perennial plant is restricted to lime-rich soils near Bristol. It has upright hairy stems with few or no branches and grows to a height of 20 cm. There are few stem leaves, which are smaller than the basal leaves and less deeply toothed. **Rosette leaves are lanceolate with a deeply toothed margin they may grow up to 3 cm long.** The glossy, thick leaves have long forked hairs emerging from raised

bumps. The creamy petals measure 3–8 mm long and are roughly twice as long as the sepals. The flowers contain 6 yellow anthers. The seedpods are long and slightly flattened, measuring up to 4 cm and fanning out away from the stem.

A. alpina (Alpine Rock-cress)
This rare mat-forming perennial plant is restricted to two rocky sites in Yorkshire and Somerset and a couple of remote locations in Scotland. It has upright hairy stems with few or no branches and grows to a height of 15 cm. There are many stem leaves, which are similar to but smaller than the basal leaves. **Rosette leaves are oval with a wavy toothed margin, they may grow to up to 4 cm long. The leaves are covered in long forked hairs, almost appearing white.** The petals measure 5–8 mm long and are roughly twice as long as the sepals. The flowers contain 6 yellow anthers. The seedpods are long and slightly flattened, measuring up to 6 cm, and fan out away from the stem.

A. caucasica (Garden Arabis)
This introduced perennial is a common garden plant and frequent escape, except in southern Ireland and Scotland. **It is rather similar to Alpine rock-cress, except it has larger flowers in a showier inflorescence, and its rosette leaves have longer stalks and fewer teeth.** It has upright hairy stems with few or no branches and grows to a height of 25 cm. There are many stem leaves, which are similar to but smaller than the basal leaves. Rosette leaves are oval with a wavy toothed margin, they may grow to up to 6 cm long. The leaves are covered in long forked hairs, almost appearing white. The petals measure 1–1.5 cm long and are roughly twice as long as the sepals. The flowers contain 6 yellow anthers. The seedpods are long and slightly flattened, measuring up to 6 cm, and fan out away from the stem.

Nasturtium officinale (Water-cress)
This perennial plant is common across most of Britain and Ireland except the mountains of Scotland. It grows in ditches, streams and wetlands. Has upright and creeping stems which branch and lack hairs, and grows to a height of 60 cm. The plants lack rosettes, and root along their stems. Stem leaves are similar to basal leaves but are smaller and narrower. **Lower stem leaves have 1–5 pairs of round leaflets and a slightly larger terminal leaflet, they measure up to 15 cm long.** The leaves lack hairs. The leaves of non-flowering plants have rounder leaflets. The petals measure 3.5–6.6 mm long and are roughly twice as long as the sepals. The flowers contain 6 yellow anthers. The seedpods are long and slightly flattened, measuring up to 2 cm, and become more horizontal as they mature. **The seedpods contain a double row of seeds.**

N. microphyllum (One-rowed Water-cress)
This perennial plant is widespread across most of Britain and Ireland except the mountains of Scotland and Wales, growing in ditches, streams and wetlands.

Seedpods and basal leaves of cresses with long seedpods

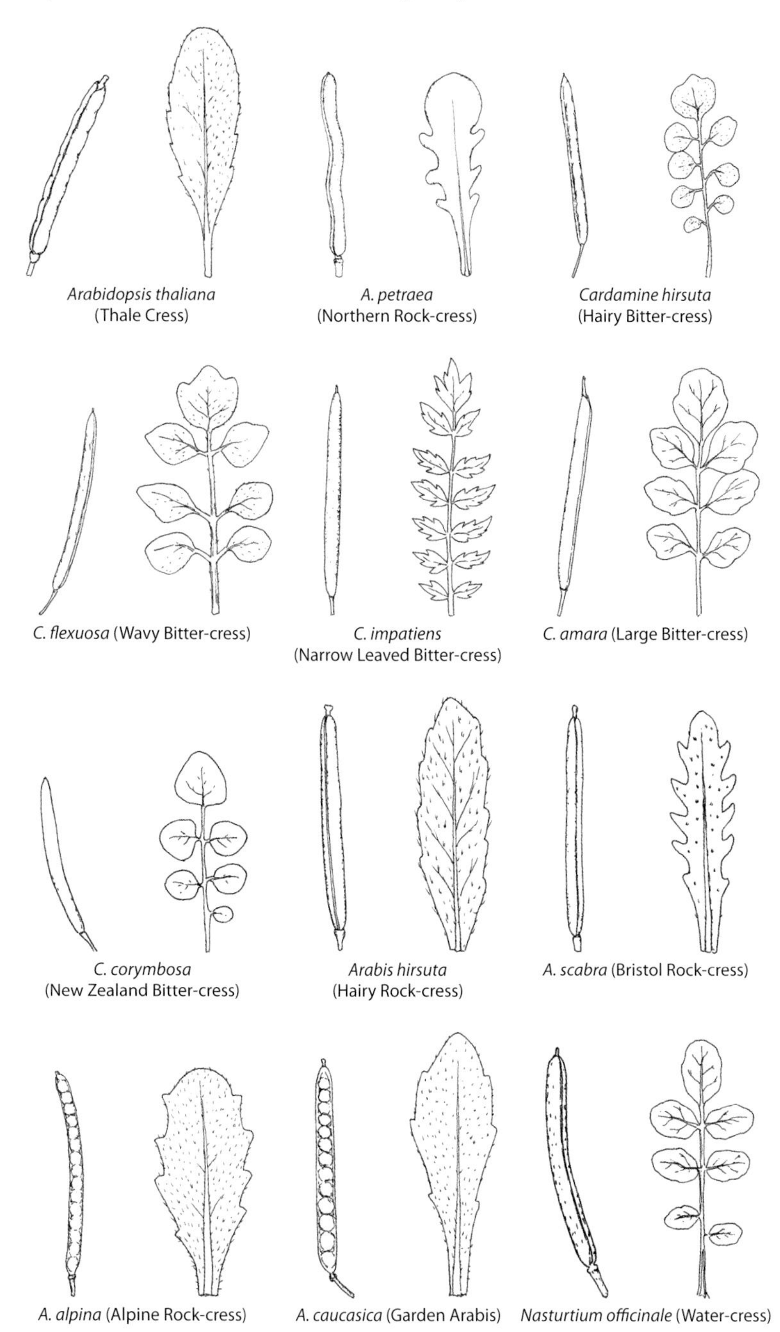

It is morphologically almost identical to Water-cress, except its seedpods contain only a single row of seeds. Has upright and creeping stems which branch and lack hairs, and grows to a height of 60 cm. **Stems turn purple-brown in autumn.** The plants lack rosettes, and root along their stems. Stem leaves are similar to basal leaves but are smaller and narrower. **Lower stem leaves have 1 to 5 pairs of round leaflets and a slightly larger terminal leaflet, they measure up to 15 cm long.** The leaves lack hairs. The leaves of non-flowering plants have rounder leaflets. The petals measure 4–6 mm long and are roughly twice as long as the sepals. The flowers contain 6 yellow anthers. The seedpods are long and slightly flattened, measuring up to 2.3 cm, and become more horizontal as they mature.

Seedpods of both species of water-cress

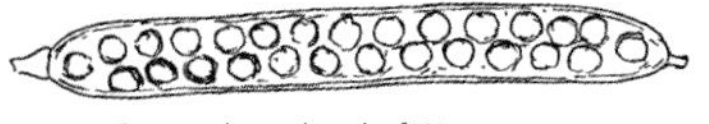

Opened seedpod of Water-cress

Opened seedpod of One-rowed Water-cress

In addition to being a wild plant, Double-rowed Water-cress is cultivated and has half the number of chromosomes of the uncultivated One-rowed species. However, the two species are known to hybridise – producing a sterile hybrid with three sets of chromosomes. This sterile hybrid, known as Brown Water-cress (*Nasturtium x sterile*), was historically also found in cultivation and is sometimes recorded in the wild.

White-flowered cresses with rounded seedpods (scurvygrasses and Awlwort)

Cochlearia danica (Danish Scurvygrass)

This annual species has a coastal distribution in southern Ireland, Wales and northern Scotland. In other regions it has also widely colonised road verges were salt has been used as a de-icer. Its **unbranched stems** grow to a height of 25 cm. Its stem leaves are small and rather ivy-like. All leaves are semi-succulent in nature. **Upper stem leaves lack stalks.** Its rosette leaves are variable, but typically heart-shaped, hairless and carried on a stalk that is twice the length of the leaf. **The rosette leaves (not including their stems) grow to a length of up to 1 cm and their leaf stalks to 4 cm.** The petals measure 2.5–4.5 mm long and are roughly 1.5 times as long as the sepals. The petals sometimes have a hint of pink. The flowers contain 6 yellow anthers. The seedpods are droplet-shaped and slightly flattened, measuring up to 5.5 mm; they are usually carried at an angle of less than 45 degrees to the stem.

Characteristics of white-flowered cresses with rounded seedpods

Species	Stems unbranched	Stem leaves: simple	Stem leaves: clasping stem	Rosette leaves < 1 cm	Seedpod droplet shaped
Cochlearia danica (Danish Scurvygrass)	■			■	■
C. officinalis (Common Scurvygrass)			■		
C. pyrenaica (Pyrenean Scurvygrass)		■			▒
C. micacea (Mountain Scurvygrass)				■	■
C. anglica (English Scurvygrass)*		■	■		
Subularia aquatica (Awlwort)	■	absent	absent		
Your specimen					

* See section below for description of English Scurvygrass

Leaves and seedpods of cresses with round seedpods

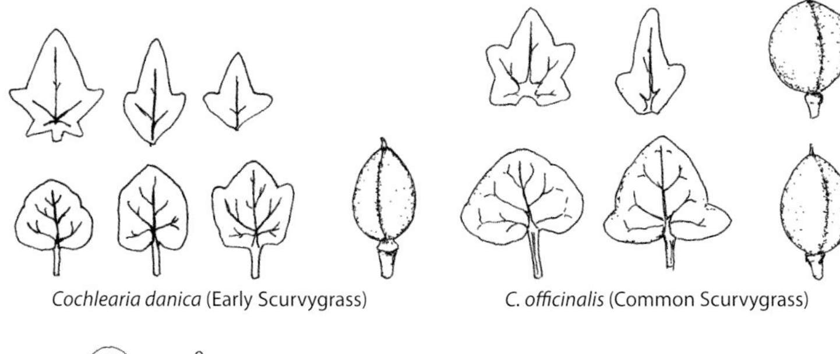

Cochlearia danica (Early Scurvygrass)

C. officinalis (Common Scurvygrass)

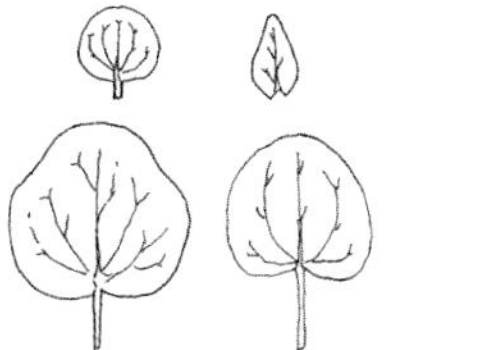

C. pyrenaica (Pyrenean Scurvygrass)

C. micacea (Mountain Scurvygrass)

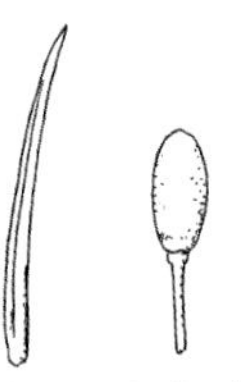

Subularia aquatica (Awlwort)

Stem leaves are shown above rosette leaves, with seedpods appearing on the right, except Awlwort which does not produce stem leaves

C. officinalis (Common Scurvygrass)
This perennial has a coastal distribution, although it may also occur inland particularly in Devon, Cornwall and northern England. **Common Scurvygrass is a highly variable species,** sometimes divided into several subspecies. Some of this variation may be linked to populations with different numbers of chromosomes, which have become adapted to local environmental conditions. A number of these ecotypes/subspecies have been recognised by some as species in their own right. **Its branched stems grow to a height of 40 cm.** All leaves are semi-succulent in nature. Its stem leaves are small and rather ivy-like. **Upper stem leaves clasp the stalk.** Its rosette leaves are variable but typically heart- or bell-shaped, hairless and carried on a stalk that is twice the length of the leaf. **The rosette leaf blades grow to a length of up to 4 cm and their leaf stalks to 10 cm.** The petals measure 3.5–9.5 mm long and are roughly twice as long as the sepals. The petals sometimes have a hint of lilac. The flowers contain 6 yellow anthers. The **seedpods are almost spherical to oval** in shape, and measure up to 7 mm long; they are held more horizontally lower down the stem.

C. pyrenaica (Pyrenean Scurvygrass)
This perennial species is not coastal in distribution – it grows in rocky upland sites and spoil heaps, mostly in northern England and Scotland. Similar to Common Scurvygrass and sometimes considered a subspecies thereof. **Its branched stems grow to a height of 30 cm.** Its stem leaves are smaller than the rosette leaves and rather ivy-like. **Upper stem leaves lack stalks but don't clasp the main stem.** Its rosette leaves are variable, but typically rounded, hairless and carried on a slender stalk that is generally twice the length of the leaf. **The rosette leaf blades grow to a length of up to 1.5 cm and their leaf stalks to 5 cm.** The petals measure 5–8 mm long and are roughly twice as long as the sepals. The flowers contain 6 yellow anthers. The seedpods are oval to droplet-shaped, and measure up to 7 mm long, and are carried fairly erect.

The alpine subspecies of Pyrenean Scurvygrass is highly distinctive. It has horizontal spreading stems and vertical flowering stems. Its leaves are small and simple in shape.

C. micacea (Mountain Scurvygrass)
This perennial species is restricted to the mountains of Scotland on base-rich soils. Again, this species is similar to Common Scurvygrass and sometimes considered as a subspecies thereof. Its stems tend to spread horizontally to 20 cm long, and grow vertically to flower to a height of 10 cm. Its stem leaves are smaller than the rosette leaves and rather ivy-like. **Upper stem leaves have short stalks or no stalks and are arrowhead-shaped.** Rosette leaves are variable, but typically rounded with a flat base, hairless and carried on a slender stalk. **The rosette leaf blades grow to a length of up to 1 cm and their leaf stalks to 4 cm.** The petals measure 5–8 mm long and are up to three times as long as the sepals. The flowers

contain 6 yellow anthers. The seedpods are an elongated droplet shape, measuring 3–6 mm long, and are carried fairly erect.

Subularia aquatica **(Awlwort)**
This aquatic annual occurs in acid lakes in northern Scotland, and less frequently in north-west England, North and West Wales and western Ireland. **This species is unlike any other cress, but when not flowering it is easily mistaken for a number of unrelated but similar-looking aquatic plants.** Its short stems grow to a maximum height of just 8 cm. These support only 2–4 flowers. Its leaves are spiky and grass-like, and grow in a rosette. They lack hairs and grow to a length of 6 cm. The petals measure 2–8 mm long and are up to twice as long as the sepals. The flowers contain 6 yellow anthers (**2 sets of 2 pairs, and 2 shorter singles**). The seedpods are oval, measuring up to 3 mm long, and are carried perpendicular to the stem.

Cresses with flattened seedpods

Capsella bursa-pastoris **(Shepherd's-purse)**
This annual plant is a common weed of gardens and disturbed ground across the whole of Britain and Ireland except the highest mountains. **It is morphologically highly variable.** It has upright stems with few branches that usually lack hairs, and grows to a height of 40 cm. The plants produce rosettes. The stem leaves lack stalks and clasp the stem. These are a simple lanceolate shape, but may be toothed, particularly lower down the stem. **The rosette leaves are highly variable; they are usually lobed, sometimes deeply so**, they may be hairy and grow to a length of 15 cm. The rosette leaves lack stems, having a broad base. The petals measure up to 3 mm long and are up to twice as long as the sepals. Each flower contains 6 pale yellow anthers, but these may be absent in winter flowers. **The seedpods are distinctively triangular-heart-shaped, with convex sides.** Seedpods measure up to 1 cm long and are carried on stalks almost perpendicular to the stem.

C. rubella **(Pink Shepherd's-purse)**
This introduced annual is a weed of gardens and disturbed ground, with a scattered distribution across southern England. **It may be considered as a variant of Shepherd's-purse** and the two are known to hybridise. It is morphologically similar to that species, differing in that its petals may be pink and are similar in length to the sepals. **The seedpods are distinctively triangular-heart-shaped, with concave sides.**

Thlaspi arvense **(Field Penny-cress)**
An annual weed of waste ground across the whole of Britain and Ireland but less common in the west and north. It has upright stems with only few short branches that lack hairs, and grows to a height of 50 cm. The plants lack a tight

rosette. The stem leaves clasp the stem. They are a simple lanceolate shape and may be toothed. All leaves lack hairs. Basal leaves may grow to a length of 7 cm, they tend to be rounder than the stem leaves and may have a short stalk. The petals measure up to 6 mm long and are twice as long as the sepals. Each flower contains 6 yellow anthers. **The seedpods are round and winged, with a distinctive notch at their apex.** Seedpods measure up to 2.2 cm long and are held more or less vertically, on stalks that are 45–90° to the main stem.

T. alliaceum **(Garlic Penny-cress)**
This rather rare annual plant is a weed of arable fields mostly in south-east England. It has upright stems with only few short branches. The stems are hairy near their base and grow to a height of 70 cm. **It has a distinctive faint smell of garlic.** The plants lack a tight rosette. The stem leaves clasp the stem. They are a simple lanceolate shape and may be toothed. All leaves lack hairs. Basal leaves may grow to a length of 10 cm, they tend to be rounder than the stem leaves and have a short stalk. The petals measure up to 5.3 mm long and are twice as long as the sepals. Each flower contains 6 pale yellow anthers. **The seedpods are swollen, slightly winged and have a distinctive small notch at their apex.** Seedpods measure up to 1.2 cm long and are carried on perpendicular stalks, the seedpods curve upwards.

Noccaea caerulescens **(Alpine Penny-cress)**
This short-lived perennial plant is uncommon and **mostly found growing on old metal mines** in the Pennines or as an alpine. It has upright stems with very few short branches. The stems lack hairs and grow to a height of 40 cm. The stem leaves clasp the stem. They are a simple lanceolate shape. All leaves lack hairs. Rosette leaves may grow to a length of 5 cm, they tend to be rounder and have a long stalk. **The inflorescence forms a tight, rounded cluster of flowers.** The petals measure up to 2.5 mm long and are up to twice as long as the sepals. The petals may be slightly pink. **Each flower contains 6 dark-purple anthers.** The seedpods are slightly winged and have an elongated heart shape. Seedpods measure up to 8 mm long and are carried on perpendicular stalks, the seedpods curve upwards.

Microthlaspi perfoliatum **(Perfoliate Penny-cress)**
This annual plant is rare (and has declined a great deal), mostly confined to limestone grasslands in the Cotswolds. It has upright stems with very few short branches. The stems lack hairs and grow to a height of 15 cm. The stem leaves clasp the stem. They are a simple lanceolate shape. All leaves lack hairs and **are a distinctive blue-green colour.** Rosette leaves may grow to a length of 4 cm, they tend to be broad and toothed. The petals measure up to 3 mm long and are about twice as long as the sepals. Each flower contains 6 yellow anthers. **The seedpods are a rounded heart shape, winged and somewhat spoon-like.** Seedpods measure up to 6 mm long and are carried on perpendicular stalks, the seedpods curve upwards.

Characteristics of cresses with flattened seedpods

Species	Plant height <30 cm	Stems		Stem leaves	Rosette leaves		Petals notched or split	Anthers 6 (or 4) yellow/purple	Seedpods winged
		hairy	branched		hairy	lobed or deeply toothed			
Capsella bursa-pastoris (Shepherd's-purse)									
Capsella rubella (Pink Shepherd's-purse)									
Thlaspi arvense (Field Penny-cress)									
Thlaspi alliaceum (Garlic Penny-cress)		at base							
Noccaea caerulescens (Alpine Penny-cress)									
Microthlaspi perfoliatum (Perfoliate Penny-cress)									
Teesdalia nudicaulis (Shepherd's Cress)									
Erophila verna (Common Whitlowgrass)		at base							
Erophila glabrescens (Glabrous Whitlowgrass)									
Erophila majuscula (Hairy Whitlowgrass)									
Drabella muralis (Wall Whitlowgrass)								4	
Draba norvegica (Rock Whitlowgrass)									
Draba incana (Hoary Whitlowgrass)									
Cochlearia anglica (English Scurvygrass)									
Hornungia petraea (Hutchinsia)						leaflets			
Iberis amara (Wild Candytuft)		at base			absent				
Your specimen									

Teesdalia nudicaulis (Shepherd's Cress)

An annual that grows in open sandy sites scattered across most of the British Isles, but it is not found in southern Ireland. Has upright stems which branch near the base. The stems usually lack hairs and grow to a height of 20 cm. There are few or no stem leaves, these clasp the stem and are a simple lanceolate shape. All leaves usually lack hairs. **Rosette leaves are fleshy and may grow**

to a length of 5 cm, they are lobed with a rounded apex. The outer flowers within the tight inflorescence tend to be larger and asymmetrical **(the outer petals being larger).** The petals measure up to 2 mm long and are up to twice as long as the sepals. Each flower contains 6 yellow anthers. The seedpods are a rounded heart shape, and narrowly winged. Seedpods measure up to 4 mm long and are carried on perpendicular stalks near the top of the stem, the seedpods curve upwards.

Erophila verna **(Common Whitlowgrass)**
This very common overwintering annual species grows in walls, waste ground, pavements, calcareous sites and open grasslands scattered across most of England and Wales, but it is less frequent in Scotland and Ireland. **Common Whitlowgrass is a highly variable species**, sometimes divided into several subspecies. As with the scurvygrasses, some of this variation may be linked to populations with different numbers of chromosomes, but this is not yet clear. A number of these ecotypes/subspecies have been recognised as species in their own right. **This species produces several upright stems which do not branch. The stems are hairy near their base** and grow to a height of 20 cm. There are no stem leaves. Rosette leaves may grow to a length of 1.5 cm, they are oval, **hairy**, with a short stalk and may be toothed. The petals measure up to 2.5 mm long and up to twice as long as the sepals. **Each petal is split into 2 deep lobes more than halfway to their base.** Each flower contains 6 yellow anthers. The seedpods are oval to droplet-shaped. Seedpods measure up to 9 mm long and are carried on upright stalks and held almost vertically.

E. glabrescens **(Glabrous Whitlowgrass)**
This overwintering annual species grows in walls, calcareous sites and open grasslands scattered across most of Britain and Ireland. It is less frequent than Common Whitlowgrass, of which it is sometimes considered a subspecies. **It has several upright stems which do not branch. The stems lack hairs** and grow to a height of 10 cm. There are no stem leaves. Rosette leaves may grow to a length of 1.5 cm, they are oval, **only slightly hairy**, with a stalk and may be toothed. The petals measure up to 2.5 mm long and up to twice as long as the sepals. **Each petal is split into 2 deep lobes to about halfway.** Each flower contains 6 yellow anthers. The seedpods are a rounded droplet shape. Seedpods measure up to 6 mm long and are carried on upright stalks, held almost vertically.

E. majuscula **(Hairy Whitlowgrass)**
This uncommon overwintering annual grows in open dry ground across most of Britain and Ireland. It is much less frequent than Common Whitlowgrass, of which it is sometimes considered a subspecies. **It usually has several upright stems which do not branch. The stems are hairy** and grow to a height of about 9 cm. There are no stem leaves. Rosette leaves may grow to a length of 1.6 cm; they are oval, **very hairy**, with a short stalk and may be toothed. The petals measure up

Leaves and seedpods of cresses with flattened seedpods

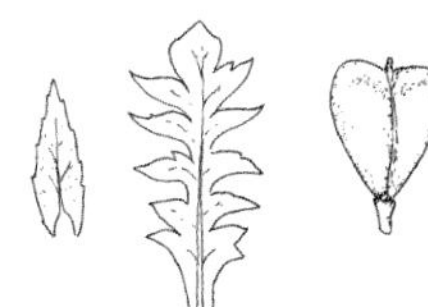

Capsella bursa-pastoris
(Shepherd's-purse)

C. rubella
(Pink Shepherd's-purse)

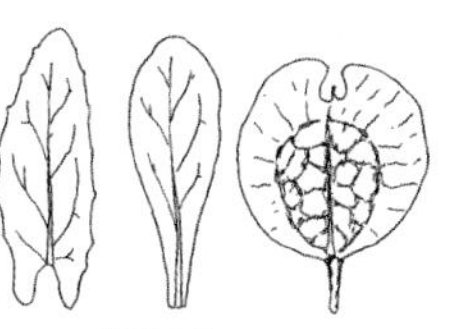

Thlaspi arvense
(Field Penny-cress)

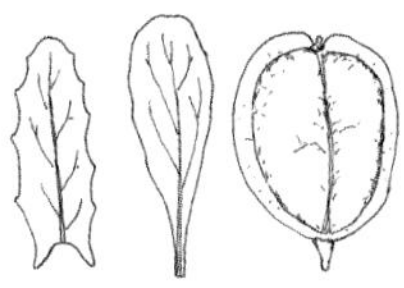

T. alliaceum
(Garlic Penny-cress)

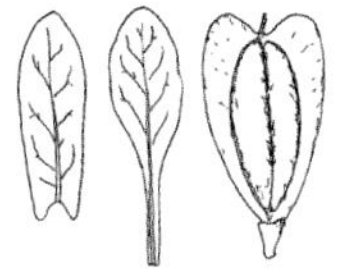

Noccaea caerulescens
(Alpine Penny-cress)

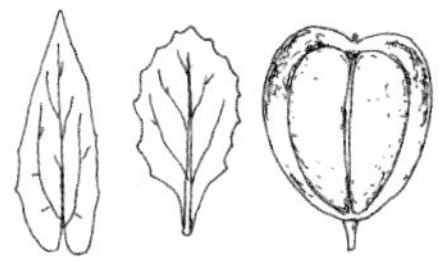

Microthlaspi perfoliatum
(Perfoliate Penny-cress)

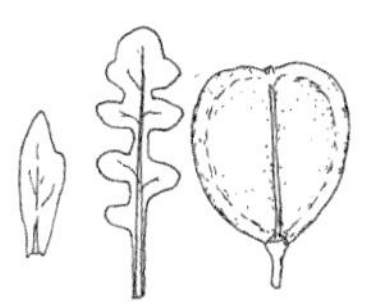

Teesdalia nudicaulis
(Shepherd's Cress)

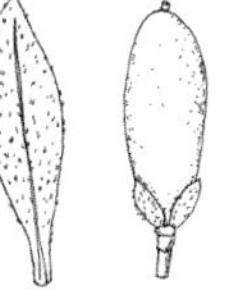

Erophila verna
(Common Whitlowgrass)

E. glabrescens
(Glabrous Whitlowgrass)

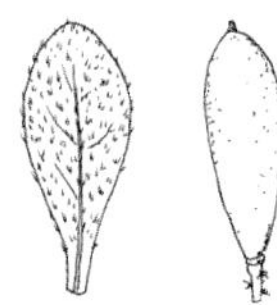

E. majuscula
(Hairy Whitlowgrass)

Drabella muralis
(Wall Whitlowgrass)

Draba norvegica
(Rock Whitlowgrass)

D. incana
(Hoary Whitlowgrass)

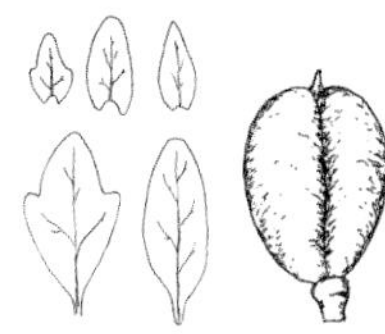

Cochlearia anglica
(English Scurvygrass)

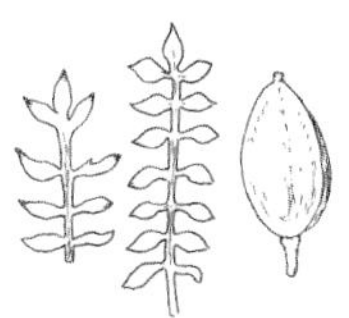

Hornungia petraea
(Hutchinsia)

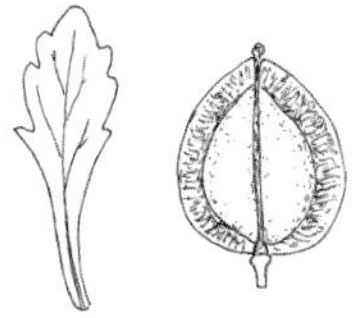

Iberis amara
(Wild Candytuft)

Stem leaves on the left, basal leaves on the right.
NB *Erophila* species have no stem leaves
Iberis amara has no basal leaves
Cochlearia anglica stem leaves above

to 2 mm long and are up to twice as long as the sepals. **Each petal is split into 2 deep, widely separated lobes to about halfway.** Each flower contains 6 yellow anthers. The seedpods are an elongated droplet shape. Seedpods measure up to 6 mm long and are carried on upright stalks and held almost vertically.

Drabella muralis **(Wall Whitlowgrass)**
This short-lived plant grows scattered across Britain and Ireland on open ground on limestone. It has upright stems, sometimes with short branches. The stems are hairy and grow to a height of up to 30 cm. **The stem leaves are triangular, toothed and clasp the stem.** All leaves are hairy. Rosette leaves may grow to a length of 4 cm, they tend to be oval in shape with a short stalk and toothed. The petals measure up to 3.2 mm long and are about twice as long as the sepals. **Each flower contains 4 yellow anthers.** The seedpods are shaped like a short sausage. Seedpods measure up to 6 mm long and are carried on perpendicular stalks.

Draba norvegica **(Rock Whitlowgrass)**
A short-lived, uncommon perennial of the mountains of Scotland. Has upright stems which lack branches. **The stems are hairy and grow to a height of up to 5 cm. Stem leaves are rare.** All leaves are hairy. Rosette leaves may grow to a length of 1.5 cm, they tend to be oval in shape with a short stalk. **The petals are slightly notched**, measure up to 3 mm long and are about 1.5 times as long as the sepals. Each flower contains 6 yellow anthers. **The seedpods are oval and hairy.** Seedpods measure up to 6 mm long and are carried on short stalks close to the stem.

D. incana **(Hoary Whitlowgrass)**
This biennial species grows on calcareous rock ledges in the uplands of northern England, Scotland, North Wales and northern and western Ireland. It has upright stems sometimes with branches. The stems are hairy and grow to a height of up to 35 cm. There are many stem leaves which are oval and sharply toothed and clasp the stem. All leaves are hairy. **Rosette leaves may grow to a length of 3 cm, they tend to be lanceolate with a short stalk and are sharply toothed.** The petals are slightly notched, measure up to 4.3 mm long and are up to twice as long as the sepals. Each flower contains 6 yellow anthers. **The seedpods are an elongated oval shape and slightly hairy.** Seedpods measure up to 9 mm long and are carried on erect short stalks.

Cochlearia anglica **(English Scurvygrass)**
This biennial or short-lived perennial grows in estuaries and on muddy shores around the whole of Britain and Ireland, except northern Scotland. Its stems have few or no branches and grow to a height of 40 cm. Its stem leaves are smaller than the rosette leaves and often a simple, elongated oval shape. All leaves are semi-succulent in nature. Upper stem leaves clasp the stalk. **Rosette leaves are variable but typically oval**, hairless and carried on a stalk that is usually more

than twice the length of the leaf. The rosette leaf blades grow to a length of up to 3 cm and their leaf stalks to 10 cm. The petals measure 5–10 mm long and are roughly twice as long as the sepals. The petals sometimes have a hint of pink. Each flower contains 6 yellow anthers. **Its seedpods differ from the other scurvygrasses in being flattened, with 2 distinct halves**; they measure up to 15 mm long, and are carried erect on the stem.

Hornungia petraea (Hutchinsia)

A rare, often very small winter annual of shallow calcareous soils in England, Wales and the Channel Islands; absent from Ireland and Scotland. Its upright stems sometimes branch. The stems are slightly hairy and grow to a height of up to 10 cm. There may be many stem leaves, which have about 5 pairs of leaflets. All leaves lack hairs. **Rosette leaves do not persist, but may grow to a length of 8 cm, they have many pairs of leaflets and a short stalk.** The short petals are very slightly notched, measure up to **1 mm long and are as long as the sepals.** Each flower contains 6 yellow anthers. The seedpods are an oval shape and somewhat spoon-like. Seedpods measure up to 4 mm long and are carried on short stalks perpendicular to the stem.

Iberis amara (Wild Candytuft)

This annual species has a highly scattered distribution, growing in arable fields and grasslands on calcareous soils mostly in England; absent from Ireland. Its upright stems are branched, slightly hairy at the base and grow to a height of up to 35 cm. There are many stem leaves, which lack hairs. The stem leaves are oval, with rounded teeth and a stalk. **No rosette leaves. The outer flowers are larger than the inner flowers, and the outer petals larger than the inner petals.** The inflorescence forms a dense flat-topped cluster. The larger outer petals measure up to **5 mm long and are up to four times as long as the sepals.** Each flower contains 6 yellow anthers. Seedpods are droplet-shaped, winged, up to 5 mm long and are carried on short stalks that cluster around the top of the stems.

Pepperworts and swine-cresses with flattened flowers and seedpods

Lepidium coronopus (Common Swine-cress)

This annual species is a common weed of waste ground and paths across much of England. It is less frequent in northern England and Scotland, while in Wales and Ireland it is largely restricted to coastal areas. Its stems are prostrate, branched and lack hairs. Stems grow to 30 cm long. **The leaves are divided into leaflets; these may be feathery, sometimes being deeply toothed or divided.** Leaves measure up to 10 cm. Stem leaves are similar to rosette leaves and lack hairs. Flowers lack stalks and are produced in small clusters along the stem. Petals measure up to 2 mm long and are between 1 and 2 times the length of the sepals. There are 6 purple anthers. Seedpods are heart-shaped, deeply veined and lumpy, measuring up to 4 mm.

Characteristics of cresses with flattened seedpods and flowers

Species	Plant height <40 cm	Stems		Basal leaves lobed or deeply toothed	Anthers 6 yellow	Seedpods winged
		hairy	branched from base			
Lepidium coronopus (Common Swine-cress)						
L. campestre (Field Pepperwort)						
L. ruderale (Narrow-leaved Pepperwort)					2	
L. didymum (Lesser Swine-cress)						
L. heterophyllum (Smith's Pepperwort)						
L. virginicum (Least Pepperwort)					2	
L. sativum (Garden Cress)					4	
L. draba (Hoary Cress)						
Your specimen						

L. campestre (Field Pepperwort)

This annual or biennial species is a common weed of open grassland and paths across much of England, but it is less common in Wales, Scotland and Ireland. Its stems are upright, hairy and highly branched above halfway. The stems grow to 50 cm long. The leaves are grey-green in colour. **Upper stem leaves clasp the stem, they are lanceolate and slightly toothed, measuring up to 4 cm. Lower stem leaves are more oval, with a short stalk.** There are no real rosettes. Flower stalks have long hairs; flowers are spread along the top quarter of the stems. Petals measure up to 2.6 mm long and are between 1.5 and 2 times the length of the sepals. There are 6 yellow anthers. **Seedpods are pear-shaped, winged and have a warty surface**; they and measure up to 4 mm.

L. ruderale (Narrow-leaved Pepperwort)

This annual species is a common weed of waste places across England, but it is less common elsewhere. Its stems are upright, hairless and highly branched. **The plants have a distinctive fetid smell.** The stems grow to 40 cm long. Upper stem leaves are simple and lanceolate; lower stem leaves are feathery with 4 or 5 lanceolate paired leaflets, the leaves measuring up to 8 cm. There are no real rosettes. Flower stalks are hairy, and flowers are spread along the top section of the stems. **Petals are often absent or 0.5 mm long and shorter than the sepals.**

There are 2 yellow anthers. Seedpods are oval, with a terminal notch, divided into two and measuring up to 2.5 mm.

L. didymum **(Lesser Swine-cress)**
Lesser Swine-cress is a very common annual weed of waste ground across much of England, Wales and the south of Ireland. Its stems are prostrate, branched and lack hairs, growing to 35 cm long. **The plants have a potent and distinctive fetid, peppery smell. The leaves are feathery, with several paired small leaflets that are toothed.** Leaves measure up to 5 cm. Stem leaves are similar to the rosette leaves but tend to be less toothed and lack hairs. Flowers are carried in inflorescences that arise along the stem, with many flowers in a spike. Petals may be absent or measure up to 0.5 mm long and are shorter than sepals. There are 6 yellow anthers. Seedpods comprise two separate lobes, and measure up to 2.5 mm.

L. heterophyllum **(Smith's Pepperwort)**
This short-lived perennial species grows in open grassland scattered across most of Britain and Ireland. **Its stems are upright, but creeping at their base**; they are hairy and branched at ground level. The stems grow to 50 cm. The leaves are grey-green in colour. **Upper stem leaves clasp the stem, they are lanceolate and toothed, and measure up to 5 cm. Lower stem leaves may be toothed and are lanceolate, with a short stalk.** There are no real rosettes. Flower stalks have long hairs, and flowers are spread along the top section of the stems. Petals measure up to 3.6 mm long and are 1.5 times the length of the sepals. There are 6 yellow anthers. **Seedpods are pear-shaped, winged, slightly notched and may have a slightly warty surface**; they and measure up to 6 mm.

L. virginicum **(Least Pepperwort)**
This introduced annual species has a scattered distribution, mostly across England, probably as a birdseed contaminant. **Its stems are upright, usually hairless, with the top half of the plants being branched.** The stems grow to 40 cm long. Upper stem leaves are simple, lanceolate and toothed. **Lower stem leaves are feathery and deeply toothed, sometimes to the point of forming leaflets**; the leaves measure up to 9 cm. There are no real rosettes. Flower stalks have fine hairs, and flowers are spread along most of the top section of the branches. Petals measure up to 1.5 mm and are 1 to 2 times the length of the sepals. **There are 2 yellow anthers.** Seedpods are round, with a terminal notch, divided into two parts and measure up to 3 mm.

L. sativum **(Garden Cress)**
An annual which is grown as a salad vegetable, it may occasionally turn up almost anywhere as a contaminant of birdseed. Its stems are erect, highly branched and lack hairs; stems grow to 100 cm long. Upper stem leaves are lanceolate or only

Leaves and seedpods of cresses with flattened flowers

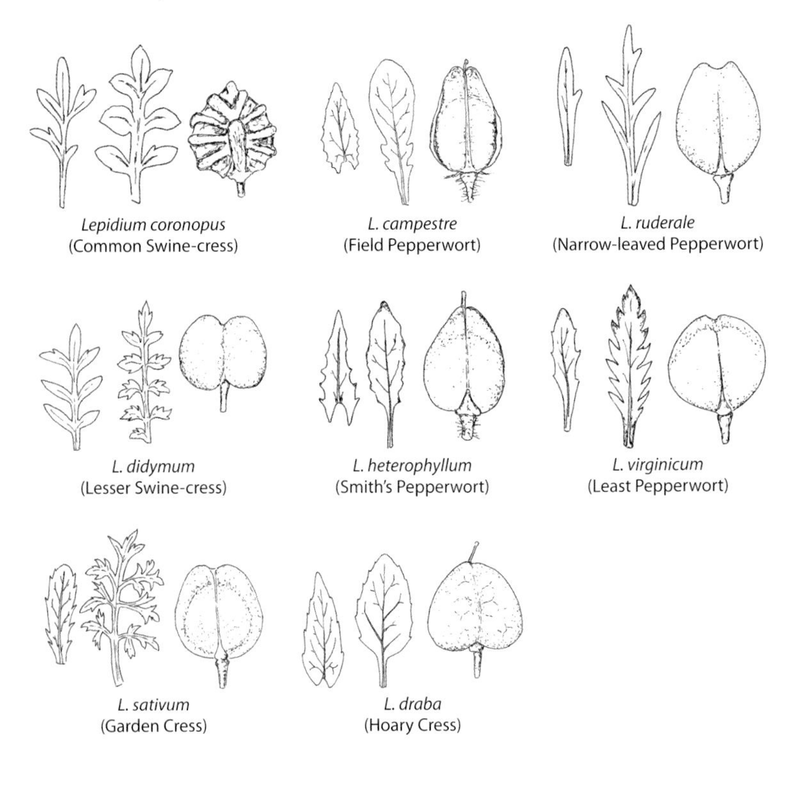

slightly lobed, with small marginal teeth. **The lower leaves are feathery, with 3 or 4 paired leaflets that are themselves deeply toothed or divided.** Leaves measure up to 8 cm and lack hairs. There are no true rosettes. Flowers are only slightly compressed; they have short stalks and are produced on terminal spikes. Petals measure up to 3.8 mm long and are about 2 times the length of the sepals. **There are 4 purple anthers.** Seedpods are a round pear shape, winged deeply, with a terminal notch, and measure up to 3 mm.

L. draba (Hoary Cress)

An introduced perennial that grows in waste places across England, but it is less common elsewhere. Its stems are upright, may be hairy or hairless and highly branched. The stems grow to 60 cm long. Upper stem leaves are lanceolate, finely toothed and clasp the stem. Lower stem leaves also lanceolate, toothed and stalked, the leaves measure up to 15 cm. Flowers are spread along the top section of the stems. Petals measure up to 4.5 mm long and are 1.5 to 2 times as long as the sepals. There are 6 yellow anthers. **Seedpods are invert heart-shaped, measure up to 4 mm, and retain a distinctively long style.**

HAVE OTHERS RECOGNISED THIS LEVEL OF VARIATION?

The taxonomy of this group is complex and historically it has changed on several occasions, to the extent that you may struggle to find two wildflower books that use the same scientific names. The more distinctive examples have been well recognised and documented for centuries. Culpeper in the seventeenth century only mentions a handful of species; much of the high resolution of morphological variation is a more recent phenomenon. For example, in the Victorian period Bentham and Hooker included only a single species of scurvygrass and they did not subdivide the *Erophila* whitlowgrasses. Interestingly, however, they did include four species of water-cress that don't neatly correspond with the two recognised today.

HOW FAR SHOULD I GO?

As always, the answer to this question depends on your motivation. If you are simply interested in safely identifying water-cress to make soup, it is more important to worry about the water quality at the site than being concerned with counting the rows of seeds. However, the knowledge that there are two species of water-cress which are almost identical except one produces seedpods containing a double row of seeds and the other has only a single row of seeds, may be appealing.

Historically, prospectors of metals such as lead and zinc were highly motivated to be able to correctly identify Alpine Penny-cress. This is because its ecological association with toxic metal-bearing rocks could lead them to profitable new sites for mining.

You may be content to identify the dark green plant by the road verge simply as a scurvygrass. Alternatively, you may need to stop if you wish to divide these plants crudely into two types: the mostly unbranched winter annuals and the longer-lived, more robust branched perennials. Others will be driven to investigate the relationships between chromosome number, morphology, distributions and tolerances to salinity or alpine conditions in the scurvygrasses. How this variation is documented is important if we are to have an accurate understanding of where these different forms occur and if this is changing over time. This knowledge has revealed the spread of Danish Scurvygrass from the coast, inland along our roads following the introduction of de-icing salt. For those wanting to tackle more unusual introduced species, some are included in Tim Rich's BSBI Handbook No. 6 *Crucifers of Great Britain and Ireland*, and other guides may be helpful with birdseed contaminants.

CHAPTER 17

Tall yellow-flowered cabbages, mustards, rapes and rockets

Members of the cabbage family are usually easy to identify to family level, with their cross-shaped arrangement of four petals and six anthers. However, the tall yellow-flowered species rather divide botanists. Some regard them as being virtually impossibly difficult to identify, while others maintain that this reputation is undeserved. They are divided into about 20 genera, which appear to be not closely related – so at least some are relatively distinct. This chapter covers species that reach taller than 50 cm when fully grown. Many of these species have had long associations with humans, either as crops or crop weeds. They are often found growing near to sites human activity and are renowned for their peppery taste. Several have been domesticated as vegetables, for their leaves, roots, stems, unopened flower buds, and as oil crops, condiments, even for making dyes. The boundary between crop and wild plant is not always clear, and many species occur both as wild and cultivated forms.

WHY IS THIS GROUP OF PLANTS COMPLEX?

A good insight into the complexity of this group is provided by the scientific names of the Wild Turnip (*Brassica rapa*) and Oilseed Rape (*Brassica napus*) because confusingly in the Latin *rapa* means 'rape' while *napus* means 'turnip'. You may wonder how this mix-up occurred. The answer lies in the fact that *Brassica napus* is not only the wild ancestor of Oilseed Rape but also the ancestor of the Swede, which is often called Turnip. Meanwhile, *Brassica rapa*, as well as being the ancestor of the Turnip, has also been domesticated as an oil crop. Several of these domestication events have involved hybridisation between different species and subsequent doubling of chromosome numbers. In fact, these hybridisations have occurred across at least three genera, too. All this suggests that the plants themselves may sometimes struggle to recognise taxonomic boundaries. Many of these different hybridisation events seem to have occurred on several occasions across both Europe and Asia, involving radically different ecotypes of the parent species. Since those early efforts at domestication, there appears to have been significant geneflow back into wild populations, through both pollen and seed dispersal, which may have further blurred species boundaries.

Given the above, you may be wondering why this group has been included in this section on inbreeders, rather than with the hybridising species. This is particularly the case when you realise that in the wild many of these species are incompatible and are obligate outbreeders. However, during the process of domestication or while evolving to become agricultural weeds, many of these species have not only become self-fertile but have also divided into many morphologically distinct and genetically narrow inbred lines. Extreme examples of this are the different forms of cabbage: kales, broccoli, cauliflower, kohlrabi and Brussels sprouts. These vegetables may look radically different from each other, but they are all members of the same species. It is unclear if there are truly wild populations of cabbage that have not been contaminated by genes from domesticated varieties, but wild cabbage does certainly contain an amazing array of forms and colours.

A final level of complexity has been added again by human activity. There are vast numbers of introduced species we might need to consider. About 40 species in total are included here, but the list could easily be extended by covering rare casual weeds that occasionally turn up in birdseed or as contaminants of imported fodder crop seed.

HOW CAN I TELL THEM APART?

This is another group that contains so many species that it is helpful to divide them into smaller chunks to make them manageable. As with the white cresses of the previous chapter, **seedpod shape** is a good place to start. But as so many of the tall yellow-flowered species have long thin pods, we need other subdivisions. The next division is provided by looking at the **stem leaves** and deciding if they **clasp the stem or not**. Having considered these two features, there are still lots of species with long seedpods and stalked leaves. To tackle these, they are further split into species **with or without lobed leaves**. This gives us five, more manageable groups of plants to examine in more detail.

1. Long thin seedpods, with divided stem leaves that do not clasp the stem.
2. Long seedpods, with undivided stem leaves that do not clasp the stem.
3. Long seedpods, with clasping stem leaves.
4. Shorter, more rounded seedpods, with stem leaves that do not clasp the stem.
5. Shorter, non-cylindrical seedpods, with clasping stem leaves.

As with all such taxonomies the boundaries of these groups are somewhat fuzzy, but hopefully it is usually clear where to start looking for your specimen.

Once you have considered the shape of the seedpods and leaves, you will need to closely examine the flowers. Key features here are the relative size of the petals and sepals. It is often helpful to note if the sepals are held close to the petals or stick out at an angle. There are always six yellow anthers, so you don't need to worry about counting them. It can be important to look at the angle that seedpods are held and their arrangement along the stem. Many floras mention the presence and structure of hairs on the leaves – it is worth looking out for these, but in a lot of species these are variable and so not always reliable for identification on their own.

Let's start with the largest group: species with long cylindrical pods, stem leaves that have lobes and do not clasp the stem

***Raphanus raphanistrum* (Wild Radish)**
An annual species of waste and cultivated ground across most of Britain and Ireland. Grows to a height of 100 cm. Its stems are erect and branched, with simple hairs near the base. Stem leaves tend to have few lobes or teeth and are sometimes lanceolate. Basal/rosette leaves are deeply divided into up to 8 paired lobes, their margins are toothed. The basal leaves measure up to 38 cm long. **Flowers may be white, mauve or yellow, often with distinctive prominent veins.** Petals are twice as long as sepals, measuring up to 2.4 cm. The sepals lie adjacent to and touching the petals. **The seedpods have a highly distinctive shape, with constrictions between each seed**, typically with 3 to 8 seeds per pod. Seedpods measure up to 8.5 cm long.

***Raphanus raphanistrum* ssp. *maritimus* (Sea Radish)**
As the name suggests, this subspecies has a coastal distribution, growing in maritime shingle, dunes and open coastal grassland. It is rarer in the north and in Ireland. A biennial or short-lived perennial, grows to a height of 130 cm. Its stems are erect and branched, they are sometimes purple with rough hairs near the base. Stem leaves tend to have few lobes or teeth and are sometimes lanceolate. Basal/rosette leaves are deeply divided into lobes, their margins toothed. The basal leaves measure up to 40 cm long. **Flowers may be deep yellow, pale yellow or sometimes white.** Petals are twice as long as sepals, measuring up to 2.2 cm. The sepals lie adjacent to and touching the petals. **The seedpods have a highly distinctive shape, appearing like a line of linked balls**, typically with 2 to 6 seeds per pod. Seedpods measure up to 5.5 cm long.

***Sinapis alba* (White Mustard)**
An annual species common on waste ground and as an arable weed across most of England, less frequent elsewhere. Its stems are upright, hairy and branched above, growing to a height of 1 m. Its stem and basal leaves are similar, being divided into a larger terminal lobe and 2 or 3 paired side lobes. The leaves have a wavy margin and short stem, they measure up to 15 cm. There are often no basal leaves. The petals measure up to 1.5 cm. and are about twice as long as the sepals.

Characteristics of species with long cylindrical pods and divided leaves that do not clasp the stem

Species	Height >75 cm	Leaf hairy	Petals pale yellow	Sepals					Seedpods		
				parallel	adjacent to petals	at an angle	between petals	hairy	>3 cm	hairy	next to stem
Raphanus raphanistrum (Wild Radish)	■	■	■	■	■			■	■		□
Sinapis alba (White Mustard)	■					■	■		□	■	
*Sinapis arvensis** (Charlock)	■					■		□	■	■	
Brassica nigra (Black Mustard)	■	■	□			■					
Sisymbrium officinale (Hedge Mustard)	■	■		■	■			■		■	■
Sisymbrium orientale (Eastern Rocket)	■			■	■			■	■	■	
Sisymbrium altissimum (Tall Rocket)	■		■			■	■		■	■	
Sisymbrium irio (London Rocket)					■				■		
Sisymbrium loeselii (False London Rocket)	■					horned			□		
Rorippa sylvestris (Creeping Yellow-cress)							■				
*Rorippa palustris** (Marsh Yellow-cress)	■		□				■				
Eruca vesicaria (Garden Rocket)	■		■	■	■			■	□		□
Diplotaxis tenuifolia (Perennial Wall-rocket)	■		□		■				■		
Diplotaxis muralis (Annual Wall-rocket)						■		■	■		
Hirschfeldia incana (Hoary Mustard)	■	□				■		■		□	□
Erucastrum gallicum (Hairy Rocket)		■	■			■		■	■		
Coincya monensis (Isle-of-Man Cabbage)				■	■				■		
Coincya wrightii (Lundy Cabbage)	■			■	■				■		
Descurainia sophia (Flixweed)	■						■				
Your specimen											

* For descriptions of *Sinapis arvensis* (Charlock) and *Rorippa palustris* (Marsh Yellow-cress) see below.

The sepals are held at an angle from the petals. **The seedpods are hairy, with a beak which has shorter hairs and is as long as the pod.** Seedpods measure up to 3 cm long, they are carried on a short stalk and are held at an angle from the stem.

Leaves, flowers and seedpods of species with long cylindrical pods and divided leaves that do not clasp the stem

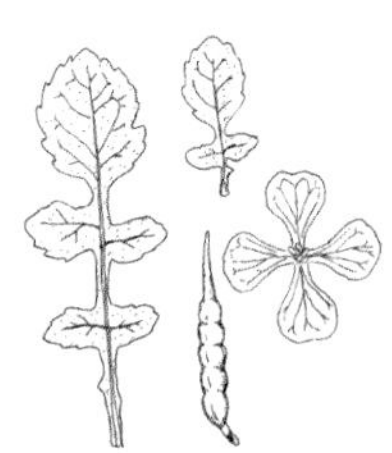

Raphanus raphanistrum (Wild Radish)

Sinapis alba (White Mustard)

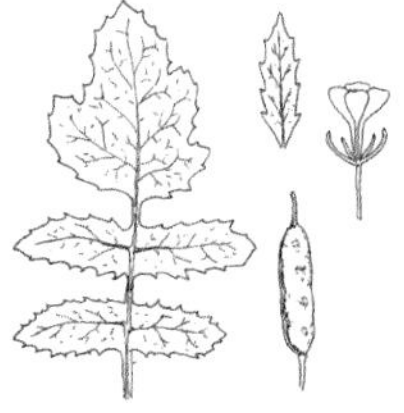

Brassica nigra (Black Mustard)

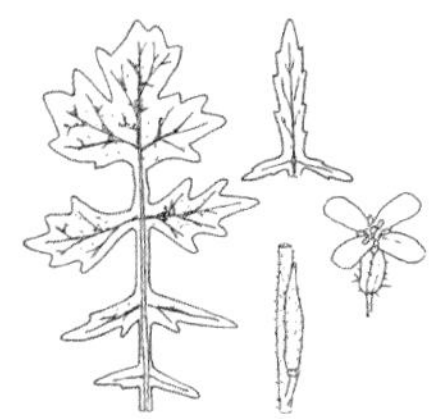

Sisymbrium officinale (Hedge Mustard)

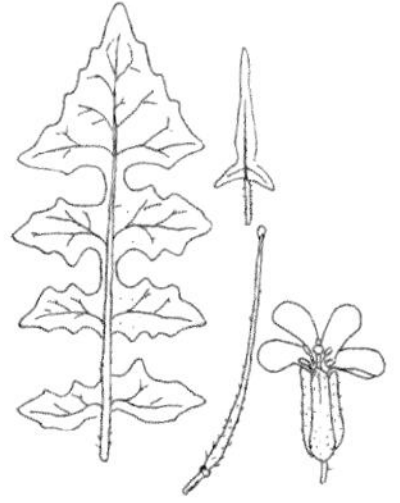

S. orientale (Eastern Rocket)

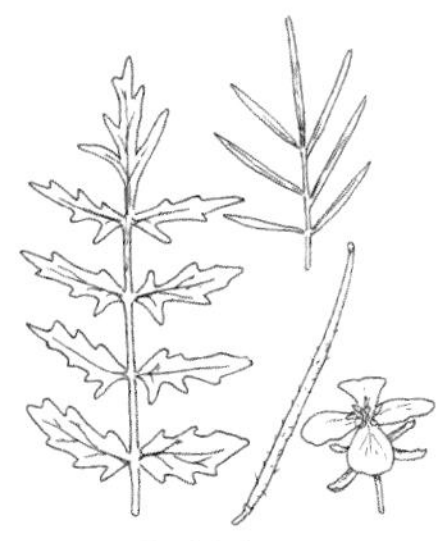

S. altissimum (Tall Rocket)

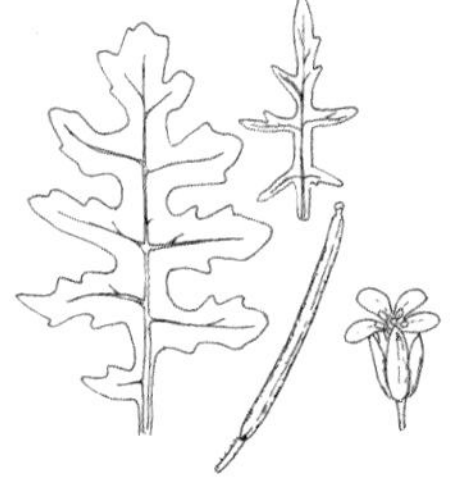

S. irio (London Rocket)

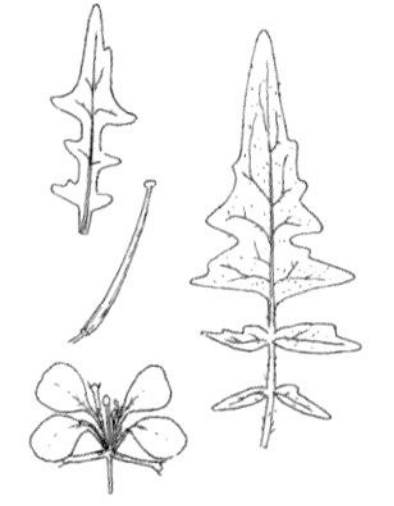

S. loeselii (False London Rocket)

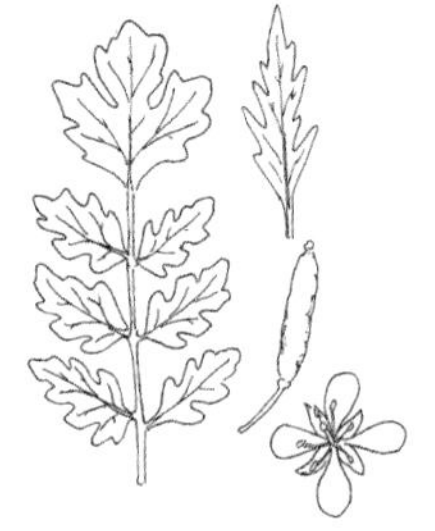

Rorippa sylvestris (Creeping Yellow-cress)

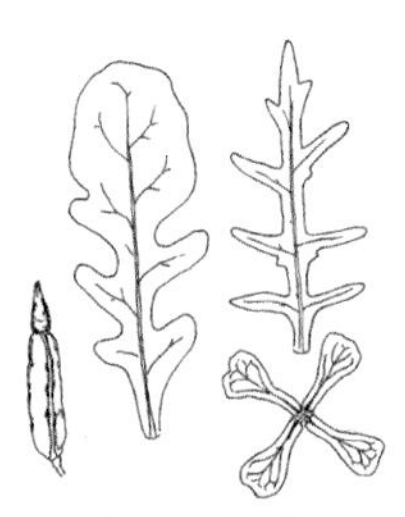

Eruca vesicaria (Garden Rocket)

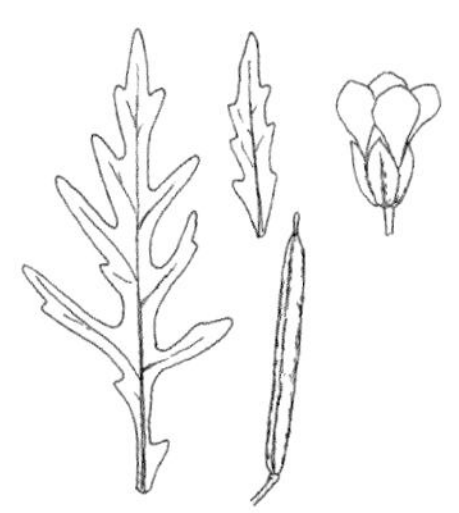

Diplotaxis tenuifolia (Perennial Wall-rocket)

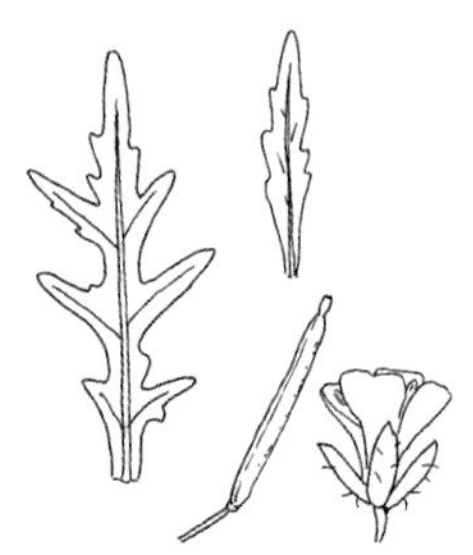

Diplotaxis muralis (Annual Wall-rocket)

Leaves, flowers and seedpods of species with long cylindrical pods and divided leaves that do not clasp the stem (*continued*)

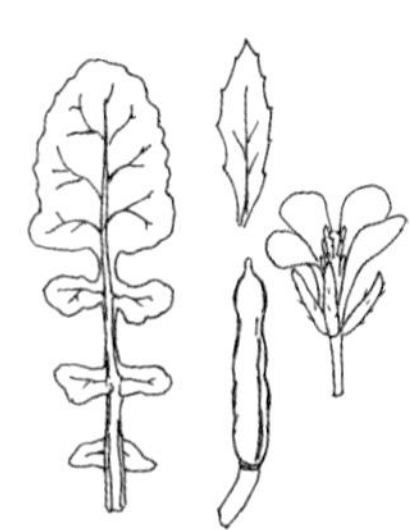

Hirschfeldia incana (Hoary Mustard)

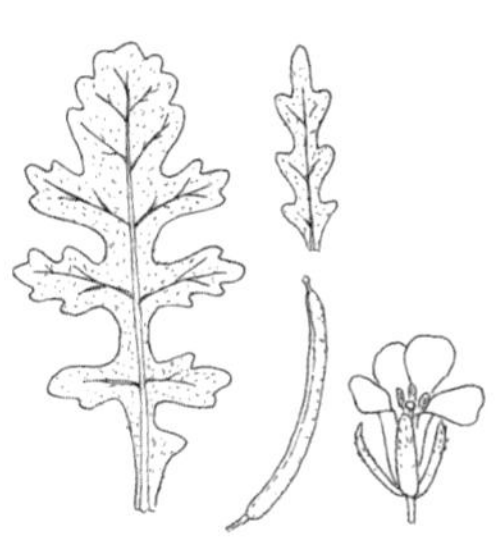

Erucastrum gallicum (Hairy Rocket)

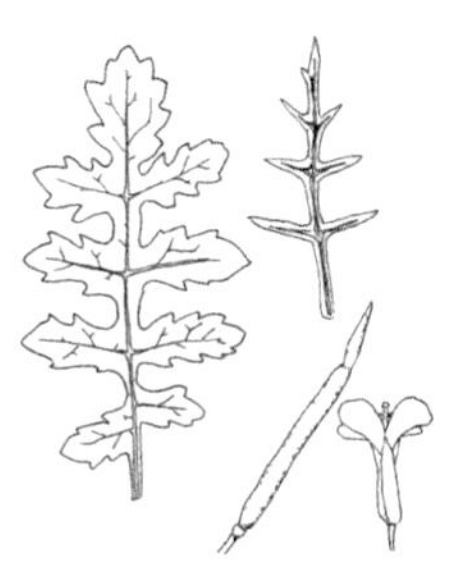

Coincya monensis (Isle of Man Cabbage)

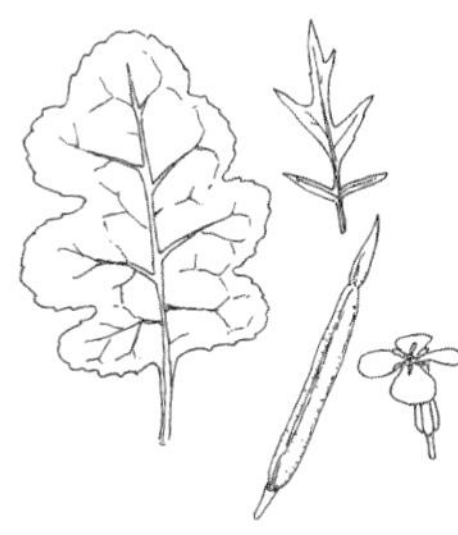

Coincya wrightii (Lundy Cabbage)

Descurainia sophia (Flixweed)

Brassica nigra (Black Mustard)

A common annual weed of open ground across much of England and Wales, much less common in Ireland and Scotland. Can grow to a height of 2 m. Its stems are erect and branched in the top half. Usually lacks hairs, but stems may have a few hairs near their base. Stem leaves tend to be simple, often lanceolate, while **basal leaves are lobed with a larger terminal lobe and a few paired side lobes.** Leaf margins are toothed, **the leaf surface appears crinkled.** Basal leaves measure up to 20 cm. Petals are yellow or pale yellow and measure up to 1.3 cm. Petals are about twice as long as sepals. Sepals are held at an angle from the petals. **The seedpods measure up to 2.5 cm and have a slender beak, they are held vertically, close to the stem.**

Sisymbrium officinale (Hedge Mustard)

This annual or biennial is very common in rough ground and roadsides across most of Britain and Ireland. Grows to a height of 1 m. Its stems are erect, branched, dark green and may have hairs. Both the stem leaves and basal leaves are lobed, with a toothed margin, and have a **rough, often hairy appearance. There are many stem leaves.** Basal leaves measure up to 15 cm. Petals are yellow and measure up to 5 mm. Petals are about 1.5 times as long as sepals. Sepals are hairy and held close to the petals. **The seedpods are hairy, measure up to 2 cm and have pointed apex, they are held adjacent to the stem.**

Sisymbrium orientale **(Eastern Rocket)**
This annual species grows in waste places across much of Britain and Ireland, often in urban areas. It grows to a height of 80 cm. Its stems are erect and hairy, with many branches. **Both stem leaves and basal leaves are lobed, and the leaf margin is shallowly toothed.** The terminal lobe is triangular and shaped like an arrowhead. **Leaf veins in lobes are not central but closer to the base.** Basal leaves measure up to 15 cm. Petals are yellow and measure up to 7 mm. Petals are about 1.5–2 times as long as sepals. Sepals are hairy and held close to the petals. **The very long seedpods are hairy near their base, narrow, curved,** measuring up to 12 cm, and are held away from the stem at an angle of approximately 45°, and extend as far as the flowers.

Sisymbrium altissimum **(Tall Rocket)**
This annual is found in waste places scattered across central England but is less common elsewhere. Grows to a height of 1 m. Its stems are erect and hairy with many branches. Both stem leaves and basal leaves are deeply lobed. **The stem leaves have distinctively narrow lobes.** The basal leaves have many paired lobes and measure up to 30 cm. Petals are pale yellow and measure up to 1 cm. Petals are about twice as long as the sepals. **Sepals sometimes have a distinctive tooth at their tip and are held away from the petals.** The seedpods are hairy, narrow, curved, measure up to 9 cm, and are held away from the stem at angle, as far as being almost perpendicular, and extending as far as the flowers.

Sisymbrium irio **(London Rocket)**
An annual species of rough ground and roadsides widely scattered across England, and very rare in Ireland, Wales and Scotland. Grows to a height of 60 cm. **Its stems are erect, highly branched,** bright green and may have hairs. Both stem leaves and basal leaves are lobed, the leaf margin is sometimes toothed. Basal leaves measure up to 16 cm. Petals are yellow and measure up to 4 mm. Petals are about 1–2 times as long as sepals. Sepals are not close to the petals. **The seedpods are long, narrow, measure up to 5.5 cm, and are held away from the stem, the lower pods being perpendicular to the stem, extending further than the flowers.**

Sisymbrium loeselii **(False London Rocket)**
An annual species of waste places and tips in the English Midlands and around London, but rare elsewhere. It grows to a height of 1.5 m, its stems are erect, with few branches, green in colour and hairy. Both stem leaves and basal leaves are lobed and the leaf margin is toothed, although the leaves are variable. **The terminal lobe is triangular.** Basal leaves measure up to 8 cm. Petals are yellow and measure up to 8 mm. Petals are about 1.5–2 times as long as sepals. **Sepals**

are not close to the petals and have two distinct horns at their tip. The seedpods are narrow, curve upwards, measure up to 3 cm, and are held away from the stem, extending to the flowers.

Rorippa sylvestris **(Creeping Yellow-cress)**
This native perennial plant is common in wet, disturbed sites across the British Isles but is less common in northern Scotland and southern Ireland. Grows to a height of 60 cm. Its stems are erect, branched and may have a few short hairs. The upper and lower leaves are similar, being lobed with toothed margins. The terminal lobe is largest. **Rosette leaves don't persist.** Lower leaves are stalked and grow to 15 cm in length. The petals are yellow and up to 6 mm long. Petals are about 1.5–2 times the length of the sepals. **Sepals are held so that they are visible between the petals.** Seedpods measure up to 2 cm, they are held perpendicular to the stem and curve upwards.

Eruca vesicaria **(Garden Rocket)**
This annual species is now widely grown as a salad vegetable and occurs as a casual, widely scattered across Britain and Ireland. Grows to a height of 80 cm. Its erect stems are branched near the top and hairy near their base. Upper and lower leaves are lobed. **The lower leaves are more rounded, and the stem leaves are more angular and narrow.** Lower leaves grow to 20 cm long. **The petals are cream to pale yellow, with dark veins, and up to 2.6 cm in length.** Petals are about twice the length of the sepals. Sepals are hairy and held close to the petals. **Seedpods have a broad beak**, and measure up to 3 cm, they are **held vertically, close to the stem.**

Diplotaxis tenuifolia **(Perennial Wall-rocket)**
This perennial species is occasionally grown as a salad vegetable and consequently turns up as a casual, widely scattered across Britain but rare in Ireland and Scotland. Grows to a height of 80 cm. Its stems are erect, branched near the top and lack hairs. **It has a blue-green appearance and a foetid smell.** Upper and lower leaves have narrow angular lobes. Lower leaves are stalked and grow to 15 cm long. The petals are pale yellow to yellow and up to 1.5 cm long. Petals are about twice the length of the sepals. **Sepals are oval and held close to the petals.** Seedpods have a short beak and measure up to 4 cm, they are held away from the stem but vertically.

Diplotaxis muralis **(Annual Wall-rocket)**
An annual weed of waste ground, widely scattered across Britain but rarer in Ireland and Scotland. Grows to a height of 60 cm. **Its stems are erect, with spreading branches from the base**; has a few hairs near the base. Upper and lower leaves have narrow angular lobes. Rosette leaves are stalked and grow to 13 cm long. The petals are bright yellow and up to 8 mm in length. Petals are about 1.5 times the length of the sepals. **Sepals are hairy and held away from the petals.** Seedpods have a short beak and measure up to 4 cm, they are held at about 45° from the stem.

Hirschfeldia incana (Hoary Mustard)
This is variable species is usually annual but may be perennial. An introduced weed of waste ground, widely scattered across Britain but rarer in Ireland and Scotland. Grows to a height of 1.3 m. Its stems are erect, with branches in the upper half. The stem may have dense hairs near the base or lack hairs. Upper leaves are lanceolate and toothed. The lower stem leaves have stalks, up to 5 pairs of side lobes and a rounder terminal lobe. The rosette leaves are similar in shape to the lower stem leaves but can have up to 9 pairs of side lobes and grow to 35 cm long. The petals are yellow and up to 1 cm long. Petals are about twice the length of the sepals. Sepals are hairy and held away from the petals. **Seedpods have a rounded beak, measure about 1 cm long; they may have a few hairs and are held parallel and close to or touching the stem.**

Erucastrum gallicum (Hairy Rocket)
This introduced annual or biennial species is a weed of waste ground, very thinly scattered across Britain (fairly common only on the Wiltshire downs) but rarer in Wales, Scotland and Ireland. Grows to a height of 60 cm. Its stems are erect, branched and covered in short hairs. **The upper and lower leaves are also hairy**, they are lobed with these lobes in turn being lobed. The terminal lobe is largest. Lower leaves are stalked and grow to 25 cm. The petals are pale yellow and up to 8 mm long. Petals are about twice the length of the sepals. **Sepals are hairy at their tips** and held away from the petals. Seedpods have a short beak and measure up to 4 cm, they are held at an angle from the stem.

Coincya monensis (Isle-of-Man Cabbage)
As the name implies this biennial species occurs in the Isle of Man, but it also grows in sandy coastal habitats in south Wales, north-west England and lowland Scotland, with scattered records elsewhere. Grows to a height of 50 cm. Its stems are upright, sparsely branched and have a few hairs near the base. Both stem and rosette leaves are deeply lobed, the lobes of rosette leaves have a toothed margin. **Upper stem leaves have rather narrow lobes.** Rosette leaves measure up to 20 cm. Petals are yellow and measure up to 2.1 cm in length. Petals are about twice as long as sepals. Sepals are held tight around the base of the petals. **The seedpods have a prominent beak**, they measure up to 7 cm long and are held at roughly 30° from the stem.

Coincya wrightii (Lundy Cabbage)
Only appears on the island of Lundy. A perennial that grows to a height of 90 cm. **Its stems are woody, upright, sparsely branched and covered with dense hairs.** Both stem and basal leaves are lobed, the lobes of basal leaves have a toothed margin. Upper stem leaves have rather narrow lobes. Basal leaves have a large basal lobe which itself may be divided into side lobes. Basal leaves measure up to 20 cm. Petals are yellow and measure up to 2 cm. Petals are about twice as long as sepals. Sepals are held tight around the base of the petals. The seedpods

have a prominent beak, they measure up to 5.5 cm long and are held almost perpendicular to the stem.

Descurainia sophia (Flixweed)
This annual grows in waste ground and on road verges in sites scattered across Britain and Ireland; common in East Anglia. Grows to a height of 1 m. Its stems are erect, branched near the top and covered in a few short hairs. **The upper and lower leaves are similar and appear feathery, having lobes with sub-lobes.** Lower leaves are stalked and grow to 10 cm. The petals are yellow and up to 2 mm long. **Petals are about as long as the sepals or shorter.** Sepals are held between the petals. Seedpods are narrow and measure up to 2.5 cm, they are held at an angle from the stem **and extend beyond the flowers.**

The second group includes species with long cylindrical pods, undivided stem leaves that do not clasp the stem

Sinapis arvensis (Charlock)
An annual arable weed and plant of waste ground that is common across the whole of Britain and Ireland, except the Scottish Highlands. It grows to a height of 1 m. Its stems are erect, branched and hairy. The upper and lower leaves are similar, being oval, having a short stalk and a toothed margin. **Some leaves may have small lateral lobes.** Lower leaves grow to 20 cm. The petals are yellow and up to 17 mm. Petals are about twice the length of the sepals. Sepals are hairy and held away from the flower. **Seedpods are hairy, measure up to 4.5 cm, and they have a beak that is about one-third to half the length of the pod.** Seedpods are held at an angle to the stem.

Erysimum cheiranthoides (Treacle-mustard)
This annual species is occasional on waste ground across most of Britain and Ireland, although less frequent in the north. Its stems are upright and roughly hairy; they are often unbranched but frequently multi-stemmed, growing to a height of 90 cm. The leaves are lanceolate, with a wavy margin and short stem, they measure up to 10 cm. There are often no basal leaves. **The petals have a flattened to notched end** and measure up to 5 mm. Petals are roughly 1.5 times as long as sepals, the petals and sepals are touching. **The seedpods are square in cross-section, and up to 2.5 cm long.** They are carried on a short stalk and held at an angle of 45° to the stem.

Brassica juncea (Chinese Mustard)
This annual species is sometimes grown as a crop and also occasionally becomes established from birdseed, in scattered sites across Britain and Ireland. It grows to a height of 1 m. Its stems are erect, green and lack hairs. Basal and stem leaves are similar and rather oval with a toothed margin, **in some plants leaf margins are almost frilly.** Basal leaves measure up to 20 cm, in some forms they have

Characteristics of species with long cylindrical pods and undivided stem leaves that do not clasp the stem

				Sepals					Seedpods		
Species	**Height >75 cm**	**Leaf hairy**	**Petals pale yellow**	**parallel**	**adjacent to petals**	**at an angle**	**between petals**	**hairy**	**>3 cm**	**hairy**	**next to stem**
Sinapis arvensis (Charlock)											
Erysimum cheiranthoides (Treacle-mustard)											
Brassica juncea (Chinese Mustard)											
B. elongata (Long-stalked Rape)											
Sisymbrium volgense (Russian Mustard)											
Your specimen											

Leaves, flowers and seedpods of species with long cylindrical pods and undivided leaves that do not clasp the stem

Sinapis arvensis
(Charlock)

Erysimum cheiranthoides
(Treacle-mustard)

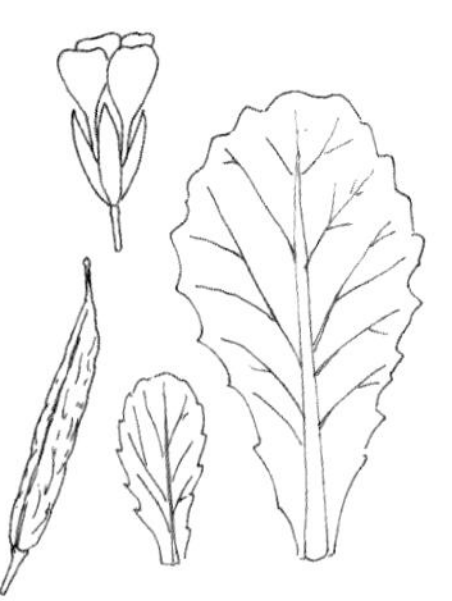

Brassica juncea
(Chinese Mustard)

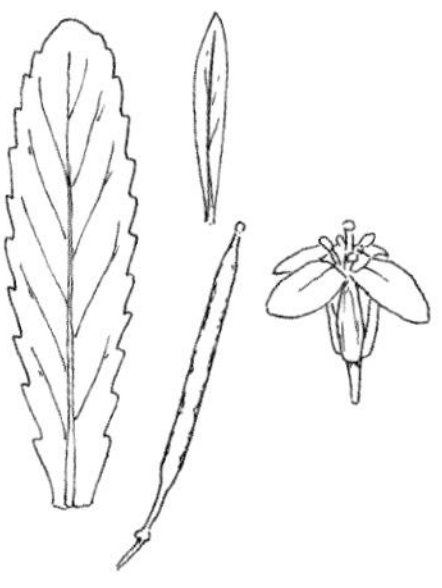

Brassica elongata
(Long-stalked Rape)

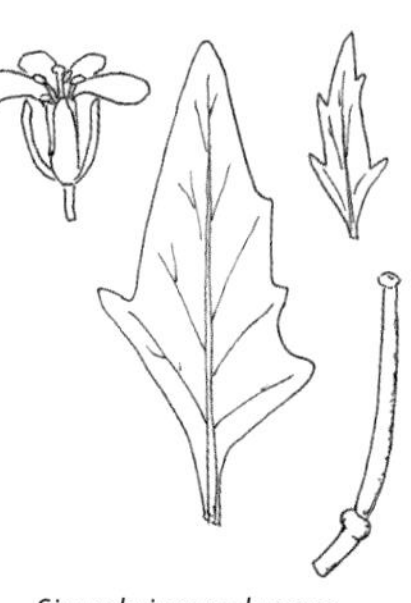

Sisymbrium volgense
(Russian Mustard)

distinctive red veins. Petals are yellow to pale yellow and measure up to 1.3 cm. Petals are about twice as long as sepals. Sepals are held at a slight angle angle to the petals. **The seedpods measure up to 4 cm and have a slender beak, they are held at an angle to the stem.**

Brassica elongata **(Long-stalked Rape)**
This short-lived perennial turns up very occasionally at sites scattered across England. Grows to a height of 1 m. Its stems are erect, branched and lack hairs. The upper and lower leaves are similar, being lanceolate and stalked. **Lower leaves may have a toothed margin.** Rosette leaves do not persist. Lower leaves grow to 30 cm. The petals are yellow and up to 6 mm. Petals are about twice the length of the sepals. Sepals are held adjacent to the flower. **Seedpods measure up to 4 cm, they have a distinct narrow lower section** and are held perpendicular to the stem and curve upwards.

Sisymbrium volgense **(Russian Mustard)**
An introduced perennial species that has been found at a few scattered locations in England, usually on roadsides. Grows to a height of 75 cm. **Its stems grow horizontally before becoming upright**, they have few branches and usually lack hairs. Both stem leaves and basal leaves are diamond to lanceolate in shape, the leaf margin is toothed; **the first tooth may be large.** Basal leaves measure up to 8 cm. **The leaves are a distinct blue-green colour.** Petals are yellow and measure up to 1 cm. Petals are about twice as long as sepals. The sepals are held close to the flower. The seedpods are narrow, measure up to 4.5 cm and are held vertically, parallel to the stem.

The third group covers species with long cylindrical pods and clasping stem leaves

Brassica oleracea **(Wild Cabbage)**
Probably native on lime-rich sea cliffs, but also occurs as an escape at many sites across Britain and Ireland, where the plants may become woody and survive for many years. The stems are robust and arch upwards, often with **many leaf scars near their base and branches in the middle section.** Grows to a height of more than 1 m. **Leaves are highly variable, reflecting the diversity of cultivated forms.** Lower leaves have a large terminal lobe and sometimes paired side lobes, the leaf margin can be highly wavy; they grow to 20 cm long. Upper stem leaves are simple and lanceolate, clasping the stem. The petals are pale yellow and up to 3 cm, about twice the length of the sepals. **The unopened buds are well above the open flowers.** Sepals are held adjacent to the flower. Seedpods measure up to 7.5 cm, have a beak and are held at an angle to the stem and curve upwards.

Characteristics of species with long cylindrical pods and clasping stem leaves

Species	Height >75 cm	Leaf hairy	>4 side lobes	Petals pale yellow	Sepals				Seedpods	
					parallel	adjacent to petals	at an angle	hairy	>3 cm	parallel to stem
Brassica oleracea (Wild Cabbage)	■			■		■			■	
Brassica rapa (Wild Turnip)	■	■					■		■	
Brassica napus (Oilseed Rape)	■		░			■			■	
Barbarea vulgaris (Winter-cress)	■		■		■	■				■
Barbarea intermedia (Intermediate Winter-cress)			░				■		■	
Barbarea verna (American Winter-cress)	■		■		■	■			■	
Barbarea stricta (Upright Winter-cress)	■				■	■		■		■
Your specimen										

Brassica rapa (Wild Turnip)

This annual/biennial species is found, probably as a relic of cultivation, at many sites across Britain and Ireland, although it is less common in Scotland. It is highly variable, with several subspecies being recognised. Its stems are erect, covered in coarse hairs and branch near the top. Plants grow to a height of over 1 m. **Lower leaves are hairy**, have a large terminal lobe and sometimes several paired side lobes. The leaf margin has blunt teeth, and the leaves grow to 20 cm in length. Upper stem leaves are simple and lanceolate, clasping the stem. **The petals are deep yellow** and up to 1.2 cm long. Petals are about 1.5–2 times the length of the sepals. **The open flowers are above the unopen buds.** Sepals are held at an angle to the flower. Seedpods measure up to 5 cm, they have a long beak and are held at an angle to the stem.

Brassica napus (Oilseed Rape)

An annual/biennial species that is found, largely as an escape from cultivation, at many sites across Britain and Ireland, although it is less common in the Highlands of Scotland and western Ireland. It is highly variable, with several subspecies being recognised. Its stems are erect, usually lacking hairs, with branches the full length of the stem. Plants grow to a height of over 1.3 m. Lower leaves are blue-green in colour, have a large terminal lobe and sometimes several paired side lobes. The leaf margin is toothed, and the leaves grow to 35 cm long. Upper stem leaves are simple and lanceolate, clasping the stem. The petals are yellow and up to 1.8 cm long. Petals are about twice the length of the sepals. **The unopened buds are just above the open flowers.** Sepals are held adjacent to the flower. **Seedpods measure up to 8.5 cm, they have a long beak** and are held at an angle to the stem.

The three *Brassica* species, Wild Cabbage, Wild Turnip and Oilseed Rape, form a complex swarm of variation. Each of them has been divided into a number of subspecies or varieties that reflect horticultural selection for different characters. It can be difficult to recognise a horticultural form if it has been allowed to flower and grow to its full potential.

Although each of these species has slightly different numbers of chromosomes, they are known to hybridise, and such hybrids are sometimes found where agricultural crops grow adjacent to wild populations. Oilseed Rape is thought to be derived from a cross between Wild Cabbage and Wild Turnip. Oilseed Rape is known to hybridise back with both its parents – so keep an eye out for plants with intermediate characteristics.

Barbarea vulgaris (Winter-cress)
This short-lived perennial species is common in watery margins and waysides across most of Britain and Ireland except the Scottish Highlands. Grows to a height of 90 cm. Its stems are erect, branched and usually lack hairs. The leaves are dark green and glossy. The basal leaves are deeply lobed, with a larger rounded terminal lobe and 3 to 5 paired side lobes, they measure up to 20 cm. **Upper stem leaves may be unlobed or have a few prominent teeth. The uppermost leaf has a broad terminal lobe.** Petals are yellow and measure up to 8 mm. Petals are about twice as long as sepals. Sepals are held tight around the petals. The seedpods measure up to 3 cm long and are held vertically on a short stalk parallel to the stem.

B. intermedia (Intermediate Winter-cress)
This biennial species is common in water margins and waysides across most of Britain and Ireland but is less common in Southern Ireland and northern Scotland. Grows to a height of 60 cm. Its stems are erect, branched and usually lack hairs. The leaves are dark green and glossy. The basal leaves are deeply lobed, with a larger oval terminal lobe and 3 or 5 paired narrow side lobes, they measure up to 12 cm. **Upper stem leaves have a few side lobes, the uppermost leaf has an oblong terminal lobe.** Petals are yellow and measure up to 6.3 mm. Petals are about twice as long as sepals. Sepals are held tight around the petals. The seedpods measure up to 3.4 cm long and are held at an angle to the stem.

B. verna **(American Winter-cress)**

This annual/biennial introduced species is mostly a plant of the south and west of Britain. It is uncommon in Scotland and western Ireland. Grows in cultivated and open ground, where it can reach a heigh of 90 cm. Its stems are erect, typically branched near the top, and usually lack hairs. The leaves are dark green and glossy. The basal leaves are deeply lobed, **with a larger rounded terminal lobe and 5 to 8 paired or staggered side lobes**; they measure up to 20 cm. Upper stem leaves have fewer narrower side lobes, and the uppermost leaf has an oblong terminal lobe. Petals are yellow and measure up to 9.5 mm. Petals are about twice as long as sepals. Sepals are held tight around the petals. **The seedpods measure up to 7 cm long** and are held at an angle to the stem.

B. stricta **(Upright Winter-cress)**

This short-lived perennial species is mostly found in central England, it is absent from Ireland and rare in both Wales and Scotland. Grows as a casual, reaching a maximum heigh of 1 m. Its stems are erect, typically branched near the top and usually lack hairs. The leaves are dark green and glossy. **The terminal lobe of the basal leaves is oval and much larger than the side lobes (there are only a few small side lobes)**, and these leaves measure up to 15 cm. Upper stem leaves are oval, with a scalloped margin. Petals are yellow and measure up to 6 mm. Petals are about 1.5 times as long as sepals. **The sepals have a distinctive tuft of hairs at**

Leaves, flowers and seedpods of species with long cylindrical pods and clasping stem leaves

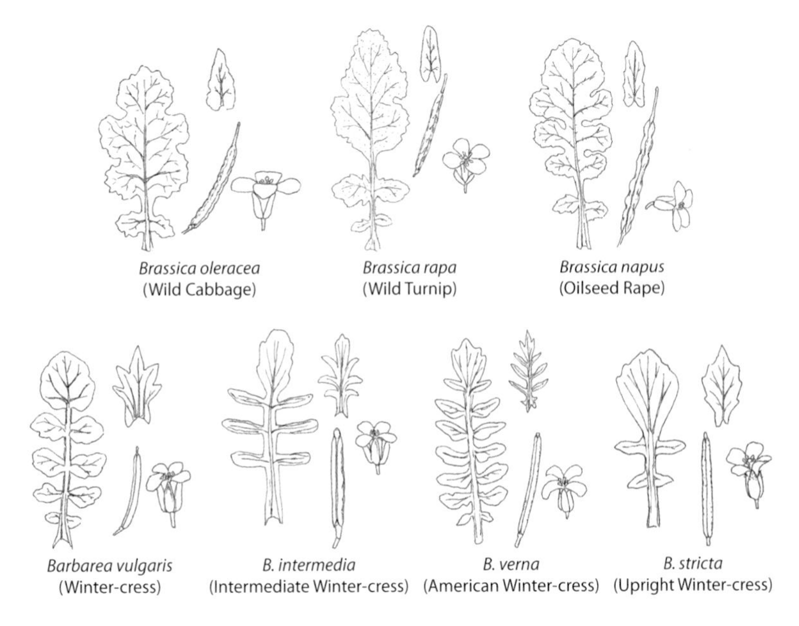

their tip, they are held tight around the petals. The seedpods measure up to 3 cm long and are held vertically.

The fourth group comprises species with shorter, rounded pods and non-clasping stem leaves

Rorippa palustris (Marsh Yellow-cress)

This common species occurs in wetlands and water margins across most of Britain and Ireland, except the north of Scotland. It is an annual or biennial that grows to a maximum height of 90 cm. Its stems are upright, branched and usually lack hairs. The stem and basal leaves are similar in form, having stalks and being divided into lobes, which are themselves lobed or deeply toothed; they measure up to 18 cm. Rosettes do not persist. **The petals and sepals are roughly the same length.** Petals are small, measuring up to 2.7 mm; they are rich yellow to pale yellow in colour. Sepals appear between petals, making it almost look like the flowers have eight petals. **The seedpods are short, fat and slightly curved like a banana.** They are held spaced along the end of stems perpendicular to the stem, with the seedpods bending upwards. Seedpods measure up to 7 mm.

R. amphibia (Great Yellow-cress)

This native perennial occurs commonly in water margins in England and Ireland but is rare in Wales and Scotland. As its scientific name suggests, may grow submerged in water. Grows to a maximum height of 1.2 m. Its stems have a spreading base, becoming more upright; they are branched above and usually lack hairs. Rosettes leaves do not persist. **The lower stem leaves are lanceolate with a toothed margin, lower leaves and submerged leaves may be lobed**; they measure up to 25 cm. The upper stem leaves are also lanceolate and less toothed. The petals are about 1.5–2 times as long as the sepals. Petals measure up to 6 mm and are yellow. **Sepals are just visible between petals.** The seedpods are oval, retain a short style and are held along the stems. Seedpods measure up to 6 mm.

Several of the yellow-cress species are known to hybridise with each other, so keep an eye out for plants with intermediate characteristics.

Rapistrum rugosum (Bastard Cabbage)

An annual that is locally abundant in many parts of England and around Dublin, and occasional in the rest of Ireland, as well as Wales and Scotland; it grows in open ground and as an arable weed. The stems are erect with wide spreading branches, they may lack hairs or have a few coarse hairs. Plants grow to a height of up to 1 m. Lower leaves are stalked and broad, with a large terminal lobe and sometimes a few paired side lobes. The leaves are often hairy, and their margin has shallow teeth. Upper stem leaves are simple and lanceolate. The petals are pale yellow and up to 1 cm. Petals are twice or more the length of the sepals. The sepals are held adjacent to the petals. **Seedpods are round and often hairy,** the style is retained. **The seedpods are held close to the stem.**

Rapistrum perenne (Steppe Cabbage)
This short-lived perennial is a rare plant found at a few open sites scattered across England. The stems are erect with spreading branches, the lower part of the **stem is covered in stiff white downward-pointing hairs, the top of the stem lacks hairs.** Plants grow to a height of up to 80 cm. Lower leaves are hairy, stalked and broad, with a larger terminal lobe and several paired side lobes. The leaf margin is toothed. Upper stem leaves are lanceolate with irregular coarse teeth. The petals are yellow and up to 1 cm. Petals are twice the length of the sepals. The sepals are held adjacent to the flower. **Seedpods are teardrop-shaped, ribbed and lack hairs. The seedpods are held close to the stem.**

Bunias orientalis (Warty-cabbage)
This short-lived perennial occurs scattered in rough grasslands across England and lowland Scotland. **The stems are erect and highly branched at the top, and covered in soft hairs and many bumpy glands.** Plants grow to a height of up to 1.2 m. Lower leaves are lanceolate, although many have a pair of prominent side lobes at their base. The leaf margins have shallow teeth. Upper stem leaves are simple and lanceolate. The petals are yellow and up to 9 mm long. Petals are twice the length of the sepals. The sepals are held away from the petals and are visible between them. The sepals may have a few hairs. **Seedpods are an irregular oval, with a bumpy surface.** The seedpods are carried on stalks 10–18 mm long.

Characteristics of species with shorter rounded seedpods and non-clasping stem leaves

					Sepals		
Species	**>3 paired side lobes**	**Leaf hairy**	**Petals pale yellow**	**Seedpods hairy**	**parallel**	**between petals**	**hairy**
Rorippa palustris (Marsh Yellow-cress)							
Rorippa amphibia (Great Yellow-cress)							
Rapistrum rugosum (Bastard Cabbage)							
Rapistrum perenne (Steppe Cabbage)							
Bunias orientalis (Warty-cabbage)							
Your specimen							

Leaves, flowers and seedpods of species with shorter, rounded pods and non-clasping stem leaves

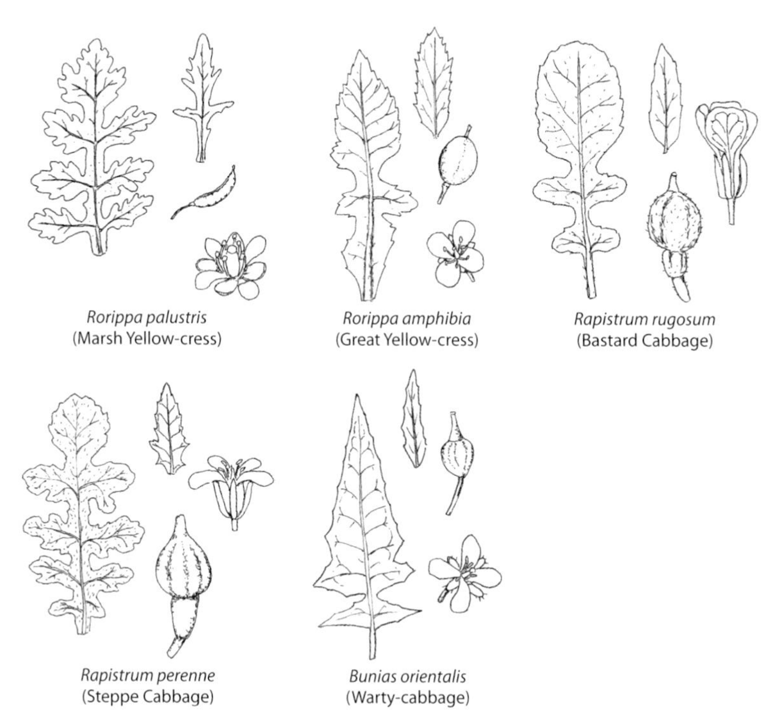

The fifth and smallest group include species with shorter pods and clasping stem leaves

Neslia paniculata (Ball Mustard)

This introduced annual species was once more common but is now only found at a few widely scattered wasteland sites. Its stems are erect with branches above, while the lower part of the stem is covered in branched hairs. Plants grow to a height of up to 80 cm. All the leaves clasp the stalk, are hairy and lanceolate with a finely toothed margin. The largest leaves reach 8 cm. The petals are yellow and small, being only 3 mm long. Petals are 1.5–2 times the length of the sepals. The sepals are hairy and are held adjacent to the flower. **Seedpods are round and slightly flattened; dried seedpods have a honeycombed surface** and retain the style. The seedpods are held at an angle to the stem.

Isatis tinctoria (Woad)

An uncommon biennial plant at sites scattered across England, where it grows to a height of 120 cm. Its stems are upright, branched at the top and have soft hairs near the base. **The leaves are a blue-green colour**. Upper stem leaves lack a stalk and clasp the stem. Basal/rosette leaves have a fairly long stem and are a simple lanceolate shape, measuring up to 30 cm long. The petals measure up to 4 mm, they are twice as long as the sepals, which are held adjacent and clearly visible between the petals. **The seedpods are flattened, pendulous, dark purple-black when ripe and measure up to 2 cm long.** The seedpods occur in clusters of roughly 10.

Rorippa austriaca (Austrian Yellow-cress)

An uncommon perennial that grows in relatively few scattered, usually wet wayside locations in southern England, South Wales and very occasionally in northern England and Scotland. Grows to a maximum height of 1 m. Its stems are upright, branched and usually lack hairs. Rosettes leaves do not persist. **The basal leaves are oval and undivided, with a toothed margin, they measure up to 15 cm. The upper stem leaves clasp the stem, are more lanceolate and also toothed.** The petals are about 1.5 times as long as the sepals. Petals are small, measuring up to 2.5 mm long and are yellow. **Sepals are visible between petals, almost looking like smaller petals.** The seedpods are round, retain the style and held along the stems. Seedpods measure up to 3 mm.

Characteristics of species with shorter pods and clasping stem leaves

Species	Height >75 cm	Leaf hairy	Pendulous seedpod	Sepals		
				adjacent	between petals	hairy
Neslia paniculata (Ball Mustard)						
Isatis tinctoria (Woad)						
Rorippa austriaca (Austrian Yellow-cress)						
Your specimen						

Leaves, flowers and seedpods of species with shorter pods and clasping stem leaves

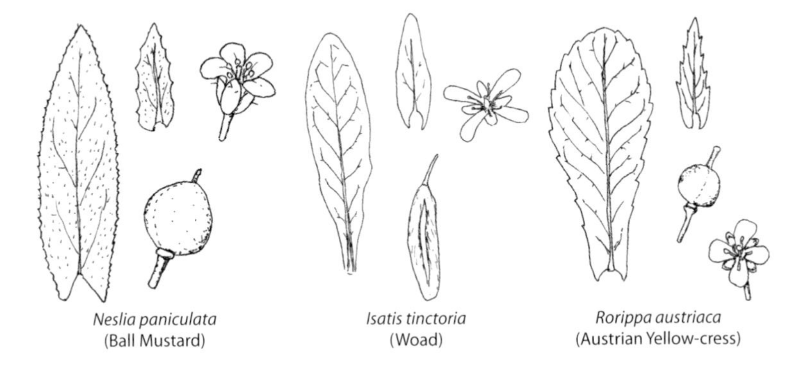

Neslia paniculata (Ball Mustard) *Isatis tinctoria* (Woad) *Rorippa austriaca* (Austrian Yellow-cress)

HAVE OTHERS RECOGNISED THIS LEVEL OF VARIATION?

The earliest records of cultivated cabbages are found in the writings of ancient Greece and Rome and date from around 600 BCE, although they were probably grown much earlier than this. By the first century AD there are accounts of cabbage varieties that resembled leafy kale, heading cabbages, kohlrabi and something like cauliflower or broccoli. It appears that these different forms were subsequently re-domesticated in Northern Europe with different progenitor populations. Their descriptions appear again in several post-Renaissance accounts. This history of domestication and associated unintentional hybridisation and geneflow between wild and cultivated populations is complex and has only recently been fully appreciated. How botanists describe and classify this complexity differs from the increasingly molecular understanding used by crop breeders.

Perhaps because these species were more frequently used as food rather than medicinally, very few crucifers are covered in early herbals. In addition to cabbages and coleworts (leafy non-head-forming cabbages), Culpeper's seventeenth-century herbal lists only black, white and hedge mustards, tall and short cresses and a single species of winter-cress. Earlier floras contained many fewer species than included here. The obvious explanation for this is that many of the species in this section, which cause modern botanists such angst, are relatively recent introductions. Thus, although historically they have been absent from British floras, these species were known and documented within their native ranges. Having said that, some of our rarer native species appear to have been genuinely overlooked. For example, the first records of Isle-of-Man Cabbage date from 300 years ago, and the endemic Lundy Cabbage was not described until 1936.

HOW FAR SHOULD I GO?

Although there are a lot of species covered in this section, most are relatively distinctive and with some practice they are easy to recognise. Hopefully this task is made more manageable by breaking down them into the above, smaller groupings. Many of these species are introduced casuals with a habit of appearing just about anywhere. So, you will always need to be on the lookout for the unexpected. This is particularly true of exotics that turn up as birdseed contaminants or escapees from the cultivation of mixed salad leaves. Tim Rich's BSBI handbook *Crucifers of Great Britain and Ireland* is a useful additional guide to some of these more unusual exotics.

The real complexity in this group arises from hybridisation between species and from the escape of genes from cultivation into wild populations. In recent decades this topic has been the focus of much research related to the perceived risk associated with the cultivation of genetically modified Oilseed Rape. The ability to recognise hybrid plants of this kind is therefore important if we are to fully appreciate the extent, and associated risks, of geneflow from crops.

SECTION IV

Polyploids and rapidly evolving species

One of the complexities in defining a species is that different individual plants within one species may sometimes contain different numbers of chromosomes. They can do this by gaining or losing a chromosome at a time, while on other occasions entire sets of chromosomes may be duplicated. Chromosome duplication is also often associated with hybridisation events. For reasons not fully understood, sometimes such chromosomal variation appears to have very little effect on the appearance of the plant. On other occasions, the duplication of chromosomes can result in profound changes in the plants' appearance, and in them being recognised as separate species. In this grouping of families, **'polyploids and rapidly evolving species'**, changes in numbers of chromosomes seem to facilitate an acceleration in the rate of evolution. Not surprisingly, recently evolved species that share much of their genome can be difficult to differentiate in the field.

In the case of broomrapes, rapid evolution is not thought to be associated with changes in the number of chromosomes but is linked with natural selection driving them to specialise on different hosts. In this instance, it is difficult to disentangle genetic differences from differences arising directly from their host species. This extreme example reminds us that phenotypic morphological variation is always the result of both genes and of the environment. Hence, observed differences between individuals with different numbers of chromosomes that may be thought to be different species may disappear if they are grown in a garden.

CHAPTER 18

Broomrapes

The broomrapes are a very unusual group. All the species that occur in Britain and Ireland are entirely parasitic on other plants. They gain nutrition by plugging their roots into the roots of other species, while their above-ground parts lack chlorophyll and are unable to photosynthesise. All our species are perennials which spend most of the year below ground. Individual plants may persist for a long time, popping up each year to flower. Although there are relatively few species of broomrape that occur in the British Isles, they have a fearsome reputation for being difficult to identify. Most are named after their usual host species, and it appears many botanists seem content to identify the adjacent plant, assume that it must be the host, and conclude the process of identification at that point. If only life could really be so simple.

WHY IS THIS GROUP OF PLANTS COMPLEX?

The most striking thing about this group is their lack of features. Being wholly parasitic, their leaves have become reduced to simple scales, and they lack photosynthetic pigments. The pigments they do contain are highly variable and most species exist in a range of colour forms. All our extant species do not have a branching growth form, and so appear as single spikes. Hemp Broomrape (*Phelipanche ramosa*) has branched stems, but it is thought to be extinct in Britain. Basically, broomrapes consist of a single upright stem, surrounded by tubular flowers. The fact that broomrapes have so few reliable features results in them being extremely difficult to identify. This problem is exacerbated because the few distinctive floral features that they do possess rapidly become indistinct as their flowers age. In addition, the structural complexity of the flowers makes it difficult to clearly preserve their diagnostic features when pressed.

The most problematic species of *Orobanche* to identify is the highly variable Common Broomrape (*O. minor*). This variability arises from it being cosmopolitan in its choice of host plants, with several subspecies being described which are each associated with different hosts. It seems likely that this species is actively in the process of evolving into several new species, each with a more restricted range of hosts – and rapidly evolving species are a taxonomist's nightmare.

HOW CAN I TELL THEM APART?

With so few features to look at, where do you start when you want to identify a broomrape, especially given that the few traits they do have are prone to being variable? The first thing to know is that we only have two species (Common Broomrape and Knapweed Broomrape) that are not described as being scarce or rare. These are your most likely contenders. Of course, it is a good idea to look at the surrounding vegetation for an indication of what the host plant might be. However, probably none of our broomrapes is entirely limited to a single host species, and several have a habit of occasionally turning up in gardens on exotic (if related) hosts. **NOTE: do NOT attempt to dig up the plant** to establish the host, because as stated above, most of our species are extremely rare.

Given that broomrapes are basically just a stalk supporting tubular flowers, you can start by measuring the length of both and consulting the table on page 182. As you will see, this can only take you so far. A plant that is 25 cm tall with flowers 2 cm long could be anything. However, if your specimen is particularly tall or short, with very long or short flowers, then you can start eliminating several of the possible species.

Beyond the very basic differences in size summarised in the table, you will need to closely examine the flowers and their interior. This will require a hand lens. Look closely at the shape of the flower tube, the number and length of the bracts at the base of the flower, and finally peer inside the flower to determine where the male stamens are attached, and to observe the colour of the female parts.

Let's start by looking at our two most widespread species

Orobanche minor **(Common Broomrape)**
This is our commonest broomrape, although it is rare in Scotland, Wales and Northern Ireland. It has been divided into four subspecies, but probably only the following two are not extinct in Britain.

Orobanche minor **spp.** *minor* **(Common Broomrape or Lesser Broomrape)**
This rather variable subspecies is erratic in where it turns up, but is **most frequently found on road verges** and even in flowerbeds. A parasite of a wide range of hosts. Its reddish-brown stems grow to a height of 10–60 cm. The stems are covered in glandular hairs and are not swollen. Its flowers are purple mixed with yellow on the outside and more yellow on the inside. They measure 10–18 mm long. The flowers occur on more than half of the stem, with the lower ones being well separated. The sepals are not usually divided into two; when they are divided, the relative lengths of the lobes are variable. **The bract is often longer than the flower tube,** measuring 7–22 mm. Inside the flower, the filaments arise less than 2–3.5 mm above the base of the flower tube. The filaments usually lack hairs, and the stigma lobe is purple.

Orobanche minor ssp. *maritima* **(Carrot Broomrape)**
This subspecies appears only to parasitise wild carrots on sea cliffs. Its stems are deep purple in colour and grow to a height of 10–30 cm. The stems are covered in white glandular hairs and the base is swollen. Its flowers are purple mixed with yellow on the outside and more yellow on the inside. They measure 10–17 mm long. The flowers occur on less than half of the stem and are fairly close together. The sepals are usually not divided into two. **The bract is shorter than the flower tube, measuring 8–16 mm.** Inside the flower, the filaments arise less than 2–3 mm above the base of the flower tube. The filaments usually lack hairs, and the stigma lobe is purple.

Orobanche elatior **(Knapweed Broomrape or Tall Broomrape)**
Primarily encountered in a line south of the Severn and the Wash, most commonly on calcium-rich soils. Its host species is Greater Knapweed (*Centaurea scabiosa*). Its stems are yellow to pale orangey brown and grow to a height of 25–75 cm. The stems are swollen at the base and covered in glandular hairs. Its flowers are yellow in colour, frequently with purple-brown veins; they measure 18–25 mm long. The flowers are densely packed on the top one-third of the stem. **On the lower side of the flower, the sepals are split into two teeth of unequal length.** Below the flower there is a bract that can be longer than the flower tube, measuring 15–25 mm. Inside the flower, the filaments arise 3–6 mm from the base of the flower tube. The filaments are hairy, and the stigma lobe is yellow.

The other broomrapes are all less common

Phelipanche purpurea **(Purple Broomrape or Yarrow Broomrape)**
This rare species is now in a different genus from our other broomrapes. Its host is Yarrow (*Achillea millefolium*) and possibly other members of the Asteraceae. Its stems are grey-blue and grow to a height of 12–45 cm. The stems are not swollen at the base, but very occasionally are branched. The stems are covered in short glandular hairs. **Flowers are lilac with darker purple veins and often yellow at the base,** they measure 18–26 mm long. The flowers are densely packed on the top third of the stem. **Its flowers are distinctly trumpet-shaped, narrowing at the base.** The greyish sepals are divided into two. Below the flower there is a bract that is similar in length to the sepals, measuring 8–15 mm. Inside the flower, the filaments arise 5–8 mm above the base of the flower tube. The filaments lack hairs, and the stigma lobe is white or pale blue.

Orobanche rapum-genistae **(Greater Broomrape)**
This scarce and much-declined species is now primarily found in Wales and the South West. It parasitises several woody members of the pea family, especially gorse. Its stems are yellow tinged with red and grow to a height of 20–90 cm. The stems are highly swollen at the base and covered in glandular hairs. Its flowers are

yellow on the outside and often red inside, they measure 20–25 mm long. **The flowers occur over most of the stem, with the lower ones being well separated.** The sepals are split into two teeth of equal length. The bract is roughly three-quarters the length of the flower tube, measuring 15–30 mm. Inside the flower, the filaments arise less than 2 mm above the base of the flower tube. The filaments lack hairs close to their base but are hairy closer to the anthers, and the stigma lobe is yellow.

Orobanche crenata **(Bean Broomrape or Carnation-scented Broomrape)**
This rare introduced species occasionally occurs in fields of peas or beans in south-east England. Its stems are reddish brown and grow to a height of 15–80 cm. The stems are slightly swollen at the base and covered in many glandular hairs. **Its flowers are white with purple veins, they measure 20–30 mm long and are somewhat trumpet-shaped with large frilly lips. The flowers are widely spaced over the top half of the stem.** The sepals are split into two teeth of unequal length. The bracts are approximately three-quarters the length of the flower tube, measuring 15–25 mm. Inside the flower, the filaments arise 2–4 mm above the base of the flower tube. The filaments are hairy, and the stigma lobe is white, yellow or pink.

Orobanche hederae **(Ivy Broomrape)**
Occurs in scattered, often coastal locations across Ireland, Wales and southern England, associated with its host plant Ivy. Can be extremely abundant where it does appear. Its stems are purple-brown (or sometimes yellow) and grow to a height of 10–60 cm. The stems are distinctly swollen at the base and covered in many white glandular hairs. Its flowers are dull cream often tinged reddish-purple above, they measure 10–22 mm long. **The flower tube tends to narrow near to its mouth. The flowers are spaced over most of the stem, often down to the ground. The sepals may be as long as the flower tube** and split into two teeth of unequal length. The bracts are often longer than the flower tube, measuring 12–22 mm. Inside the flower, the filaments arise 3–4 mm above the base of the flower tube. The filaments usually lack hairs, and the stigma lobe is yellow or occasionally pink.

Orobanche picridis **(Oxtongue Broomrape)**
This very rare species is now found at just a few sites on the south coast of England. It probably parasitises several members of the daisy family. It looks rather like Common Broomrape, to which it is closely related. Its stems are purple-brown or sometimes pale yellow, they grow to a height of 20–60 cm. The stems are slightly swollen at the base and covered in glandular hairs. Its flowers are cream sometimes with a hint of purple, they measure 14–22 mm long. The flowers are densely packed and only occur on the top half of the stem. **The sepals are split into two lobes of very unequal length, the upper one being almost as long as the flower tube.** The bract is almost as long as the flower tube, measuring

12–20 mm. Inside the flower, the filaments arise less than 3–5 mm above the base of the flower tube. The filaments are covered in white hairs at their base, and **the stigma lobe is purple.**

***Orobanche caryophyllacea* (Bedstraw Broomrape or Clove-scented Broomrape)**

This rare species is now confined to a few sites in Kent, where it parasitises members of the bedstraw family. Its stems are straw-coloured or pink and grow to a height of 15–40 cm. The stems are covered in many glandular hairs and are not swollen at the base. **The flowers smell distinctively of cloves.** Its flowers are dull cream or pink, they measure 20–32 mm long. The flowers are widely spaced over the top half of the stem, and they widen towards their mouth. **The sepals are short and split into two equal lobes.** The bracts may be as long as the flower tube, measuring 17–25 mm. Inside the flower, the filaments arise 1–3 mm above the base of the flower tube. The filaments are hairy at their base and glandular closer to the anthers, and the **stigma lobe is dark purple.**

***Orobanche alba* (Thyme Broomrape or Red Broomrape)**

This species sometimes grows as an annual. Mostly occurs in Western Scotland, Northern Ireland, North Yorkshire and scattered other locations, where it is a parasite on Thyme. Its stems are short, rich orange-red in colour, and grow to a height of 8–25 cm. The stems are slightly swollen at the base and covered in glandular hairs. **Its flowers are creamy yellow on the inside and dark red on the outside**, they measure 15–25 mm long. **The flowers occur on the top half of the stem. There are relatively few, well-separated flowers (10 to 15). The sepals are not divided into two.** The bract is similar in length of the flower tube, measuring 12–25 mm. Inside the flower, the filaments arise 1–3 mm above the base of the flower tube. The filaments are hairy near their base and glandular near the anther, and the stigma lobe is red or purple.

***Orobanche reticulata* (Thistle Broomrape)**

This rare species is now only found at a few calcareous sites in Yorkshire, where it parasitises several species of thistle. Its stems are yellow tinged with purple (sometimes almost appearing grey) and grow to a height of 15–70 cm. The stems are slightly swollen at the base and covered in glandular hairs. **Its flowers are dull yellow, but often purple on their upper surface closest to the mouth of the flower tube,** they measure 12–22 mm long. **The flower tube is distinctively curved and covered in dark glands.** The flowers occur over most of the stem, with the lower ones being well separated. The sepals are split into two, with the upper lobe longer. The bract is of similar length to the flower tube, measuring 12–25 mm. Inside the flower, the filaments arise 2–4 mm above the base of the flower tube. The filaments lack hairs close to their base but are glandular closer to the anthers, and the stigma lobe is purple.

Dimensions of broomrape flowers and plants

Species	Flower length mm											Plant height cm											
	10	12	14	15	18	20	22	25	26	30	32	8	10	15	20	25	40	45	60	70	75	80	90
O. minor (Common Broomrape)																							
O. elatior (Knapweed Broomrape)																							
P. purpurea (Purple Broomrape)																							
O. rapum-genistae (Greater Broomrape)																							
O. crenata (Bean Broomrape)																							
O. hederae (Ivy Broomrape)																							
O. picridis (Oxtongue Broomrape)																							
O. caryophyllacea (Bedstraw Broomrape)																							
O. alba (Thyme Broomrape)																							
O. reticulata (Thistle Broomrape)																							
Your specimen																							

Other characteristics of broomrapes

Species	Swollen stem base	Flowers close together	Flowers on more than half of stem	Bract longer than flower tube	Stigma lobe colour
O. minor spp. *minor* (Common Broomrape)		At top			
O. minor spp. *maritima* (Carrot Broomrape)					
O. elatior (Knapweed Broomrape)					
P. purpurea (Purple Broomrape)					
O. rapum-genistae (Greater Broomrape)					
O. crenata (Bean Broomrape)					varies
O. hederae (Ivy Broomrape)		varies			
O. picridis (Oxtongue Broomrape)				similar	
O. caryophyllacea (Bedstraw Broomrape)				similar	
O. alba (Thyme Broomrape)				similar	
O. reticulata (Thistle Broomrape)				similar	
Your specimen					

Side and front views of broomrape flowers

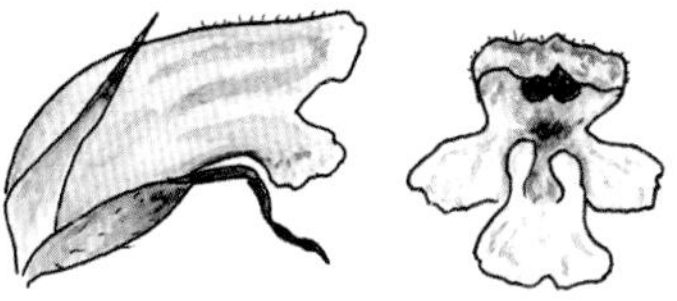

O. minor spp. *minor* (Common Broomrape)

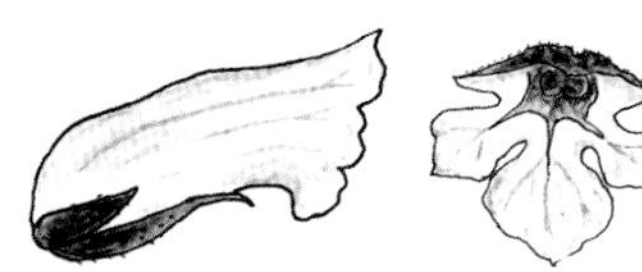

O. minor spp. *maritima* (Carrot Broomrape)

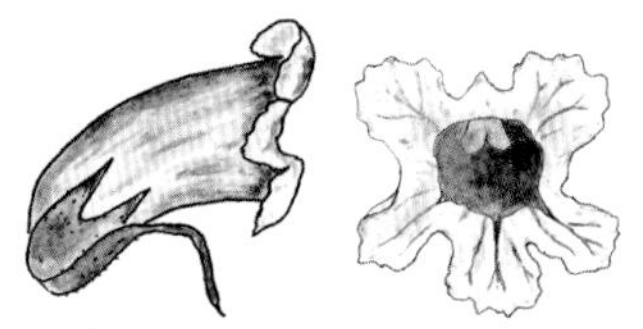

O. elatior (Knapweed Broomrape)

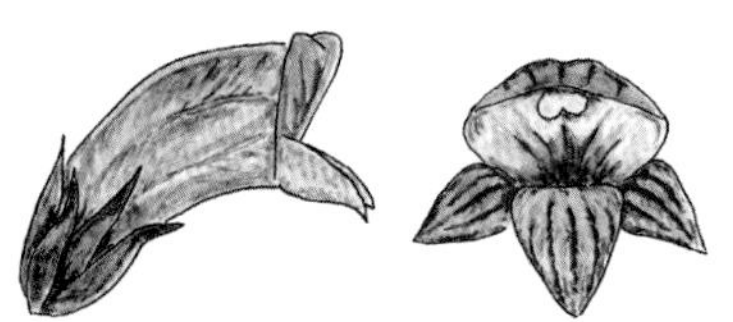

P. purpurea (Purple Broomrape)

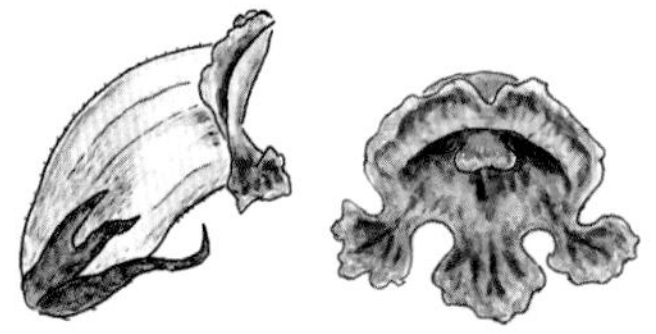

O. rapum-genistae (Greater Broomrape)

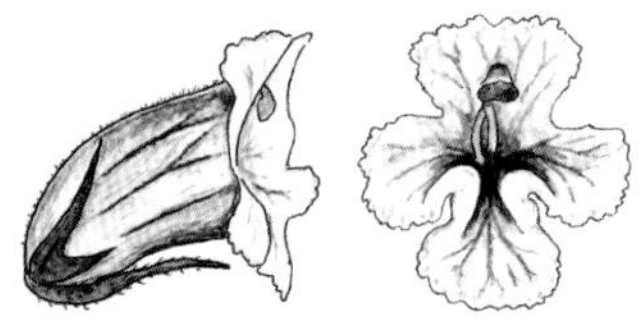

O. crenata (Bean Broomrape)

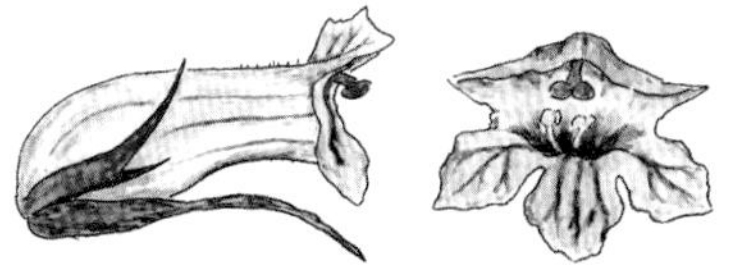

O. hederae (Ivy Broomrape)

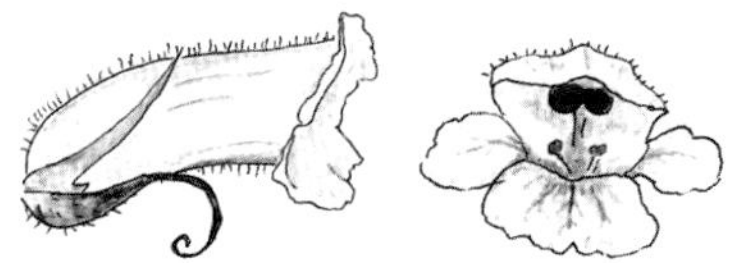

O. picridis (Oxtongue Broomrape)

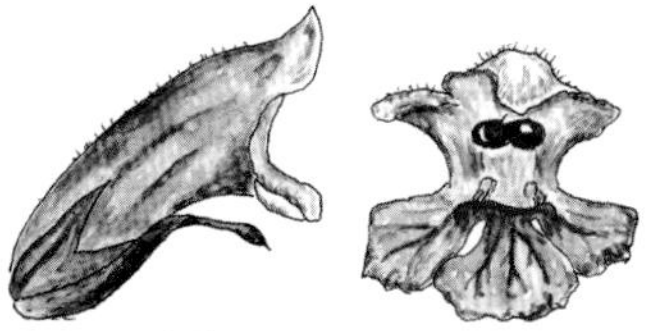

O. caryophyllacea (Bedstraw Broomrape)

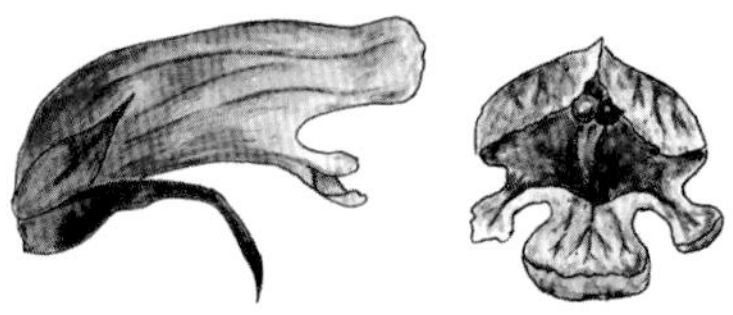

O. alba (Thyme Broomrape)

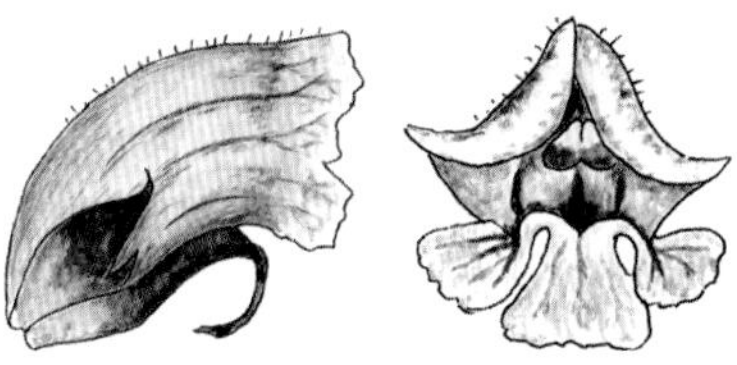

O. reticulata (Thistle Broomrape)

HAVE OTHERS RECOGNISED THIS LEVEL OF VARIATION?

Given that broomrapes are such striking and unusual plants, it is not a surprise that humans have noticed them since ancient times. Early herbals in both Western and Eastern traditions included broomrapes. They were thought to cure a vast range of medical complaints, including gangrene, impotence, kidney problems and even cancer. Several of these early botanical records mention different host species, as this was thought to influence their medicinal properties. However, there was apparently little interest in describing them as separate species until much later. Victorian floras included about half of the species covered here. This was primarily because the rarity of some resulted in them being overlooked until the twentieth century.

HOW FAR SHOULD I GO?

Broomrapes are prone to being variable in size and colour. The combination of this variation and the somewhat erratic nature of their distributions is responsible for their difficulty of identification. So, even experienced botanists tend to double check their records. Because of this, if you intend to engage in the challenge of identifying broomrapes at all, then it surely makes to go the full Monty. The rarity and curious nature of these species will always reward those prepared to find out more. For those wishing to travel down this path, the BSBI Handbook No. 22 *Broomrapes of Britain & Ireland* by Chris Thorogood and Fred Rumsey is recommended.

CHAPTER 19

Forget-me-nots

The genus *Myosotis* is a very familiar group of plants. They are easily recognised by their clusters of distinctive small blue flowers with yellow or white centres. Some species are widely cultivated in gardens, from where they regularly escape back into the wild. Their rather memorable common name seems to ensure that even non-botanists know what they are called. However, it often comes as a bit of a shock to discover that there are in fact ten different species of forget-me-nots found in the British Isles. Although some are common, others are much rarer, having more restricted habitat requirements. Thus, if you wish to see all our native species, you will need to visit both mountainous regions in the north and the Channel Islands.

WHY IS THIS GROUP OF PLANTS COMPLEX?

Forget-me-nots appear to be a genus that has rapidly evolved and appeared only fairly recently. Being relatively young species, they are not always easy to tell apart. Some of our more familiar species make the task more difficult by hybridising as they escape from cultivation, taking with them flower-colour variation that has been selected for by gardeners.

Quick evolution in forget-me-nots is associated with changes in numbers of chromosomes. Thus, although our ten species may be morphologically extremely similar, they contain very different numbers of chromosomes per cell. These differences in ploidy (to use the technical term) have acted as a barrier to crossing, ensuring that the new species became as genetically isolated from each other as if they were located on their own little islands. In addition to this, rapid evolution in forget-me-nots has been accelerated further by the possession of transposons (sometimes called jumping genes). As the name suggests, these sections of DNA have the ability to move location among other genes between generations. While doing this, they can increase the mutation rate, cause earlier mutations to revert and are associated with changes in the amount of DNA per cell.

This combination of polyploidy and jumping genes has allowed different mutations to occur in different isolated populations, so new species have arisen much more quickly than usual. No wonder they are difficult to identify.

HOW CAN I TELL THEM APART?

The flowers of different species of forget-me-not are all very similar. They are typically shades of blue, although pink and white forms are common in several species. The centre of the flowers is usually yellow, which frequently changes to white or pink with age. Flower morphology is therefore generally not the most helpful of characters when it comes to species identification.

Some field guides suggest that you use the relative length of the female stigma compared to the flower-tube to identify forget-me-nots. Although this information is included here too, these features can only be compared following a fiddly dissection of the flower. This is generally not easy in the field. Similarly, it has been claimed that in some regions Changing Forget-me-not and Early Forget-me-not can only be reliably distinguished based on pollen-grain size. This is also less than ideal as a field identification tool.

You may find the most helpful features to identify forget-me-nots include the angle of hairs on the stem and on the sepals and/or developing seedheads (are the hairs pressed close or raised at an angle?). You will of course need a hand lens to fully determine the nature of any stem hairs. A second helpful character is the length of the individual flower stalk compared to the length of the flower or seedhead itself, see page 188.

Let's start with the more widespread species

***Myosotis arvensis* (Field Forget-me-not)**

This short-lived species is common across the British Isles, where it grows in gardens and disturbed open ground. It has an upright growth form which is sometimes rather spindly. It grows to a height of 50 cm. The flowering stalks are covered in short hairs that point upwards at an angle of about 45°. The flowers are pale blue and measure up to 5 mm across. The female style is shorter than the flower tube. **The stalks of the individual seedheads are up to twice the length of the seedheads. Seedheads are open.** The lower part of the seedhead is covered in hairs which are spreading, stiff and hooked. Seeds measure 2–2.5 mm.

***M. sylvatica* (Wood Forget-me-not)**

This short-lived species is common across the British Isles, where it is native to woodlands but is now more common in gardens and as a garden escape. It has an upright growth form which is often branched. It grows to a height of 50 cm. This species is often confused with the Field Forget-me-not. The flowering stalks are covered in short hairs that point upwards at an angle of about 45°. **The flowers are rich blue (although pink and white forms are common in garden escapes) and measure 6–10 mm across.** Later flowers tend to be smaller. The female style is longer than the flower tube. **The stalks of the individual seedheads are a little longer than the length of the seedheads. Seedheads are closed.** The lower part of the seedhead is covered in hairs which are spreading, stiff and hooked. Seeds measure 1.5–2 mm.

Characteristics of species of forget-me-not

Species	Ratio of flower stem to flower length			Stem hairs		Seedhead hairs		
	shorter	similar	longer	Lying flat	at an angle	Lying flat	at an angle	hooked
M. arvensis (Field Forget-me-not)			×2					
M. sylvatica (Wood Forget-me-not)								
M. discolor (Changing Forget-me-not)								
M. ramosissima (Early Forget-me-not)								
M. secunda (Creeping Forget-me-not)			much	varied				
M. scorpioides (Water Forget-me-not)			×1.5					
M. laxa (Tufted Forget-me-not)								
M. stolonifera (Pale Forget-me-not)			×2					
M. alpestris (Alpine Forget-me-not)								
M. sicula (Jersey Forget-me-not)								
Your specimen								

M. discolor (Changing Forget-me-not)

This is a fairly common annual species that occurs in many disturbed, often dry habitats. It can grow to a height of 30 cm, but it is often much shorter. The flowering stalks are covered in short hairs that point upwards at an angle of about 45°. **Its flowers are tiny, only measuring about 2 mm across. Initially they are pale creamy yellow, but they change colour becoming white, then pink and finally blue as they age.** The female style is the same length or slightly longer than the flower tube. **The stalks of the individual seedheads are shorter than the seedheads.** The lower part of the seedhead is covered in hairs which are spreading, stiff and hooked.

There are two common subspecies of Changing Forget-me-not, although it has recently been suggested that these should be considered as two separate species. The first (sub-)species *dubia* has 12 chromosomes and can be identified because its flowers are initially white to pale cream in colour and its upper leaves are alternate. The second (sub-)species *discolor* has 72 chromosomes, and its flowers are brighter yellow and its upper leaves are opposite.

Seedheads and stalks of forget-me-nots

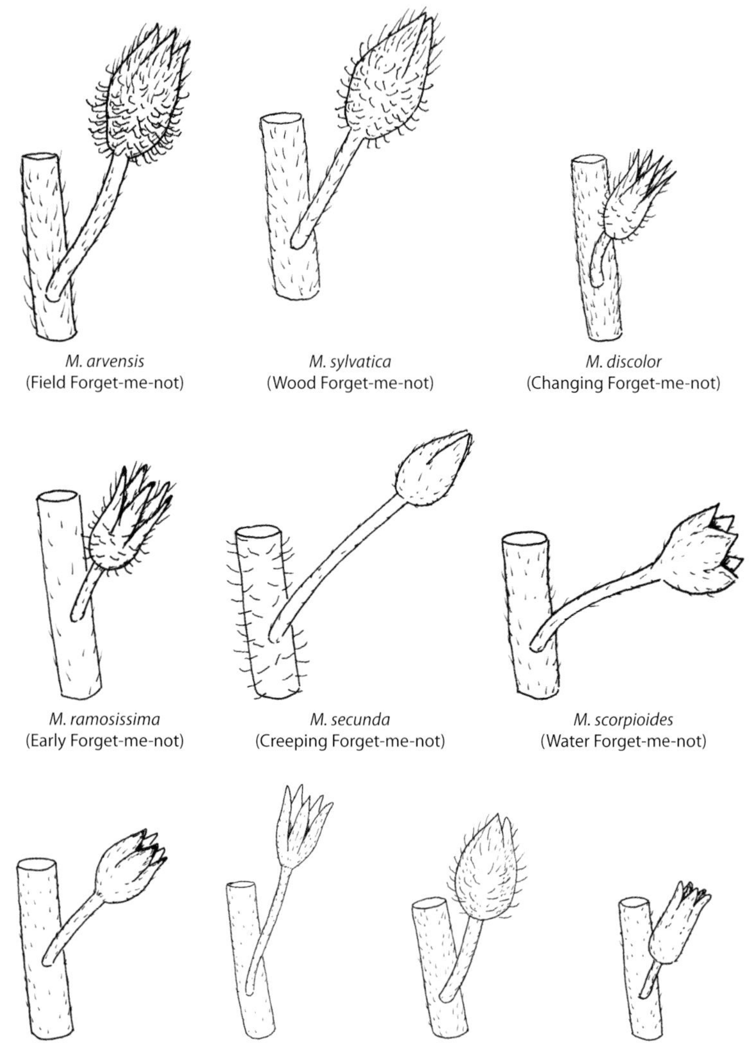

M. ramosissima (Early Forget-me-not)

This annual species is fairly common across most of lowland Britain and Ireland where it grows in dry open ground. It can grow to a height of 25 cm but **in dry sites it may only be a few centimetres tall.** Can be easily confused with Changing Forget-me not. The name implies this species flowers earlier than other species of *Myosotis*, but in fact it starts flowering in April as do most forget-me-nots. However, this species does finish flowering earlier, generally in June. The flowering stalks are covered in short hairs that point upwards at an angle of about 45°. The flowers are blue and measure up to 3 mm across. The female style similar in length to the flower tube. **The stalks of individual seedheads are shorter than the seedheads themselves.** The lower part of the seedhead is covered in hairs which are spreading, stiff and hooked. The seedheads are open.

***M. secunda* (Creeping Forget-me-not)**

This short-lived species is common in damp, largely upland areas of the British Isles, where it may grow to a height of 50 cm. It has an upright growth form but spreads vegetatively via stolons. **The tops of the stems and sepals are covered in short hairs which lie close to the plant. Lower down the hairs on the stems arise at 90° from the stem and may curve upwards.** The flowers are blue and measure up to 6 mm across. The female style is the same length or slightly longer than the flower tube. The stalks of the individual seedheads are 2.5–5 times longer than the seedheads. The seedheads are open and covered in short hairs that are pressed close to sepals.

***M. scorpioides* (Water Forget-me-not)**

A very common perennial species of wet fields and water margins across the British Isles, where it grows to a height of 70 cm. It has an upright growth form but spreads vegetatively, forming clumps. The flowering stalks are covered in short hairs that are pressed flat to the stalks. The flowers are blue (rarely white) and measure 8–13 mm across. The female style is longer than the flower tube. **The stalks of the individual seedheads are about one and a half times the length of the seedheads.** The seedheads are open and recurved when mature, and have a few short hairs which are pressed close.

***M. laxa* (Tufted Forget-me-not)**

A common, short-lived species of wet ground across the British Isles where it grows to a height of 40 cm. The plants have an upright growth form. The stalks are covered in short hairs that are closely pressed to the stem. The flowers are sky blue and measure up to 5 mm across. The female style is short and about half the length of the flower tube. **The stalks of the individual seedheads are slightly longer than length of the seedheads.** Seedheads are open and covered in short hairs which are pressed close.

Now let's look at the less common species

***M. stolonifera* (Pale Forget-me-not)**

This perennial species is restricted to wetlands in the uplands of northern England and southern Scotland. A short plant, growing to only 20 cm. Has an upright growth form and spreads vegetatively via stolons. The flowering stalks are covered in short hairs that are pressed close to the stem. The flowers are pale blue and measure up to 5 mm across. The female style is slightly longer than the flower tube. The stalks of the individual seedheads are up to twice the length of the seedheads. Seedheads are open with recurved sepals which are covered in short hairs that are pressed close.

M. alpestris (Alpine Forget-me-not)
A rare perennial that occurs at a few mountainous sites in northern England and Scotland. It grows to a height of 25 cm. Has an upright growth form and may spread vegetatively. The flowering stalks are covered in short hairs that point upwards at an angle. The flowers are blue and measure 4–8 mm across. The female style is longer than the flower tube. The stalks of individual seedheads are about the same length of the seedheads or slightly longer. Seedheads open slightly. **The lower part of the seedhead is covered in some hairs pressed close and others which are spreading, stiff and hooked.**

M. sicula (Jersey Forget-me-not)
As the name suggests this is a rare species, within the British Isles **only found at a few sites on Jersey. It is an annual of damp grasslands**, grows to a height of 20 cm. The flowering stalks are covered in short hairs that are pressed close to the stem. The flowers are blue and measure less than 3 mm across. **The stalks of the individual seedheads are generally shorter than the seedheads.** Seedheads are open and covered in short hairs which are pressed close to the surface.

HAVE OTHERS RECOGNISED THIS LEVEL OF VARIATION?

Knowing that the differences between *Myosotis* are so subtle, it is no surprise that several species have only recently been recognised. It is perhaps more of a revelation to learn that the name 'forget-me-not' dates from only the early nineteenth century. The current common name was popularised following the publication of a translation of a German folktale by Coleridge in 1802. John Gerard's herbal of 1597 refers to these plants as 'scorpion-grass', because the unfurling inflorescence is reminiscent of the tail of a scorpion. The common name used in French 'ne m'oubliez pas' also translates as forget-me-not. In Welsh, many older regional names survive: 'llygaid doli', 'llygad y ddol', 'llygaid y gors', 'llygaid yr aderyn bach', which translate as the eyes of the doll, the meadow, the bog, the little bird – but they don't seem to be associated with any particular species.

Of the ten species of *Myosotis* covered here, only two were described by Linnaeus. The Victorian flora of Bentham and Hooker included just five species. The Pale Forget-me-not was first recorded in Britain in 1918, followed by the Jersey Forget-me-not in 1922.

HOW FAR SHOULD I GO?

If definitive identification of some species of *Myosotis* is dependent on measuring their pollen size, you can be forgiven for not doing so in the field. You may wish to rely on location, habitat and several morphological traits for some trickier individuals. Generally, however, habitat and seedhead morphology will be sufficient to confidently identify all ten British and Irish species.

CHAPTER 20

Speedwells

The speedwells are probably not our most challenging plant family. These are small herbaceous annuals or perennials that are often found growing close to areas of human habitation. Although their bright blue flowers might be expected to attract attention, their ubiquitous nature seems to result in them being often overlooked. Thus, while many people may be able to identify that the delicate blue flower at their feet is a speedwell, they may struggle to tell you which species of *Veronica* it is. In contrast, arable agriculturalists are proficient at identifying the smallest of speedwell seedlings because some species are important weeds of crop fields.

WHY IS THIS GROUP OF PLANTS COMPLEX?

Newly evolved species can be more difficult to differentiate than those separated by many millennia. This seems to be the case with speedwells. As a family, they appear to have undergone a process of recent rapid evolutionary change. Many of our *Veronica*s have adapted to living in close proximity to humans. As 'opportunistic species' they have the ability to swiftly colonise disturbed or cultivated ground. This life history appears to have evolved on several occasions within this family. Rapid evolution in plants is often associated with doubling the number of chromosomes within each cell. This genetic 'trick' can occur in a single generation, and it can result in the creation of plants with different morphologies and distinct ecologies. In the British Isles there are speedwells that contain very different numbers of chromosomes. Sometimes this even occurs within a species, so that subspecies may have different numbers of chromosomes and may still be in the process of evolving to become new species in their own right.

The speedwells that live alongside humans sometimes get lucky and find themselves growing in unoccupied fertile ground, while others nearby may find themselves struggling in a crack in the pavement. This environmental stochasticity can result in dramatic differences in plant size and form, even between siblings. Such plasticity does not make their identification any easier.

One of the adaptations to life alongside humans is a change from being perennial to being an annual. Humans are an unpredictable lot, and few species survive many years in their company. The shift to reproducing and dying every year is thought to speed up the rate of evolution. Species with shorter generation times are able to adapt more rapidly to change than those that require many years to reach maturity. This provides more evidence to support the theory that speedwells have undergone a period of rapid speciation.

In addition to their recent evolutionary history, speedwells have an annoying habit that can make them tricky to identify. Once picked, their flowers drop very easily. So, if you have collected a sample while on a walk, by the time you get home, the chances are it will have shed its petals.

HOW CAN I TELL THEM APART?

The flowers of different species of speedwell are superficially rather similar. This could be daunting. However, the ease with which they can be identified to family level should be reassuring. At least you are confident you are looking in the right chapter!

Many speedwells are rather small plants and therefore a hand lens will be helpful to examine the features you need to compare species. Key characters include hairs on the stems and leaf surfaces. While the flowers themselves are somewhat similar, how they are located – on side branches, on long stalks or in terminal spikes – is helpful in differentiating species.

Finally, different species of speedwell have very different ecologies, so considering their habitat and location can be very informative too.

Let's start with the common species of speedwell

Veronica officinalis **(Heath Speedwell)**
A perennial of grasslands, open woodlands and heaths across the British Isles. The plants are semi-upright with hairy stems which grow to a height of about 40 cm and root at the lower nodes. The oval leaves occur in pairs and are hairy, with a toothed margin. **The lilac-coloured flowers measure 5–9 mm across and occur in spikes of 10 to 20 flower buds.**

V. serpyllifolia **(Thyme-leaved Speedwell)**
A perennial species that grows in lawns and cultivated ground. Common throughout the British Isles. A low-growing species with stems to a height of 30 cm, often rooting at the nodes. The stems lack hairs. **The small, oval leaves occur in pairs, are bright green and lack hairs. The leaf margin has a suggestion of small teeth.** The flowers are white to pale blue, with darker blue veins. The flowers measure 5–8 mm across and occur in spikes at the end of stems, with about 10 flower buds per spike.

Characteristics of *Veronica* species

Species	Max. height cm			Rooting stems	Hairs on stems	Hairs on leaves	Flower size <5 mm	Position of flowers			
	<20	20–40	>40					on side branch	tip of stems	along stem	spike >50 flowers
V. officinalis Heath Speedwell		■		■	■	■		■			
V. serpyllifolia Thyme-leaved Speedwell		■		■			■		■	■	
V. chamaedrys Germander Speedwell			■	■	■ *	■		■	■		
V. filiformis Slender Speedwell			■	■	■	■		■		■	
V. persica Common Field-speedwell			■	■	■	■		■	■	■	
V. hederifolia Ivy-leaved Speedwell			■	■	■	■	■	■	■	■	
V. arvensis Wall Speedwell	■				■	■	■		■	■	
V. agrestis Green Field-speedwell		■		■	■	■		■	■	■	
V. polita Grey Field-speedwell		■		■	■	■		■	■	■	
V. montana Wood Speedwell		■			■	■		■			
V. beccabunga Brooklime			■	■				■			
V. anagallis-aquatica Blue Water-speedwell			■					■	■		
V. catenata Pink Water-speedwell			■					■	■		
V. scutellata Marsh Speedwell			■	■				■			
V. longifolia Garden Speedwell			■						■		■
V. praecox Breckland Speedwell	■				■	■	■		■		
V. triphyllos Fingered Speedwell	■				■	■	■	■	■	■	
V. verna Spring Speedwell	■				■	■	■		■		
V. spicata Spiked Speedwell			■		■	■	■		■		■
V. fruticans Rock Speedwell	■			■					■		
V. alpina Alpine Speedwell	■				■				■		
Your specimen											

* Hairs in two distinct rows running along the stems

Speedwells of water and waterlogged ground

Flowers and leaves of common species of speedwell

V. officinalis (Heath Speedwell)

V. serpyllifolia (Thyme-leaved Speedwell)

V. chamaedrys (Germander Speedwell)

V. filiformis (Slender Speedwell)

V. persica (Common Field-speedwell)

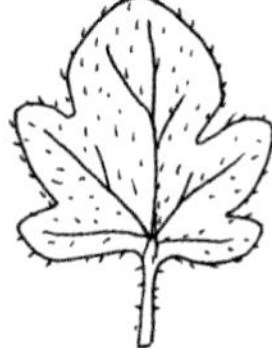

V. hederifolia (Ivy-leaved Speedwell)

V. arvensis (Wall Speedwell)

V. agrestis (Green Field-speedwell)

V. polita (Grey Field-speedwell)

V. montana (Wood Speedwell)

V. chamaedrys **(Germander Speedwell)**

A common perennial species of grasslands and wooded areas. The plants tend to be rather erect. They grow to a height of 50 cm and root at the lower nodes. **The stems are highly distinctive in having two rows of hairs that run opposite each other along their length.** The paired leaves are somewhat triangular in shape, with pronounced teeth along their margins. The flowers are bright blue and measure 8–12 mm across, they occur on short side shoots with fewer than 10 flowers per spike, and only one or two open at a time.

V. filiformis **(Slender Speedwell)**

An introduced perennial species that is now **commonly found spreading in lawns** and churchyards across the British Isles. Grows to a length of 50 cm. The stems are covered in short hairs and root at the lower nodes. Its leaves are small, heart-shaped and covered in short hairs. Only the terminal leaves grow in pairs. The flowers are pale lilac-blue, with the lowest petal palest. The flowers measure 8–15 mm. **Single flowers are found on long stalks,** so that flowers appear scattered equally over the plant.

V. persica **(Common Field-speedwell)**

An introduced annual found in cultivated and waste ground across the British Isles. A low-growing, spreading species that roots at the lower nodes and can attain a maximum height of 50 cm. The stems and leaves are hairy. The leaves vary between being oval to heart-shaped. The leaf margins are deeply toothed. Only the terminal leaves grow in pairs. **The flowers are bright blue, with the lowest petal almost white.** The flowers measure 8–12 mm, with single flowers being found on their own stalk.

V. hederifolia **(Ivy-leaved Speedwell)**

A common annual scrambling plant that grows to a height of 60 cm. Found in woodland edges and disturbed ground across the British Isles. Present in early spring, it disappears completely by summer. The stems are hairy and root at the lower nodes. As its name suggests, its leaves are somewhat ivy like, being well lobed, although unlike ivy they are covered in hairs. Leaves do not occur in pairs. **The pale blue or blue flowers are small, measuring 4–9 mm, and often appear not to be fully open. Their sepals are hairy and as long as the petals.** Single flowers are found on their own stalk.

There are two subspecies of the Ivy-leaved Speedwell in Britain. They are easily distinguished by looking at their anthers.

- Ssp. *lucorum* is considered native. Found in shady spots in less disturbed sites. **Its anthers are blue.** Cells contain 36 chromosomes.
- Ssp. *hederifolia* is introduced. Grows in more open and cultivated places. **Its anthers are white.** Cells contain 54 chromosomes.

V. arvensis **(Wall Speedwell)**
As its name suggests, this common annual is often seen growing on walls. It is also found across the British Isles growing in cultivated ground and dryish spots. A small plant, only up to 15 cm tall. Its stems and leaves covered in short hairs. It has paired, oval leaves with toothed margins. **The intensely blue flowers are tiny, only 2–3 mm across. The flowers have no stalks, single flowers are found at the base of the leaves near the end of stems.**

V. agrestis **(Green Field-speedwell)**
This annual is found in cultivated ground across most of the British Isles, although it is less common than the other species and is scarce in the north west of Scotland and in Ireland. Stems grow to a height of 30 cm and root at the lower nodes. Its stems and leaves are covered in short hairs. The paired leaves are oval and their margins are deeply toothed. Its mature leaves are longer than they are wide. The flowers are variable in colour – being white, pale blue or pale lilac. **Sepals are longer than petals.** The flowers measure 3–8 mm across. **Single flowers are found on short stalks that arise one stalk per pair of leaves.**

V. polita **(Grey Field-speedwell)**
This annual occurs in cultivated ground across most of the British Isles, although it is rare in the north west of Scotland and Ireland. Its stems grow to a height of 30 cm and root at the lower nodes. Its stems and leaves are covered in short hairs. The paired leaves are oval and their margins are deeply toothed. **The leaves are distinctly dull grey-green in colour. Can be differentiated from Green Field-speedwell because its mature leaves are wider than they are long.** The flowers are deep blue in colour and measure 3–8 mm across. **Single flowers are found on short stalks that arise one stalk per pair of leaves.**

V. montana **(Wood Speedwell)**
This perennial species grows in damp woods across the British Isles, except the north of Scotland. May be locally abundant. A rather scrambling plant that grows to only 40 cm. **Its stems are obviously hairy, while its leaves may appear hairless until you look more closely.** The leaves are paired, oval in shape with deep teeth along their margins. The leaves are carried on small stems that measure 5–15 mm long. The flowers are pale lilac blue and measure 8–10 mm across. **Flowers occur in groups of about 5 to 10 buds on delicate stalks which arise as a single stalk per pair of leaves.**

Speedwells of water and waterlogged ground

V. beccabunga **(Brooklime)**
A common perennial plant across the British Isles, growing in wetlands and streams as well as wet hollows alongside paths and tracks. Botanists joke that it is called *Veronica beccabunga* because it bungs up becks. It may grow tangled

Flowers and leaves of speedwells of water and waterlogged ground

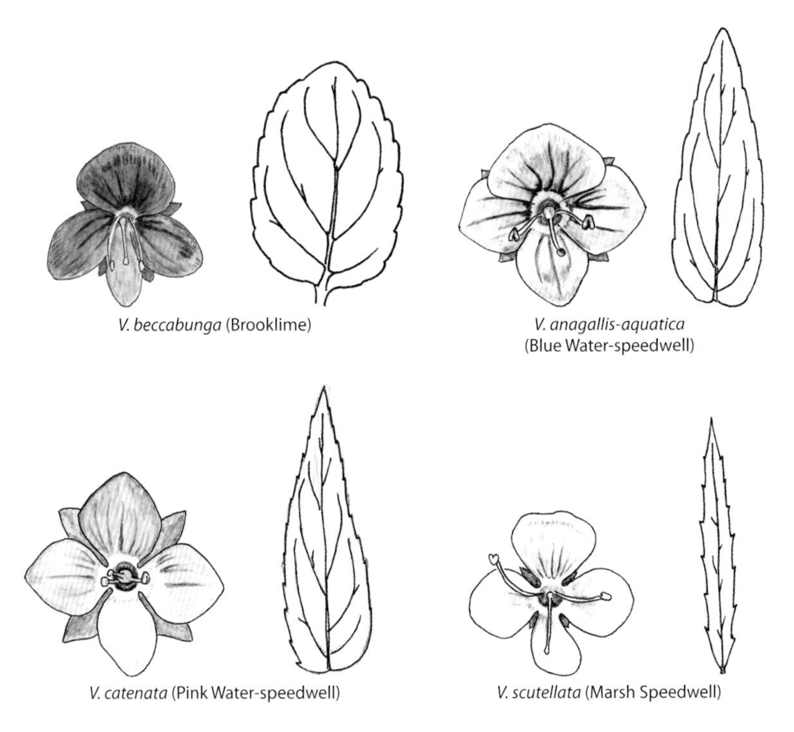

V. beccabunga (Brooklime)

V. anagallis-aquatica (Blue Water-speedwell)

V. catenata (Pink Water-speedwell)

V. scutellata (Marsh Speedwell)

with other vegetation, or to a height of 60 cm. **Its stems lack hairs and are often reddish in colour. Its oval paired leaves are somewhat fleshy, glossy bright green and hairless.** The flowers are variable in colour from pale to bright deep blue and measure 5–8 mm. The flowers are found in paired spikes which arise at the base of leaves. Each spike contains about 10 flower buds.

V. anagallis-aquatica **(Blue Water-speedwell)**
A short-lived perennial or annual in wetlands and ponds across the British Isles. Less common in the west and in northern Scotland. In areas where it does occur, it can be abundant. An upright plant, to a height of 60 cm. Stems and leaves are hairless. **Its paired leaves are long and lanceolate.** The flowers are pale blue and measure 5–7 mm. The four sepals are not well visible between the petals. The flowers are found in paired spikes which arise at the base of leaves. **Each spike contains 20 or more flower buds.**

V. catenata **(Pink Water-speedwell)**
Generally a plant of water-logged ground rather than standing water. Occurs across the British Isles but is less common in the north and west. Very similar

in appearance to Blue Water-speedwell, in all respects; as the name suggests the way to tell them apart is the flower colour. **This species has pale pink flowers, with pink veins. The four sepals are clearly visible between the petals. Blue and Pink Water-speedwells are known to hybridise.** Intermediate forms are not uncommon where the two species occur together, which is most of Britain and Ireland except for northern Scotland.

V. scutellata (Marsh Speedwell)
A rather delicate-looking plant that scrambles among other vegetation in wetlands across the British Isles, where it can sometimes be locally abundant. Grows to a height of 60 cm and roots at the lower nodes. Its stems and leaves are usually hairless, but occasionally can be hairy. Its leaves occur in pairs and are long and thin. The leaf margin has small teeth. **Flower colour is variable, including white, pale blue or pale pink to lilac. The flowers measure 5–8 mm across. Flowers occur in groups of about five buds on delicate branched stalks which typically arise as a single stalk per pair of leaves**, but occasionally there are paired flower stalks.

Next let's look at a less common species

V. longifolia (Garden Speedwell)
An introduced species, now quite regularly found on waste ground and road verges across the British Isles except for Ireland. A tall upright plant that can grow to 1.2 m tall. **The stems and leaves lack hairs**, except occasionally they may have a few small hairs. The paired leaves are lanceolate, with a saw-like toothed margin. The leaves are carried on small stems that measure 5–10 mm. **The flowers occur in densely packed spikes of 50 to 100 tiny, tubular, purplish-blue flowers. Individual flowers measure less than 4 mm across. The flower spikes occur at the tip of the plant.**

Finally let's look at the rarer speedwells

If you want to find rare British speedwells the place to look is cultivated ground in the Breckland area of East Anglia. Here you may find four species that are almost unknown outside the region.

V. praecox (Breckland Speedwell)
An introduced annual plant that is almost entirely restricted to sandy arable fields in the Brecks. Its stems are erect and grow to a height of 20 cm. Both the stems and leaves are hairy. **The paired leaves have deep irregular teeth along their margins.** The flowers are relatively small, measuring 2.5–4 mm across. **The flowers are blue with dark blue veins. Only a few flowers open at a time near the tip of the longer stems.**

Flowers and leaf shapes of less common *Veronica* species

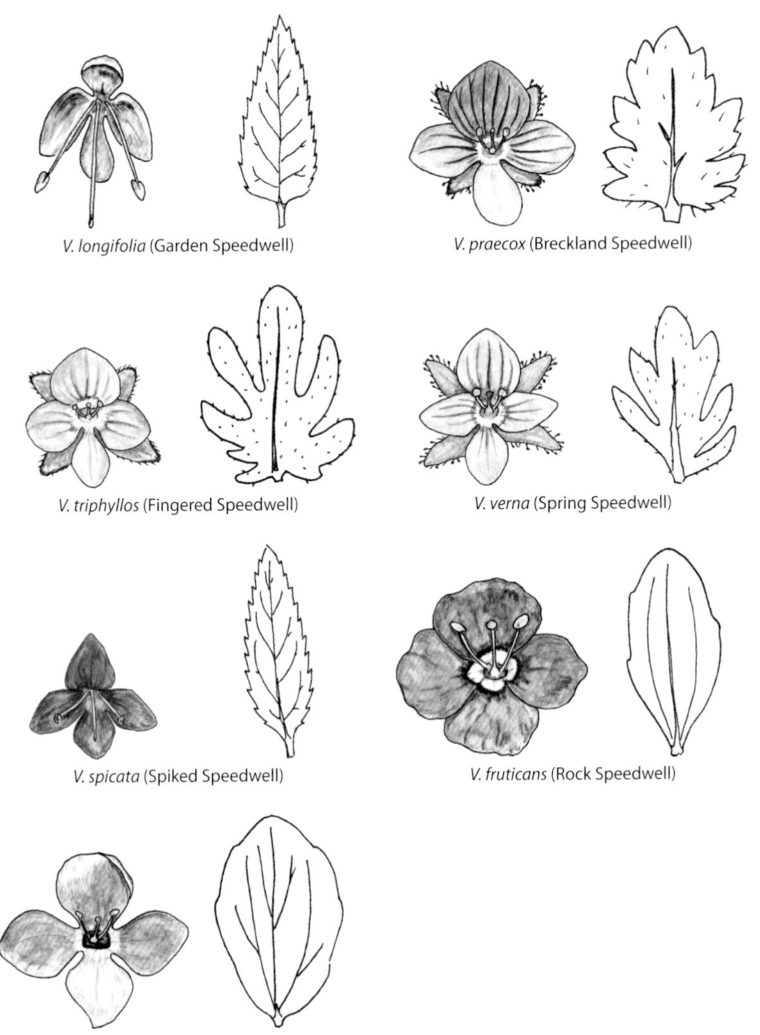
V. longifolia (Garden Speedwell)
V. praecox (Breckland Speedwell)
V. triphyllos (Fingered Speedwell)
V. verna (Spring Speedwell)
V. spicata (Spiked Speedwell)
V. fruticans (Rock Speedwell)
V. alpina (Alpine Speedwell)

V. triphyllos **(Fingered Speedwell)**

This annual species is restricted to sandy arable fields in Breckland. It is a small erect plant that grows to a height of only 20 cm. The stems and leaves are covered in glandular hairs. **The paired leaves have 3 to 7 distinctive pronounced lobes (fingers).** The deep blue flowers measure 3–4 mm across. The flowers occur near the tips of longer stems and are carried on short stalks that measure 5–8 mm.

V. verna (Spring Speedwell)
Found in grasslands on dry sandy soils in Breckland. Its erect stems grow to a height of only 15 cm. Both the stems and leaves are hairy. **The leaves are deeply lobed. The deep blue flowers are small, measuring only 2–3 mm across.** The flowers occur near the tips of the plants and lack stalks.

V. spicata (Spiked Speedwell)
Has a scattered distribution, growing on limestone and other base-rich soils across Wales, the South West and central England. A rather robust clump-forming perennial that looks somewhat like the Garden Speedwell. Its upright hairy stems grow to a height of 60 cm or more. Its paired leaves are hairy and lanceolate in shape, with a saw-toothed margin. **The deep blue flowers occur in densely packed spikes of 50 to 100 flower buds. Individual flowers measure about 4 mm across. The flower spikes occur at the tip of the plant.**

V. fruticans (Rock Speedwell)
A rare herbaceous perennial plant of the Highlands of Scotland. It grows among rocks above 500 m. The stems are upright and grow to a height of 20 cm and root at the lower nodes. The stems and leaves usually lack hairs. The paired leaves are oval in shape and the margins usually lack teeth. **The flowers are deep blue, and measure 10–15 mm across. Flowers occur in pairs on short stalks arising from the base of leaves near the tip of the plant.**

V. alpina (Alpine Speedwell)
A rare herbaceous perennial of the Highlands of Scotland. Grows among damp rocks. It has upright hairy stems that grow to a height of 15 cm. The oval leaves are paired and usually lack leaves but can have a scattering of short hairs. **The flowers are a dull deep blue and measure 5–10 mm across. The flower spikes are found at the tip of the plant.** The flowers themselves are unstalked.

HAVE OTHERS RECOGNISED THIS LEVEL OF VARIATION?

There are lots of common names for speedwells in English and in our other native languages. However, these names seem to refer to just a few of the more common species. This is perhaps no surprise as the rarer species have very restricted distributions and are unlikely to have been known outside their respective areas. The use of the scientific name *Veronica* can be traced to the late 1500s and is associated with St Veronica. Culpeper's herbal mentions just a single species of speedwell, while the Victorian flora of Bentham and Hooker includes 16 species, rather than the 21 listed here.

HOW FAR SHOULD I GO?

Speedwells tend not to hybridise very often, and there are few subspecies to worry about, so with a little effort the common species can easily be recognised. The annual species are to be found growing where the ground is regularly disturbed by humans, while the perennials tend to occur in more natural habitats.

Unless you are in the Highlands of Scotland or in arable East Anglia, you are unlikely to encounter our more unusual species – even then this guide should provide you with all the assistance you need, unless you encounter an exotic garden escape.

SECTION V

Successful families with lots of species

This final grouping, **'families with lots of species'**, are only problematic to identify because there are simply so many of them. The differences between individual species are distinct, so they can be easily recognised. However, these families contain so many different species, many of which are uncommon, that inevitably some will be unfamiliar. The vast numbers of species involved can be daunting for the novice botanist. The approach taken here involves providing a guide for dividing these families into more manageable subgroups.

For hundreds of years, botanists have struggled to understand why some plant families contain many more species than others. There is still no definitive answer to this question, because it is not open to experimentation. Analysis of characteristics associated with more successful plant families, however, suggests that the answer is related to insect pollination. It is thought that plants with highly specialised flowers attract more specialised insects, which perform more precise high-fidelity pollination. This reduces contamination of pollen from other sources and acts to genetically isolate the plants as if they were living on a remote island. This genetic isolation enables them to speciate more easily, as they are not always being swamped by genes from a larger gene pool.

It is notable that the families included here tend to have complex inflorescences which are attractive to insects. This group could also be said to include orchids and yellow members of the dandelion family, which have been covered earlier. These families also contain lots of species and have complex insect-pollinated flowers. This observation again emphasises the overlapping nature of these groupings, and the fact that frustrating flowers often have multiple layers of complexity.

CHAPTER 21

Dead-nettles, mints and woundworts

Many members of this rather large family are common and well known as annual weeds, long-lived wayside wildflowers and garden herbs. Although the family is now known as the Lamiaceae, many botanists still affectionately use their old name and refer to them as Labiates. The family is generally easy to recognise: plants typically have square stems, with pairs of opposite leaves alternating up their stems. The flowers are typically constructed of five petals that fuse into a tube that has a single, vertical line of symmetry. The flower tubes generally have four lips, giving the impression of four petals, and these may bifurcate so that flowers sometimes appear to have six petals. Many members of the family are herbs with highly aromatic leaves, which are used medicinally or in cooking. Although the smell of some species is far from appealing, it can still be a useful character when it comes to identifying them.

WHY IS THIS GROUP OF PLANTS COMPLEX?

Much of the complexity of this family arises from its success – there are just so many of them. As a result, trying to identify them usually involves working through many dichotomous questions, with descriptions and illustrations spread out over several pages of your field guide. Simply knowing where to start can be frustrating. In addition to our many native species, humans have accidentally introduced a good number of weedy species and intentionally introduced and widely cultivated many herbs and ornamental alien species. To this already long list has been added a plethora of different horticultural cultivars and hybrids. The mints are particularly prone to hybridisation. Virtually all possible parental combinations are known, as are a few three-way parental crosses. The hybrid mints appear to have been generated on several occasions, as they occur with a wide range of different numbers of chromosomes. This chromosomal diversity has made it easy for horticulturalists to select many different culinary forms. Since mints have a habit of escaping cultivation and being difficult to eradicate, their variability that delights the chef also frustrates the field botanist.

HOW CAN I TELL THEM APART?

Although these plants are usually easy to recognise to family level, identifying them to species can be more of a puzzle. The first thing to do is to assign them to one of the following smaller, more manageable groups. To do this you need to look at the upper petal in the flower. Most commonly these are hooded, but they may also curve upwards and backwards; the upper petal may be small or missing, and in the mints and some other herbs they can be similar to the other petals.

Comparison of four different flower forms within the Lamiaceae

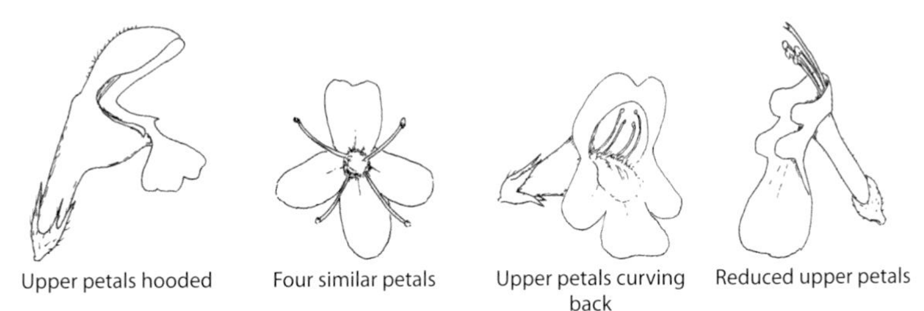

1. **Species with a hood-like upper petal: dead-nettles, woundworts …**
   This is a large and varied group of plants. The key thing to look out for is **the upper petal being hooded and shaped like an upside-down spoon**. This hood usually conceals the male anthers and female stigma. The lower part of the flower is tubular, and usually occurs in a whorl with several other flowers. The leaves and stems are often hairy. For ease here these species have been divided into those with purple or blue flowers, and those with white, pink or yellow flowers. This is a rather unsatisfactory division, as the intensity of anthocyanin pigments varies, being greatly influenced both by genetics and the environment.
2. **Species with four rather similar small petals: mints, thymes …**
   Plants in this group have **rather small flowers, which appear to have four fairly similar unfused petals and a short flower tube.** The male anthers and female stigma usually extend well beyond the flower. The flowers occur in dense whorls with many other flowers, often in a terminal flower spike. All parts of the plant are highly aromatic.
3. **Species with upturned upper petals: Betony, Wild Basil …**
   This group is somewhat similar to the mints. Their flowers have three lower lobes, which are wider than the two upper lobes. **The upper petal often has two lobes which turn upwards.** The male and female organs are usually protected by the upper lobes. The flowers occur in rather open whorls with relatively few other flowers. Many of these species are aromatic.

4. **Species with much-reduced upper petals: Bugle, Wood Sage …**
   This group is perhaps the most distinctive. **The flowers lack the upper petal, and the anthers and their stigmas protrude clearly beyond the flower tube.** The central lower lip is much larger than are the side lobes. Flowers come in a range of different colours.

1. Species with a hood-like upper petal: dead-nettles, hemp-nettles, woundworts etc.

a. With purple or blue flowers

Lamium purpureum **(Red Dead-nettle)**
This annual species is common across most of Britain and Ireland, where it occurs as a weed of disturbed ground. May grow to a height of 40 cm. Its leaves are heart-shaped with a pronounced toothed margin and may grow to 5 cm long. The lower leaves are carried on a long hairy stalk. The pink-purple flowers occur in about five tight whorls around the main stem, with leaves at the apex. The tubular flowers have a hooded upper petal and **much reduced side lobes, which may appear tooth-like.** The flower tube measures 7–12 mm and it has a ring of hairs near its base.

L. maculatum **(Spotted Dead-nettle)**
This garden escape is widespread in much of England and eastern Scotland but is less frequent elsewhere. Spreads vegetatively, the plants are rough and hairy and may grow to a height of 60 cm. Its leaves are a triangular heart shape, with a toothed margin. **Most plants have distinctive silvery blotchy leaves that grow to 5 cm long.** The pink-purple flowers occur in about five tight whorls around the main stem, usually with flowers uppermost. Flower colour is variable. The tubular flowers have a hooded upper petal and much reduced side lobes, the lower petal may be covered in purple blotches. The flower tube measures 2–3.5 cm and has a ring of hairs near its base.

L. hybridum **(Cut-leaved Dead-nettle)**
This annual species is widespread in much of England but tends to be coastal elsewhere. Has spreading stems, plants are slender and slightly hairy and may reach a height of 30 cm. **Its leaves are rather round with a deeply lobed irregular toothed margin**, growing to 5 cm long. The pink-purple flowers occur in about three tight whorls around the main stem, usually with leaves uppermost. Uppermost leaves may be purple in colour. The tubular flowers have a hooded upper petal and much reduced side lobes with a sharp tooth, the lower petal may have a few purple blotches. The flower tube measures 7–12 mm, and it has **no ring of hairs near its base.**

Characteristics of labiates with hooded upper petals and purple or blue flowers

Species	Disturbed ground	Height <50 cm	Flower spike	Leaf margin irregular	Leaves on long stalks	Flower colour	Reduced side lobes	Extended style
Lamium purpureum (Red Dead-nettle)								
Lamium maculatum (Spotted Dead-nettle)								
Lamium hybridum (Cut-leaved Dead-nettle)								
Lamium confertum (Northern Dead-nettle)								
Lamium amplexicaule (Henbit Dead-nettle)								
Galeopsis angustifolia (Red Hemp-nettle)								
Galeopsis bifida (Bifid Hemp-nettle)								
Galeopsis tetrahit (Common Hemp-nettle)*								
Stachys sylvatica (Hedge Woundwort)								
Stachys palustris (Marsh Woundwort)								
Stachys germanica (Downy Woundwort)								
Stachys alpina (Limestone Woundwort)								
Ballota nigra (Black Horehound)								
Scutellaria galericulata (Skullcap)								
Prunella vulgaris (Selfheal)								
Salvia pratensis (Meadow Clary)								
Salvia verbenaca (Wild Clary)								
Your specimen								

* See section below for description of Common Hemp-nettle

L. confertum (Northern Dead-nettle)

This annual species mostly occurs around the coast in Scotland, Ireland and the Isle of Man. May have spreading stems, plants are slender and almost hairless and may grow to a height of 25 cm. **Its leaves are a rather broad heart shape with**

a deeply lobed regular toothed margin and grow to 2.5 cm long. The leaves are carried on long stalks. The pink-purple flowers occur in about three or four tight whorls at the top of the main stem, usually with leaves uppermost. Uppermost leaves may be purple in colour. The tubular flowers have a hooded upper petal and much reduced side lobes with a small tooth, the lower petal and inner flower tube may have a few purple blotches. The flower tube measures 8 to 12 mm, and it has faint ring of hairs inside.

L. amplexicaule **(Henbit Dead-nettle)**
This annual species is widespread in much of England but tends to be more coastal elsewhere. Has somewhat spreading stems, which are covered in fine hairs. It may grow to a height of 25 cm. **Its leaves are rather small and round with a rounded, lobed margin. The leaves grow up to 2.5 cm long. The lower leaves have long stalks up to 5 cm long.** The pink-purple flowers usually occur in two or three tight whorls around the main stem, with the flowers projecting above the leaves. The tubular flowers have a hooded upper petal and much-reduced side lobes that turn outwards, the lower petal may have a few pink-purple blotches. The flower tube measures 1.4–2 cm, **it is hairless inside but covered in dense white hairs on the outside.**

Galeopsis angustifolia **(Red Hemp-nettle)**
This scarce annual species has a rather scattered distribution but is most frequent in England. It mostly occurs on open calcareous ground, where it can grow to a height of 50 cm but is often much smaller. **Its leaves are hairy, long and lanceolate, measuring less than 1 cm across and up to 8 cm long, with a margin of fine teeth.** The leaves are carried on short stalks, plants are regularly branched. The rosy-purple **flowers occur in about two or three whorls at the top of the main stem and on branches,** usually with flowers uppermost. The tubular flowers have a hooded upper petal, side lobes that project out and back; the lower petal has a white throat and may have a few purple blotches. The flower tube measures 1.5–2.5 cm, and it has a ring of hairs inside.

Galeopsis bifida **(Bifid Hemp-nettle)**
This coarse annual species occurs across most of Britain and Ireland, usually as an arable weed, where it can grow to a height of 100 cm. **It has red-tipped glandular hairs below the nodes on its stems. Its leaves are hairy, oval with a pointed tip, measuring up to 10 cm long, with a toothed margin.** Plants are regularly branched. Flowers can be a mix of purple and white with yellow markings. They occur in compact whorls at the top of the main stem and on branches, usually with flowers uppermost. **The tubular flowers have a hooded upper petal, which is covered in long hairs.** Its side lobes project out and back, the lower petal has a clear central notch and a few purple blotches. The flower tube measures 13–16 mm and has a ring of hairs inside.

Stachys sylvatica **(Hedge Woundwort)**
This perennial species is common across all of Britain and Ireland in hedges, woodland and in rough ground. Grows to a height of 100 cm. It spreads vegetatively, forming patches. The hairy leaves are nettle-like, with a serrated margin, a long leaf stalk and rather foetid smell. The leaves measure up to 9 cm long. Stems are hairy. Plants branch, with flowers carried in a spike at the end of the main stem and side branches. **The flower is a distinct dark purple-claret colour, the outside is covered in short hairs. The throat of the flower has white blotchy markings.** The side lobes project outwards and back. The flower tube measures 13–15 mm.

Stachys palustris **(Marsh Woundwort)**
This perennial species is common across all of Britain and Ireland, usually in damp or rough ground, often by waterbodies. Grows to a height of 100 cm. It spreads vegetatively, forming patches. The leaves are long, thin and lanceolate, with a serrated margin, and a short or no leaf stalk. The leaves measure up to 12 cm long. Stems are hairy. Plants branch, with flowers carried in a spike at the end of the main stem and side branches. **The flower is a distinct pink-mauve colour, the outside covered in short hairs. The lower petal of the flower has darker purple blotchy markings.** The side lobes project outwards and back. The flower tube measures 12–15 mm.

Hedge and Marsh Woundworts quite often hybridise and produce the sterile intermediate offspring *Stachys* x *ambigua*.

Stachys germanica **(Downy Woundwort)**
This rare perennial species occurs mostly in an area west of Oxford on calcareous ground. It grows to a height of 100 cm. **The entire plant is covered in distinctive dense white silky hairs.** The leaves are lanceolate, deeply veined, with a serrated margin and lack a leaf stalk. The leaves grow up to 12 cm long. Plants branch, with flowers carried in a spike at the end of the main stem and side branches. The flowers are pink-purple in colour, **the upper petal is only shallowly concave (so it might be considered to have upper petals turned up)** and the outside of the upper petal is covered in long hairs. The lower petals are paler. The side lobes project outwards. The flower tube measures 12–20 mm.

Stachys alpina **(Limestone Woundwort)**
This extremely rare, perhaps introduced perennial species occurs in just a small number of woodlands in Denbighshire and Gloucestershire. Grows to a height of 100 cm. The plant is covered in soft short hairs. **The leaves grow to a length of 16 cm, they are an elongated heart shape, deeply veined and with a serrated margin. The lower leaves have longer stalks which can be up to 10 cm.** Flowers are carried in a spike at the end of the main stem in loose whorls. The flowers are a dull pink with red to purple markings and are covered in glandular hairs. The upper petal is shallowly concave. The side lobes project outwards. The flower tube measures 15–20 mm.

Ballota nigra (Black Horehound)
This perennial species is widespread in Wales and England, but much less common in Scotland and Ireland. Grows to a height of 100 cm in hedgerows and in rough ground. A robust, branched plant covered in hairs. The rough leaves, which grow to a length of 5 cm, are rather nettle-like with a serrated margin. The lower leaves are broader and have longer stalks. Flowers are carried in a spike at the end of the main stem and branches in loose whorls, usually with leaves uppermost. The flowers are mauve-purple in colour, and are covered in hairs. The upper petal is shallowly concave. The side lobes project outwards and backwards. **The lower and side lobes are often crossed with white veins, which emerge from the throat of the flower.** The flower tube measures 15–20 mm. The whole plant is pungently foetid.

Scutellaria galericulata (Skullcap)
This creeping perennial is widespread across most of Britain and Ireland. Occurs in wetlands, often near waterbodies, where it may grow to 50 cm tall. The lanceolate leaves grow to 5 cm long, they are usually hairless and have a short stalk. **The curved, tubular rich blue-violet flowers usually arise in pairs, facing the same direction.** The upper petal is deeply hooded, almost closing the mouth of the flower. The side lobes are much reduced, the lower petal is paler. The flower tube measures 1–1.8 cm.

Skullcap and Lesser Skullcap are known to hybridise, producing sterile intermediate but vigorously growing offspring. Lesser Skullcap is covered in the next section.

Prunella vulgaris (Selfheal)
This common perennial species occurs across the whole of Britain and Ireland in old grasslands, rough ground and lawns. Grows to a height of 30 cm. The leaves and stems often have a purple tinge and are slightly hairy. The leaves are oval with a pointed tip and short stalk. **The blue-purple flowers are located in terminal spikes at the end of the main stem and branches.** The upper petal is deeply concave, the side lobes project outwards and back. The flower tube measures 1–1.4 cm long.

Salvia pratensis (Meadow Clary)
This rare perennial species is known from a small number of calcareous grasslands in southern England, and very occasionally Wales and Scotland. May grow to a height of 80 cm. Its stems are branched and are covered in glandular hairs. The leaves measure up to 15 cm long; they are elongated ovals, deeply veined with a wavy margin. Upper leaves have long stalks, while lower leaves are stalkless. **The flowers are an intense deep blue colour** and occur in open whorls in a terminal flower spike. The upper petal is a closed hood, which is flattened vertically. **The female stigma projects beyond the upper petal on an elongated style.** The flower tubes measure 1.5–2.5 cm.

***Salvia verbenaca* (Wild Clary)**

This uncommon plant occurs in calcareous grasslands, mostly in southern England but occasionally in other parts of the British Isles. May grow to a height of 80 cm. Its stems are tinged blue-purple, branched and covered in hairs. The leaves measure up to 12 cm long; they are oval, deeply veined with a very wavy margin. Upper leaves have stalks, while lower leaves are stalkless. **The flowers are an intense blue colour** and occur in open whorls in a terminal flower spike. The upper petal is a closed hood, which is flattened vertically. **The female stigma projects beyond the upper petal on an elongated style.** The flower tubes measure up to 1.5 cm.

Purple- or blue-flowered labiates with hooded upper petals

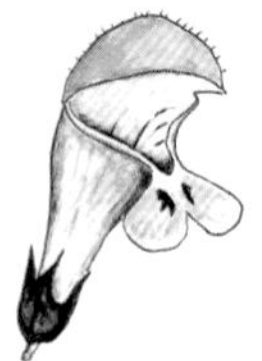

Lamium purpureum (Red Dead-nettle)

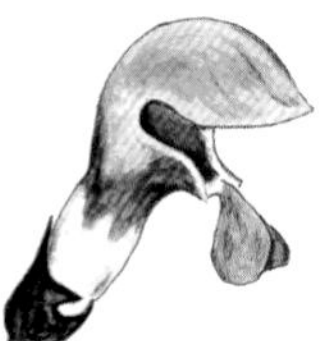

L. maculatum (Spotted Dead-nettle)

L. hybridum (Cut-leaved Dead-nettle)

L. confertum (Northern Dead-nettle)

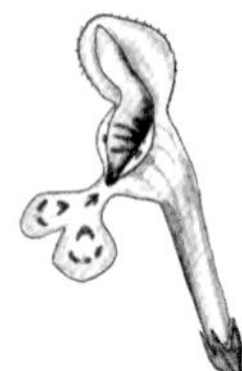

L. amplexicaule (Henbit Dead-nettle)

Galeopsis angustifolia (Red Hemp-nettle)

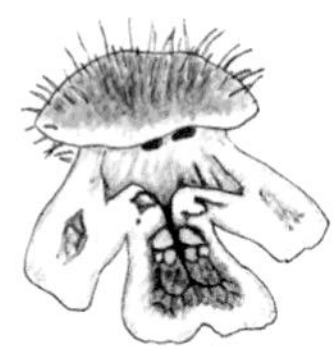

Galeopsis bifida (Bifid Hemp-nettle)

Stachys sylvatica (Hedge Woundwort)

Stachys palustris (Marsh Woundwort)

Stachys germanica (Downy Woundwort)

Stachys alpina (Limestone Woundwort)

Ballota nigra (Black Horehound)

Scutellaria galericulata (Skullcap)

Prunella vulgaris (Selfheal)

Salvia pratensis (Meadow Clary)

Salvia verbenaca (Wild Clary)

1. Species with a hood-like upper petal: dead-nettles, hemp-nettles, woundworts etc.

b. With white, pink or yellow flowers

Lamium album (White Dead-nettle)

This perennial occurs in rough ground and hedges almost everywhere, but it is uncommon in the west of Ireland and western Scotland. Grows to a height of 60 cm. Its stems are rough and hairy. The leaves are rather nettle-like, with a toothed margin; the upper leaves are stalkless, while those lower down the stem are carried on short stalks. The leaves measure up to 7 cm long. **The white flowers occur in dense whorls of 6 to 10 flowers around the stem. The upper petals are hooded, shielding four black anthers and a white stigma.** The side lobes are highly reduced to a curled lip, sometimes with a fine projecting tooth. The lower petal has two lobes which curve downwards. The flower tube measures 1.8–2.5 cm.

Galeopsis tetrahit (Common Hemp-nettle)

This annual species occurs in rough and disturbed ground almost everywhere, but it is less common in the west of Ireland and the Scottish Highlands. Grows to a height of 100 cm. Its stems are branched and hairy with red-tipped glandular hairs below the nodes. The rather nettle-like leaves are slightly hairy with a toothed margin; the upper leaves are stalkless, while those lower down the stem are carried on short stalks. The leaves measure up to 10 cm long. **The flowers are variable in colour from white through pale pink to purple, with yellow and purple patches in the throat of the flower.** Near the top of the stem the flower whorls are densely packed but they become more spaced out lower down the stem. The upper petals are hooded and hairy. The side lobes are folded downwards. The lower petal curves downwards and usually has purple markings in the throat. The flower tube measures 1.5–2 cm.

Galeopsis speciosa (Large-flowered Hemp-nettle)

This annual species grows in disturbed ground and arable land often associated with root crops. It has a scattered distribution, potentially turning up anywhere, but is more frequent in the north. Grows to a height of 100 cm. Its stems are uniformly hairy. The leaves are oval with a pointed apex, the leaf margin is toothed, the leaves are carried on short stalks. The leaves measure up to 10 cm long. **The flowers are highly distinctive: pale yellow, with a mauve-purple lower petal, while the throat is yellow with purple markings.** Near the top of the stem the flower whorls are densely packed, but they become more spaced out lower down the stem. Flowers usually project above the top of the plant. The upper petals are hooded and hairy. The side lobes project outwards. The lower petal projects forwards. The flower tube measures 2.2–3.4 cm.

Characteristics of labiates with hooded upper petals and white, pink or yellow flowers

Species	Disturbed ground	Height <50 cm	Flower spike	Leaf margin irregular	Middle leaf with stalks	Flower colour	Reduced side lobes	Extended style
Lamium album (White Dead-nettle)								
Galeopsis tetrahit (Common Hemp-nettle)								
Galeopis speciosa (Large-flowered Hemp-nettle)								
Galeopsis segetum (Downy Hemp-nettle)								
Melittis melissophyllum (Bastard Balm)								
Lamiastrum galeobdolon (Yellow Archangel)								
Stachys recta (Yellow Woundwort)								
Stachys annua (Annual Yellow Woundwort)								
Stachsy arvensis (Field Woundwort)								
Scutellaria minor (Lesser Skullcap)								
Nepeta cataria (Cat-mint)								
Leonurus cardiaca (Motherwort)								
Your specimen								

Galeopsis segetum (Downy Hemp-nettle)

This extremely rare annual species occurs in disturbed ground and arable land is now only found at a few sites. Grows to a height of 50 cm. Its stems branch and lack hairs. **The leaves are downy, particularly on the undersurface.** The leaf shape is a narrow oval with a pointed apex, the leaf margin is toothed, and the leaves are carried on short stalks. The leaves measure up to 8 cm long. **The flowers are white with a pale lemon throat, the flowers stand erect.** Near the top of the stem the flower whorls are densely packed, but they become more spaced out lower down the stem. Flowers usually project above the top of the plant. The upper petals are hooded and slightly hairy. The side lobes project outwards. The lower petal projects downwards. The flower tube measures 2–3 cm.

Melittis melissophyllum (Bastard Balm)

This perennial occurs in woodlands and hedgerows in the South West of England and south-west Wales. It may occur occasionally elsewhere as a garden escape. **The plant has a distinctive smell of lemons.** The stems are hairy and grow to

70 cm tall. The leaves grow to a length of 8 cm, they are oval in shape with a pointed tip and serrated margin. **The flowers are distinctive, being white with a pink-purple lower lip.** The upper petal is only shallowly hooded. The side lobes project outwards and slightly backwards. The flowers occur in whorls spaced along the stems with few flowers per whorl. The flower tube measures 2.5–4 cm.

Lamiastrum galeobdolon (Yellow Archangel)
The perennial occurs in woodlands and hedgerows across the Britain and Ireland, although it is less common in Scotland and Ireland. It spreads vegetatively, forming patches, and its stems grow to a height of 60 cm. Its leaves grow to a length of 7 cm, they are rather nettle-like with a deeply irregularly toothed margin. **Its flowers are distinctively lemon yellow in colour.** The upper petal is deeply hooded, **the side and lower lobes have orange-brown markings.** The lower lobe is slightly longer than the side lobes. Flowers are carried in several whorls around the stems. The flower tube measures up to 2 cm.

There are three subspecies of Yellow Archangel in the British Isles. They are distinguished by looking at their stems, leaves and flowers.

- Ssp. *galeobdolon* is native, sometimes has white blotches on its leaves early in the year, its stems have hairs only along their edges. Its hooded upper petal is less than 8 mm wide. Its cells contain 18 chromosomes. Very rare.
- Ssp. *montanum* is native, has green leaves and hairs on all faces of its stems. Its cells contain 36 chromosomes. The usual subspecies, common.
- Ssp. *argentatum* is an introduced garden escape, its leaves have distinctive silver-white blotches all year-round, sometimes with purple markings in the centre. Its stems have hairs only along their edges. Its hooded upper petal is more than 8 mm wide. Its cells contain 36 chromosomes.

Stachys recta (Yellow Woundwort)
A very rare, introduced perennial species that is found at a handful of scattered disturbed locations, where its hairy stems grow to a height of 70 cm. Its leaves are lanceolate with a toothed margin. Mid-leaves lack stalks. **Its flowers are pale yellow with purple marks in their throat.** The side lobes are much reduced, the lower petal projects downwards. The flowers are carried in a terminal spike. The flower tube measures 1.5–2 cm.

Stachys annua (Annual Yellow Woundwort)
This rare, introduced annual occurs at a small number of scattered disturbed sites, where its hairy stems grow to a height of 30 cm. Its leaves are lanceolate with a toothed margin. Mid-leaves have short stalks. **Its flowers are white with a hint of yellow with purple marks on the lower lobes.** The side lobes project sidewards,

the lower petal projects downwards. The flowers are carried in whorls towards the top of the stems. The flower tube measures 1–1.6 cm long.

Stachys arvensis **(Field Woundwort)**
This annual plant occurs across Britain and Ireland but is rarer in Scotland and Northern Ireland. It tends to be a weed of non-calcareous arable land, where it grows to a height of 25 cm – notably smaller than the other woundworts. Its stems are hairy. Its leaves are oval with a blunt end and serrated margin. The leaves grow to 3 cm long and have short stems. **The flowers are a very pale purple-white colour. The sepals are darker purple. The upper petal is a shallow hood.** The side lobes project sidewards and the lower lobe projects forwards. There are pink blotches on the lower lobes. The flowers are carried in whorls near the ends of the stems. The flower tube measures up to 8 mm.

White, pink or yellow flowered labiates with hooded upper petals

Lamium album (White Dead-nettle)

Galeopsis tetrahit (Common Hemp-nettle)

G. speciosa (Large-flowered Hemp-nettle)

G. segetum (Downy Hemp-nettle)

Melittis melissophyllum (Bastard Balm)

Lamiastrum galeobdolon (Yellow Archangel)

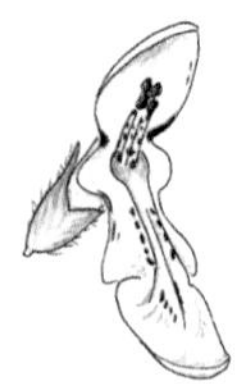
Stachys recta (Yellow Woundwort)

S. annua (Annual Yellow Woundwort)

S. arvensis (Field Woundwort)

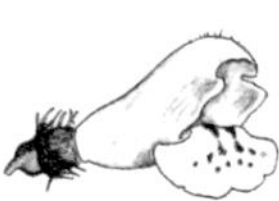
Scutellaria minor (Lesser Skullcap)

Nepeta cataria (Cat-mint)

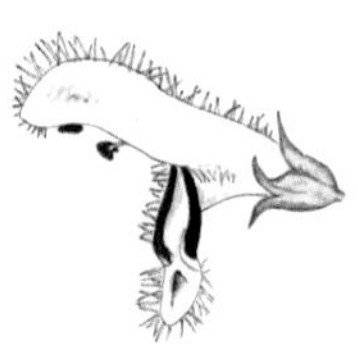
Leonurus cardiaca (Motherwort)

Scutellaria minor **(Lesser Skullcap)**
Most common in Wales, southern England, the west of Scotland and southern Ireland. A low-growing perennial plant of wet acid ground, where it grows to a height of 15 cm. Its stems lack hairs. Its leaves are narrow, lanceolate and lack teeth. The leaves have no stalk and grow to a length of 3 cm. The flowers are pale pink. **The upper petal is deeply hooded but may turn upwards once open.** The side lobes are much reduced. The lower petal is often divided into three lobes which are covered in small purple spots. **The flowers are often found in pairs on the same side of the stem.** The flower tube measures up to 1 cm.

Nepeta cataria **(Cat-mint)**
This perennial species is famous for its strong and distinctive smell. Mostly restricted to England where it grows in calcareous grasslands, to a height of 100 cm. Its leaves are somewhat triangular in shape with a deeply toothed margin. The leaves have stalks, and they may grow to 7 cm long. **The flowers are highly variable, because there are many horticultural varieties that may escape into the wild.** The flowers are pale mauve, the upper petal is usually hooded. The side lobes are much reduced. The lower lobe is variable but usually covered in purple spots. The flowers occur in terminal spikes. The flower tubes measure up to 1.2 cm.

Leonurus cardiaca **(Motherwort)**
Motherwort is an uncommon hairless perennial, introduced in the Middle Ages, which occurs in disturbed scattered sites mostly restricted to England. It grows to a height of 120 cm. **Its lower leaves are distinctive in being palmate and having irregular deep lobes.** The leaves grow to a length of 12 cm. The flowers are white with a hint of purple, **the upper petal is hooded and covered in dense, long hairs – giving a fluffy appearance.** The side lobes are much reduced and may be more deeply pigmented. The lower petal is smaller than the upper and also covered in long hairs. The flowers are carried in whorls at the top of the stem. The flower tubes measure up to 1.2 cm.

2. Species with four rather similar small petals: mints, thymes …

Mentha aquatica **(Water Mint)**
This perennial species occurs commonly across the whole of Britain and Ireland, although it is less frequent in northern Scotland. Grows in waterlogged ground often by water, where it can reach a height of 90 cm. Like all mints, it has a strong and distinctive smell. A highly variable species, sometimes being highly branched. **Its leaves and stems can be tinted purple in high light or stressed conditions.** Its leaves are broadly oval with a pronounced tip and serrated margin. Mid-leaves have a short stalk and can grow up to 4 cm long. Both leaves and stems may be slightly hairy. **The flowers are carried in a spherical terminal spike,** with 1 to 3 more whorls of flowers lower down the stem. The flowers are pale pink and appear to have four rather similar lobes. The stigma and anthers protrude beyond the flower tube. The flower tubes measure up to 4.5 mm.

Characteristics of labiates with upper petals similar to lower petals

Species	Leaves				Terminal flowers	Leaf margin serrated
	hairy	rounded	crinkled	stalkless		
Mentha aquatica (Water Mint)						
Mentha arvensis (Corn Mint)						
Mentha spicata (Spearmint)						
Mentha suaveolens (Round-leaved Mint)						
Mentha pulegium (Pennyroyal)						
Origanum vulgare (Wild Marjoram)						
Your specimen						

M. arvensis (Corn Mint)

This perennial occurs fairly commonly across the whole of Britain and Ireland, although it is less frequent in Ireland and Scotland. It grows in arable fields and damp meadows, where it can grow to a height of 60 cm. Like all mints it has a strong **distinctive smell, sometimes described as sickly**. A highly variable species, sometimes being branched. Its stems die back in winter. The leaves are broadly oval with a pronounced tip and serrated margin. Mid-leaves have a short stalk and can grow up to 6 cm long. Both leaves and stems may be hairy. **The flowers are carried in about 3 or 4 whorls that decrease in size ascending the stem. The flower whorls are interspersed with leaves, which also decrease in size towards the top, but are always longer than the flower whorls. Stems terminate in leaves.** The flowers are pale pink and appear to have four rather similar lobes, although sometimes the upper one can bifurcate, so the flowers appear to have five petals. The stigma and anthers protrude beyond the flower tube. The flower tubes measure up to 2.5 mm.

M. spicata (Spearmint)

A perennial that occurs relatively commonly across the whole of Britain and Ireland, often as a garden escape, although it is less frequent in Ireland and Scotland. Often appears in rough ground close to human habitation, where it can grow to a height of 90 cm. **It has a distinctive, familiar spearmint smell.** However, it can be very difficult to differentiate from some hybrid mints. Its leaves are lanceolate, with serrated margins; they usually lack a stalk and can grow up to 9 cm long. **Its leaves usually lack hairs and have a bright green, crinkled appearance. The flowers are carried in a conical terminal spike of closely packed whorls of**

flowers. Terminal flower spikes are also found on side branches. The flowers are pale pink to white and appear to have four rather similar lobes. The stigma and anthers protrude beyond the flower tube. The flower tubes measure up to 3 mm.

M. suaveolens (Round-leaved Mint)

This perennial species occurs in scattered sites across the whole of Britain and Ireland in damp ground and verges, where it can grow to a height of 90 cm. It spreads vegetatively, forming patches. Has a strong aromatic smell. **As its name suggests, the leaves are round, with a crinkled appearance** and serrated margin, they lack a stalk and can grow up to 4 cm long. **The leaves and stems are covered in dense hairs, often giving the plant a whitish appearance. The flowers are carried in a conical terminal spike of closely packed whorls of flowers.** Terminal flower spikes are also found on side branches. The flowers are pale pink to white and appear to have four rather similar lobes. The stigma and anthers protrude beyond the flower tube. The flower tubes measure up to 3 mm.

M. pulegium (Pennyroyal)

This perennial species grows in scattered sites across England, but is rare elsewhere. Grows in damp meadows and heaths often near the sea, where it can get to a height of 30 cm. It has a pungent smell. The stems are often pigmented red and are highly branched. Its leaves are narrowly oval with a pronounced tip, and serrated margin. Mid leaves have short stalks and can grow up to 2 cm long. Both leaves and stems are hairy. **The flowers are carried in many dense but widely spread whorls, that decrease in size ascending the stem.** The flower whorls are interspersed with leaves, which also decrease in size towards the top but are always longer than the flower whorls. Stems terminate in leaves. **The flowers are pale pink and appear to have four rather similar lobes, although the upper one is slightly notched. The flowers are hairy on the outside.** The stigma and anthers protrude beyond the flower tube. The flower tubes measure up to 3.5 mm.

Reported mint hybrids with two parents. Each of these crosses has been given a name in its own right, but it is more informative to refer to their parentage.

Parent species	*M. aquatica* (Water Mint) **T**	*M. arvensis* (Corn Mint)	*M. spicata* (Spearmint) **T**	*M. suaveolens* **T** (Round-leaved Mint)
M. aquatica (Water Mint)		*M.* x *verticillata* (Whorled Mint)	*M.* x *piperita* (Peppermint) **T**	*M.* x *suavis* (Sweet Mint) **T**
M. arvensis (Corn Mint)	*M.* x *verticillata* (Whorled Mint)		*M.* x *gracilis* (Bushy Mint)	*M.* x *carinthiaca* (Austrian Mint)
M. spicata (Spearmint)	*M.* x *piperita* (Peppermint) **T**	*M.* x *gracilis* (Bushy Mint)		*M.* x *villosa* (Apple Mint) **T**
M. suaveolens (Round-leaved Mint)	*M.* x *suavis* (Sweet Mint) **T**	*M.* x *carinthiaca* (Austrian Mint)	*M.* x *villosa* (Apple Mint) **T**	

T = Plants with terminal flower spikes. *M. arvensis* and all its hybrid progeny have flowers in whorls, but not in terminal spikes.

***Origanum vulgare* (Wild Marjoram)**

This perennial species is common across Britain and Ireland but is less frequent in Scotland and Ireland. Grows in dry grasslands and verges, typically on calcareous soils, where it can attain a height of 80 cm. Has a strong aromatic smell. The stems are upright and much branched. **Its leaves are oval with a smooth margin and often two smaller leaflets at their base.** Mid-leaves have short stalks and can grow up to 4.5 cm long. Both leaves and stems are slightly hairy. **The flowers are carried in terminal clusters, side branches also terminate in flower clusters. The flowers are pale pink or white and appear to have four rather similar lobes, although the lower one is slightly longer and the side ones project downwards.** The stigma and anthers protrude beyond the flower tube. The flower tubes measure up to 9 mm. The sepals are darkly pigmented purple in colour, except in white-flowered forms.

***Thymus drucei* (Wild Thyme)**

A low-growing mat-forming perennial common across Britain and Ireland. Grows in short grasslands and in rocky places, to a height of 7 cm. **The stems are covered in hairs of varying lengths on two opposite sides only.** The leaves are usually held horizontally. The leaves are small and oval, growing to a length of 8 mm and often lacking hairs. Flowers occur in terminal clusters. The flowers are pink, with four lobes. The lower three lips project downwards. Two anthers project sidewards beyond the flower tube, while two remain covered by the upper petal. The flower tubes measure up to 5 mm.

***T. pulegioides* (Large Thyme)**

A low-growing perennial that is largely restricted to southern England. Found in short dry calcareous grasslands and rocky places, where it grows to a height of 25 cm. **The stems are covered in short hairs on two opposite sides only and longer hairs on the corners of the stems.** The leaves are usually held horizontally. The leaves are small and oval, growing to a length of 12 mm and often lacking hairs. Flowers occur in terminal clusters. The flowers are pink, with four lobes. The lower three lips project downwards. Two anthers project sidewards beyond the flower tube, while two remain covered by the upper petal. The flower tubes measure up to 8 mm.

***T. vulgaris* (Garden Thyme)**

This perennial is a garden escape that occurs in scattered locations across Britain and Ireland. Grows in rocky places such as walls, where it can reach a height of 30 cm. **All four sides of the stems are covered in short hairs.** The leaves are usually upright. The leaves are grey-green in colour, with very short hairs, narrowly oval in shape and grow to a length of 8 mm. Flowers occur in terminal clusters. The flowers are pink to pale purple with four lobes. The lower three lips project downwards. Two anthers project sidewards beyond the flower tube, while two remain covered by the upper petal. The flower tubes measure up to 8 mm.

T. serpyllum **(Breckland Thyme)**

This rare species is restricted to a few heathland sites in Suffolk, Norfolk and Kent. Grows to a height of less than 7 cm. **All four sides of its stems are equally covered in hairs.** The leaves are usually upright. The leaves are small and oval, growing to a length of 5 mm and lack hairs. Flowers occur in terminal clusters. The flowers are pink, with four lobes. The lower three lips project downwards. Two anthers project sidewards beyond the flower tube, while two remain covered by the upper petal. The flower tubes measure up to 5 mm.

Thymes are readily identified by looking at the distribution of hairs on their stems – see the illustrations below.

Morphologies of labiates with four similar petals

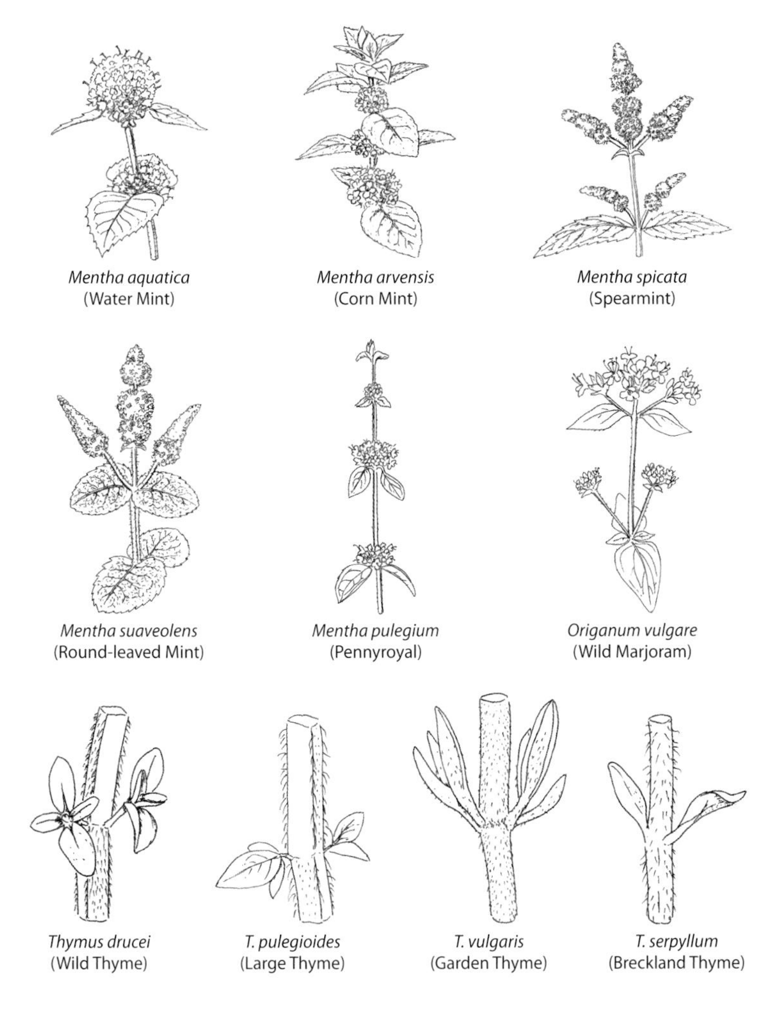

3. Species with upturned upper petals: Betony, Wild Basil, calamints etc.

Glechoma hederacea (Ground-ivy)
Common across most of Britain and Ireland, except the extreme north of Scotland and west of Ireland. It grows in rough ground, hedges and woodlands, forming straggly patches as it spreads vegetatively, reaching a height of 30 cm. Its trailing stems may range from being hairy to almost lacking hairs. **Its leaves are a rounded heart shape with a deeply serrated margin.** The leaves have long stalks and grow to up to 3 cm. The flowers occur in loose whorls along the stem, roughly three or four flowers per whorl. The flowers are a bluey violet colour, the lower lip bifurcates and usually projects horizontally. The throat of the flower is paler, hairy and blotched purple. The upper petal is usually the smallest and curves upwards. The flower tube is long, measuring up to 2 cm.

Betonica officinalis (Betony)
A perennial plant widespread in England and Wales but rare in Ireland and Scotland. Grows in grasslands, heaths and verges and can reach a height of 60 cm. When not flowering it forms rosettes. The stems may be slightly hairy, but the leaves lack hairs. **The leaves are an elongated heart shape, with a deeply serrated margin** and rather crinkly surface. The leaves grow to a length of 7 cm and have stalks, which are shorter further up the stem. **The flowers are carried on a distinctive terminal spike of compressed whorls on a long stalk.** The flowers are deep pink to reddish-purple. The flower tube is held almost horizontally; the lower lip is broad and partly bifurcated, the upper lip projects upwards. The flower tubes measure up to 18 mm.

Clinopodium vulgare (Wild Basil)
This perennial species is widespread across most of Britain, except northern Scotland, and uncommon in Ireland. Occurs in rough grasslands and verges, where it grows to a height of up to 80 cm. Has a faint smell. Its stems are very hairy, its leaves less so. The leaves are oval and tapered, with shallow serrations; they are carried on a short stalk and grow to a length of 5 cm. The flowers are pinkish-purple, the lower and upper lips are broad and slightly bifurcated. The side lobes are smaller and project downwards. **The sepals are very hairy. The flowers are usually carried in two whorls, one of which is terminal.** The flower tubes measure up to 2 cm.

Clinopodium acinos (Basil-thyme)
Usually an annual, widespread across most of England but less frequent elsewhere. Occurs in open ground and arable fields usually on calcareous soils, where it grows to a height of up to 25 cm. Stems are hairy, the leaves less so. The leaves are oval and tapered, with shallow serrations; they are carried on a short stalk and grow to a **length of only 1.5 cm.** The flowers are usually carried in loose whorls near the

Characteristics of labiates with upturned upper petals

Species	Disturbed ground	Height <50 cm	Flower spike	Leaf margin irregular	Leaves downy	Mid leaves with stalks	Flower colour	Bifurcated upper lobe
Glechoma hederacea (Ground-ivy)								
Betonica officinalis (Betony)								
Clinopodium vulgare (Wild Basil)								
Clinopodium acinos (Basil-thyme)								
Clinopodium ascendens (Common Calamint)								
Clinopodium nepeta (Lesser Calamint)								
Clinopodium menthifolium (Wood Calamint)								
Melissa officinalis (Lemon Balm)								
Satureja montana (Winter Savoury)								
Lycopus europaeus (Gypsywort)								
Marrubium vulgare (White Horehound)								
Your specimen								

end of stems, with only 2 or 3 flowers per whorl. The stems terminate in leaves. **The flowers are violet with a white throat which is blotched purple. The flower tube tends to point upwards.** The lower and upper lips are broad and slightly bifurcated, the side lobes are smaller and project downwards. The sepals are very hairy. The flower tubes measure up to 1 cm.

Clinopodium ascendens (Common Calamint)

This perennial species is widespread across much of southern England, south Wales and southern Ireland. Occurs in rough grasslands mostly on calcareous soils, where it grows to a height of up to 60 cm. Stems and leaves are hairy. The leaves are oval, sometimes almost a diamond shape, with shallow serrations; they are carried on a stalk which may be up to 1 cm long, while the leaf blade can grow to a length of 4 cm. The flowers are pale mauve, with a white throat blotched purple. The lower and upper lips are broad and slightly bifurcated. The side lobes are smaller and project downwards. **The flowers are usually carried on short, branched stems in whorls with about four or five flowers per branch.** The flower tubes measure up to 1.5 cm.

Flowers of labiates with upturned upper petals

Glechoma hederacea
(Ground-ivy)

Betonica officinalis
(Betony)

Clinopodium vulgare
(Wild Basil)

Clinopodium acinos
(Basil-thyme)

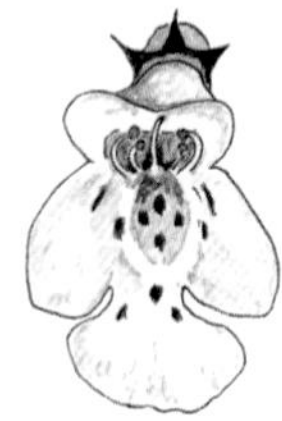

Clinopodium ascendens
(Common Calamint)

Clinopodium nepeta
(Lesser Calamint)

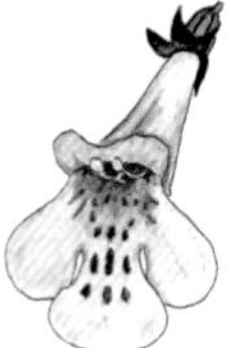

Clinopodium menthifolium
(Wood Calamint)

Melissa officinalis
(Lemon Balm)

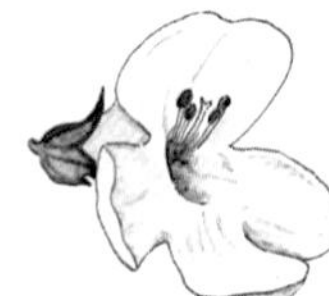

Satureja montana
(Winter Savoury)

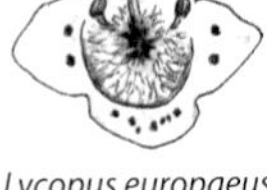

Lycopus europaeus
(Gypsywort)

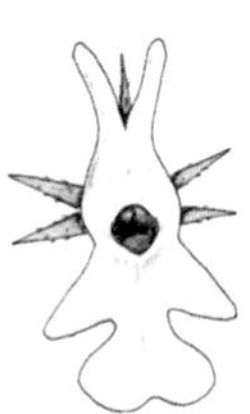

Marrubium vulgare
(White Horehound)

Clinopodium nepeta (Lesser Calamint) (previously *Calaminta nepeta*)

This perennial shrubby species most frequently occurs in south-east England, with a few other scattered records; unknown in Scotland and Ireland. **Smells like a mix of basil and mint.** Occurs in rough grasslands mostly on calcareous soils, where it grows to a height of up to 60 cm. Stems and leaves are hairy. The leaves are rather round with a pointed tip, with shallow serrations; they are carried on a stalk which may be up to a 5 mm long, while the leaf blade can grow to a length of 2 cm. The flowers are pale mauve, sometimes with a white throat blotched purple. The upper lips are slightly bifurcated. The lower and side lobes are similar and project downwards. The flowers are carried in a terminal flower spike of loose whorls of about four or five flowers per whorl. The flower tubes measure up to 1.5 cm.

Clinopodium menthifolium (Wood Calamint)

This perennial erect herb is known from a single site on the Isle of Wight on chalk and from a site in Powys; grows to a height of up to 60 cm. Stems and leaves are hairy. The leaves are rather round with a pointed tip, with shallow serrations; they are carried on a stalk which may be up to a 1.5 cm long, while the leaf blade can grow to a length of 4.5 cm. The flowers are mauve, sometimes with a white throat blotched with purple. The upper and lower lips are slightly bifurcated, the lower petal being the widest. The side lobes project sidewards and down, sometimes even back. **The flowers are carried in a terminal flower spike of loose whorls of short branches carrying about four or five flowers each, usually with only one flower open per branch.** The flower tubes measure up to 2 cm.

Melissa officinalis **(Lemon Balm)**
This introduced perennial herb is a common garden escape in England but is less frequent in northern Scotland and the north of Ireland. Forms patches which can grow to a height of 60 cm. **Has a distinctive sweet lemony smell.** Stems and leaves may be slightly hairy. The leaves are oval with a broader base and a pointed tip. The leaves are bright green and deeply veined with shallow serrations; they are carried on a stalk which may be up to 2 cm long, while the leaf blade can grow to a length of 5 cm. There may be smaller leaflets at the base of the leaf stalks. The flowers are white. The upper lips are slightly bifurcated. The lower lips are broad and project downwards. The side lobes are somewhat reduced and project sidewards. The flowers are carried in five or six loose whorls near the ends of stems, with leaves uppermost and three or four flowers per whorl. The flower tubes measure up to 1.2 cm.

Satureja montana **(Winter Savoury)**
This perennial **evergreen shrub** is a very occasional garden escape in the British Isles, can grow to a height of 50 cm. Stems and leaves lack hairs. The leaves are narrow and lanceolate. They are leathery, dark green, lack serrations and grow to a length of 2 cm. The flowers are white or pale pink. **The upper lips turn upwards but may sometimes be slightly hooded.** The lower lips are broad and somewhat spoon-like. The side lobes project sidewards. The flowers are carried in compressed whorls near the ends of stems, with leaves uppermost. The flower tubes measure up to 1 cm.

Lycopus europaeus **(Gypsywort)**
This perennial species is common across the whole of Britain and Ireland but less so in north-east Scotland. Occurs in wet grasslands and near waterbodies, where it grows to a height of up to 100 cm. Stems and leaves may be slightly hairy. **The leaves are highly distinctive, being lanceolate with deep irregular teeth.** The leaves are carried on a short stalk and can grow to a length of 10 cm. The flowers are white, sometimes with purple spots on the lower and side lobes. All the flower lobes are rather short. The flowers are carried near the ends of stems in up to eight dense whorls with many flowers per whorl and leaves uppermost. The flower tubes measure up to 5 mm.

Marrubium vulgare (White Horehound)
Widespread but has a scattered distribution, being most frequent in England where it occurs in rough grasslands and waste places. Grows to a height of up to 60 cm. Stems and leaves are hairy, **its leaves appearing white as they are covered in downy hairs.** The leaves are round and deeply veined, appearing crinkly, with rather rounded teeth. The leaves are carried on a short stalk and can grow to a length of 4 cm. **The flowers are white and distinctive: the upper lobes are deeply bifurcated and extend upwards, the lower lip is broad and rounded, the side lobes are smaller.** The flowers are carried near the ends of stems in up to ten dense whorls with many flowers per whorl and leaves uppermost. The flower tubes measure up to 1.5 cm.

4. Species with a much-reduced upper petal: bugles, Wood Sage, germanders etc.

Ajuga reptans (Bugle)
Bugle is a common perennial plant that occurs almost all over Britain and Ireland. It is found in woods and damp grasslands, where it grows to a height of 30 cm. However, be aware there are many unusual horticultural varieties that occur as garden escapes. The plant spreads vegetatively, forming clumps. The stems are often hairy on opposite sides. **The leaves are oval in shape, often slightly purple, deeply veined and grow to a length of 7cm. The leaves lack hairs, are carried on short stalks and distinctively become regularly smaller up the stem.** The flowers are blue, occasionally pink. The upper petal is much reduced, the lower petal may be horizontal projecting away from the stem. **The flowers are carried in a showy flower spike in which the flowers are longer than the leaves.** The flower tubes measure up to 1.8 cm.

Ajuga pyramidalis (Pyramidal Bugle)
This rare perennial only occurs in northern Scotland and a few sites in Ireland and northern England. A species of rocky ground at altitude, where it grows to a height of 30 cm. The plant spreads vegetatively, forming small clumps. The stems have hairs on all sides. The leaves are oval in shape, often slightly purple, deeply veined and grow to a length of 6cm. **The leaves are usually covered in hairs,** are carried on short stalks and become regularly smaller up the stem. The flowers may be pale blue to deep purple. The upper petal is much reduced, the lower petal may be horizontal projecting away from the stem. **The flowers are carried in a showy flower spike, in which the leaves are longer than the flowers.** The flower tubes measure up to 1.8 cm.

Ajuga chamaepitys (Ground-pine)
This rare annual only occurs in southern England. **Has a distinctive smell of pine when crushed.** Grows on calcareous arable soils where it can get to 20 cm tall. The leaves and stems are hairy. **The leaves are divided into**

three, finger-like lobes. The leaves grow to a length of 4 cm. **The flowers are a distinctive yellow colour with deep-orange splodges.** The upper petal is much reduced; the lower petal may be horizontal, projecting away from the stem. The flowers are carried in loose whorls on the stem. The flower tubes measure up to 2 cm.

Teucrium scorodonia (Wood Sage)

A common perennial across most of Britain and Ireland (although it is rare in central Ireland), where it occurs in dry woodlands. Grows to a height of 30 cm. Its stems are erect, branched and hairy. Its leaves are an elongated heart shape and grow to a length of 7 cm. The leaves are deeply veined, crinkly and have a short stalk. **The flowers are a distinctive pale lime green colour, the filaments within the flower are purple.** The upper petals are much reduced, the side lobes are small, the lower petal projects downwards. The flowers are carried in a terminal spike. The flower tube measures up to 8 mm.

Teucrium scordium (Water Germander)

This rare perennial species is found at a few scattered wet calcareous sites in Britain and Ireland, where it can grow to 50 cm tall. Stems and leaves are hairy. The leaves are lanceolate with a scalloped edge and grow to 5 cm in length. **The flowers are mauve in colour.** The upper petals are much reduced, the side lobes rather small, the lower petal is rather spoon-like. The flowers are carried scattered around the main stem. The flower tubes measure up to 12 mm.

Characteristics of labiates with much-reduced upper petals

Species	Flower spike	Leaves			Flower colour
		deeply lobed	hairy	stalkless	
Ajuga reptans (Bugle)					
Ajuga pyramidalis (Pyramidal Bugle)					
Ajuga chamaepitys (Ground-pine)					
Teucrium scorodonia (Wood Sage)					
Teucrium scordium (Water Germander)					
Teucrium chamaedrys (Wall Germander)					
Teucrium botrys (Cut-leaved Germander)					
Your specimen					

Flowers of labiates with reduced upper petals

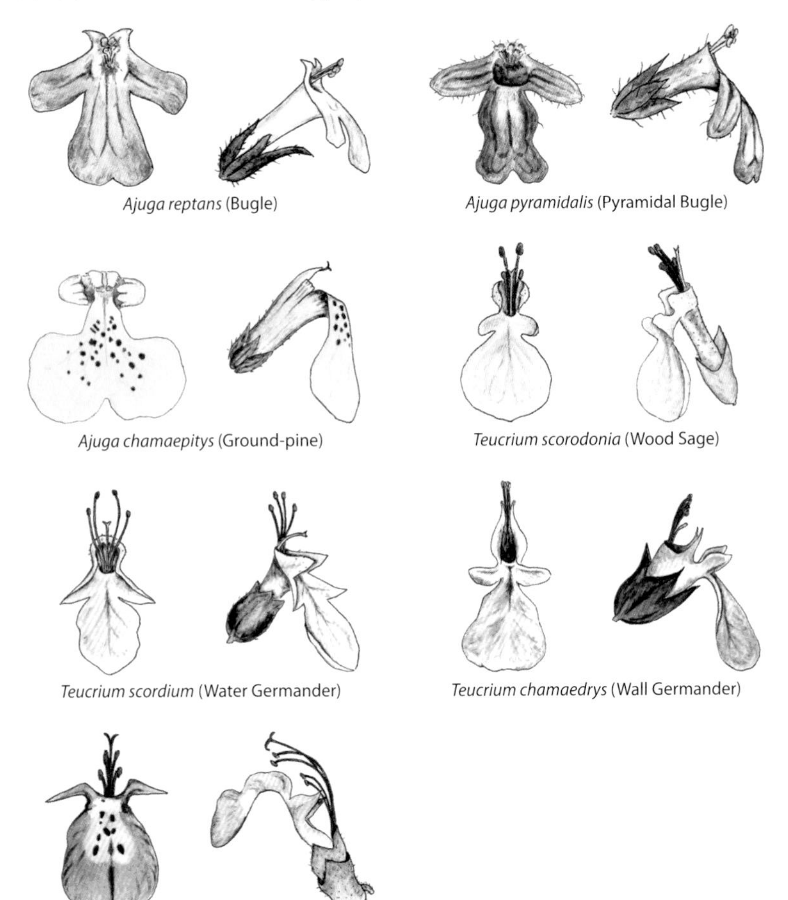

Ajuga reptans (Bugle)

Ajuga pyramidalis (Pyramidal Bugle)

Ajuga chamaepitys (Ground-pine)

Teucrium scorodonia (Wood Sage)

Teucrium scordium (Water Germander)

Teucrium chamaedrys (Wall Germander)

Teucrium botrys (Cut-leaved Germander)

Teucrium chamaedrys (Wall Germander)

This rare native perennial is found at a few very scattered calcareous grassland sites mostly in England, where it can grow to 40 cm tall. However, introduced plants occasionally occur as garden escapes. **Has a rather horizontal growth form.** The stems and leaves are hairy. The leaves are lanceolate with a deeply serrated edge and grow to 3 cm. The flowers are pink in colour. The upper petals are much reduced, the side lobes quite small, the lower petal rather spoon like. The flowers are carried in a terminal spike. The flower tubes measure up to 15 mm.

Teucrium botrys (Cut-leaved Germander)

This rare annual species is found in a few scattered calcareous sites in southern England, where it can grow to 30 cm tall. The stems and leaves are hairy.

The leaves are oval and highly dissected into deep lobes. The leaves grow to 2.5 cm and have stalks. The flowers are pink in colour. The upper petals are much reduced, the side lobes rather small, the lower petal has a white throat with purple blotches. The flowers are carried in an open terminal spike. The flower tube measures up to 6 mm.

HAVE OTHERS RECOGNISED THIS LEVEL OF VARIATION?

The majority of species in this family were described by Linnaeus, with the herbs in particular having been very well known and used by herbalists for centuries. However, several species are relatively recent introductions to our shores. The complexity arises from horticulturalists, who have generated a plethora of new varieties and hybrids. Frustratingly, this vast array of genetic diversity is often inconsistently documented. For example, Peppermint is the most widely cultivated mint; its use was recorded by the Roman writer Pliny the Elder (AD 23–79). Yet it seems unlikely that the Peppermint familiar to the Ancient Greeks and Romans is the same plant we know today. The English botanist John Ray mentions Peppermint in his 1696 herbal. He described it at the time as a new hybrid between Spearmint and Apple Mint. This situation is further complicated by the fact that European Peppermint is different to Chinese Peppermint. Since that period, this horticultural confusion has only deepened as ever more new varieties have been created and given exotic-sounding names.

HOW FAR SHOULD I GO?

Once you have got to grips with the number of species, most members of this family are relatively easy to identify. As we have seen, the complexity arises because of the generation of a vast array of horticultural varieties. If you really want to fully get to grips with the many mints, thymes and other herbs, you may need to delve into the specialist horticultural literature. It is worth being aware, though, that botanical and horticultural scientific names do not always correspond, and that the common names and colloquial names given to cultivars can be more of a hinderance than help.

CHAPTER 22

Blue and purple vetches and peas

These plants are a loosely defined set of about 30 members of the pea family. Although they are united by having instantly recognisable blue-purple pea-like flowers, they belong to several different but closely related genera. They are annuals or low-growing perennials which usually grow through other herbaceous vegetation, often supported by tendrils (a spring-like extension found at the end of their leaves, with which they attach themselves to other plants). Several of these species are common and widespread, and grow in tall grasslands, roadsides and hedge bottoms. There are several other similar species that are choosier about where they occur, and these tend to have more scattered distributions. Into this mix, there are also a number of introduced species. These include plants cultivated for forage, for amenity horticulture or introduced accidentally. These non-natives may turn up almost anywhere.

WHY IS THIS GROUP OF PLANTS COMPLEX?

Experienced botanists might consider that the plants in this group are probably not as difficult to identify as are some others. However, modern DNA analysis has revealed that many of these species that were previously thought to be closely related are in fact only distant cousins. In other cases, the reverse has turned out to be true. Appearances can be deceptive, and here lies some of the complexity of this group.

There are two main reasons why these plants can frustrate the novice. Firstly, there are lots of them. The pea family (the Fabaceae) is the third-largest plant family on the planet. It contains almost 20,000 known species. Of course, not all of these have blue-purple flowers or are found in the Britain and Ireland. But there are enough of them here to make the task challenging for beginners. This situation has been made more difficult because over the years humans have introduced several other similar species, as well as importing new subspecies of native species. Many of our less common species have widely scattered, almost random distributions, so you always need to be on the lookout for the unexpected.

The second factor that makes these species challenging is their morphological variability. Many are associated with heterogeneous habitats. If they find themselves growing in rich, fertile soil, they have the capacity to grow tall and produce numerous flowers. By contrast, individuals in shallow, infertile conditions may only grow a few centimetres tall and produce just one or two flowers. Even flower colour is variable. In species with blue-purple flowers, floral colour can vary from intense purple through to pale pink, and even white forms are regularly found. Although flower colour variation has a genetic basis, pigmentation is known to be more intense in more stressful, suboptimal environments.

HOW CAN I TELL THEM APART?

When trying to identify members of this group of plants, flower colour may both help and deceive. Once you are confident about their identity, you will recognise the unique and subtle colour tones that are typically associated with each of these species. Unfortunately for the beginner, there is a vast array of colour variation displayed by most, which means that flower colour can be an unreliable attribute upon which to entirely base your identification.

The process of identifying your specimen will be made easier if you can allocate it to one of the following four groups, which are dealt with separately below: vetches, peas, forages and milk-vetches.

1. Vetches are typically tall, straggly, climbing plants, with leaves that comprise many narrow leaflets, and which frequently terminate in a tendril. Their flowers are relatively small, measuring less than 1 cm.

2. Vetchlings and peas are usually more robust, with thicker often winged stems. Their leaves generally comprise a few broader leaflets. Their flowers are larger, measuring more than 1 cm; they occur in clusters on long flower stalks, and tend to be more pink-purple than blue.

3. The forage species are a rather miscellaneous group of not very closely related plants. They tend to be taller (usually over a metre), more upright species. Being upright, they have no need for tendrils. Their flowers are small and occur in dense clusters.

4. The milk-vetches, our final group, contains just three closely related, short species. They usually grow to a height of less than 30 cm. Their leaves comprise many narrow leaflets, which lack tendrils. Their flowers are generally small and occur in dense clusters.

Key features to look out for are the number of leaflets per leaf and whether the leaves terminate in a tendril. When looking at the flowers, count the number per cluster, measure the length of a sample of individual flowers, measure the length of any seedpods and finally count the number of seeds they contain.

Let's start with the commoner vetches and vetch-like species

Vicia sativa **(Common Vetch)**

This familiar annual species (which occurs as three subspecies) is found climbing through other vegetation, where it can reach a heigh of up to 1.5 m. It has leaves

Characteristics of purple vetches

Species	Tendrils			Number of leaflets		Flower stalks >2 cm	Flowers per cluster			Seeds per pod	
	none	simple	branched	<10	>10		<2	<10	>10	<5	>5
Vicia sativa (Common Vetch)											
Vicia cracca (Tufted Vetch)											
Vicia sepium (Bush Vetch)											
Ervilia hirsuta (Hairy Tare)										2	
Ervum tetraspermum (Smooth Tare)										4	
Ervilia sylvatica (Wood Vetch)											
Vicia orobus (Wood Bitter-vetch)											
Vicia lathyroides (Spring Vetch)				<4							
Vicia tenuifolia (Fine-leaved Vetch)											
Vicia villosa (Hairy Vetch)											
Ervum gracile (Slender Tare)				<4				<4			
Vicia bithynica (Bithynian Vetch)				<3		vary					
Your specimen											

Less common species

with 3 to 8 pairs of leaflets, which terminate in tendrils that are usually branched. At the end of each leaflet is a short needle-point. **Its bright pink-purple flowers are usually solitary or found in pairs. Flowers measure 1–2.5 cm.** It produces 4 to 12 brown seeds per pod. The pods measure 2–4 cm long.

Vicia cracca (Tufted Vetch)

A common native perennial that grows in many grassy habitats all over Britain and Ireland, reaching a length of up to 2 m. It has leaves with 5 to 15 pairs of leaflets, which terminate in branched tendrils. At the end of each leaflet is a short point. **Its blue-violet flowers are usually found in upright clusters of 10 to 30, each individual flower measures 8–12 mm.** It produces 2 to 6 brown seeds per pod. The pods measure 1–2.5 cm long.

Vicia sepium (Bush Vetch)
This common native perennial occurs in a wide range of grassy habitats across Britain and Ireland, where it may grow to 1 m tall. It has leaves with 5 to 9 pairs of leaflets, which terminate in tendrils. The lower tendrils are branched. **Leaflets are a distinctive egg shape.** At the end of each leaflet is a short needle-point. **Its blue-purple flowers are usually found in clusters of 2 to 6, each flower measuring 1.2–1.5 cm.** Produces 3 to 10 brown (speckled with black) seeds per pod. The pods measure 2–3.5 cm long.

Ervilia hirsuta (Hairy Tare)
A common annual species of rough grasslands across most of lowland Britain and Ireland. A delicate scrambling plant that grows to 80 cm. Its leaves are divided into 4 to 10 pairs of narrow leaflets, which end in branched tendrils. Its flowers are very pale lilac and may appear white from a distance. The flowers occur in clusters of 2 to 9, each flower measuring 3–5 mm long. **Seedpods are covered in fine hairs and typically contain just 2 seeds.**

Ervum tetraspermum (Smooth Tare)
This is a widespread native annual species of dry grasslands across most of Britain, but it is rarer in Scotland and Ireland. A delicate scrambling plant that grows to 80 cm. Its leaves are divided into 3 to 6 pairs of narrow leaflets, which end in branched tendrils. Its flowers are very pale lilac in colour with darker veins. **The flowers usually occur in pairs on a long flower stalk.** Each flower measures 4–8 mm long. **Seedpods are hairless and typically contain 4 seeds.**

Ervilia sylvatica (Wood Vetch)
This climbing perennial species occurs occasionally in open woods and coastal habitats across Britain and Ireland, where it can grow to 2 m in height. Its leaves are divided into 4 to 12 pairs of leaflets, which end in branched tendrils. **Its beautiful flowers are white with distinctive purple veining**; they appear in clusters of 4 to 15 flowers, with each flower measuring 12–20 mm long. Seedpods measure 2.5–3 cm long and typically contain 4 or 5 seeds.

Vicia orobus (Wood Bitter-vetch)
This climbing perennial has a distinctly westerly distribution in Britain and Ireland, only common in Wales. Occurs in rocky and grassy places, where it can grow to 60 cm in height. Its leaves are divided into 6 to 15 pairs of leaflets, **which lack tendrils. Its flowers are white in colour with distinctive purple veining**; they occur in clusters of 6 to 20 flowers, with each flower measuring 12–15 mm long. Seedpods measure 2–3 cm long and typically contain 4 or 5 seeds.

Vicia lathyroides (Spring Vetch)
A short weakly climbing, uncommon annual species which has a distinctly eastern distribution in Britain and Ireland, although it also occurs around some western

coasts. It occurs in sandy heaths and coastal habitats where it grows to only 20 cm in height. Its leaves are divided into **2 to 4 pairs of hairy leaflets, with a short unbranched tendril.** Its flowers are dull purple in colour, although **often the lower wing petals are pale pink. The flowers are solitary** and measure 6–9 mm long. Seedpods measure 1.5–3 cm long and typically contain 6 or 12 seeds.

Now for the less common vetches and vetch-like species

Vicia tenuifolia (Fine-leaved Vetch)
This introduced species occurs in scattered locations across the south of England. **Similar to Tufted Vetch, except its flowers are more lilac in colour and its lower wing petals are usually white.** It has leaves with between 5 to 15 pairs of **narrow leaflets,** which terminate in branched tendrils. At the end of each leaflet is a short point. **Its flowers are usually found in upright clusters of 10 to 30,** each individual flower measuring 12–18 mm. It produces 2 to 6 brown seeds per pod. The pods measure 1–2.5 cm long.

Vicia villosa (Hairy or Fodder Vetch)
This introduced species occurs in scattered locations across the south of England. **Similar to Tufted Vetch, except its flowers are an intense purple colour and its lower wing petals are paler. As the scientific name suggests, its stems and leaves are hairy.** It has leaves with 4 to 12 pairs of leaflets, which terminate in branched tendrils. At the end of each leaflet is a short point. **Its purple flowers are usually found in upright clusters of 10 to 30,** each individual flower measures 10–20 mm. It produces 2 to 8 bark-brown seeds per pod and the pods measure 1–2.5 cm long.

Ervum gracile (Slender Tare)
A rare annual plant of chalk grasslands in the south of England. Grows to a height of 80 cm. Its leaves comprise 2 to 4 pairs of fine leaflets. The leaves terminate in **tendrils which are usually not branched. Its flowers are pale lilac in colour and usually occur in clusters of 1 to 4 flowers.** Flowers are found on long flower stalks. Each flower measures 6–9 mm long. Seedpods are hairless and contain 4 to 8 seeds.

Vicia bithynica (Bithynian Vetch)
A scrambling annual plant of rough grasslands and scrub. Occurs at scattered sites, mostly in southern England and Wales, and very occasionally further north; grows to 60 cm tall. Its leaves comprise just 2 or 3 pairs of leaflets, terminating in branched tendrils. Its **distinctive two-tone flowers are purple, with white lower wing petals.** The flowers are often solitary or occur in pairs on **flower stalks which vary in length from 0.5 to 4 cm.** Each flower measures 1.6–2 cm long. Seedpods measure 2.5–5 cm and contain 4 to 8 seeds.

Leaves, inflorescences and flowers of blue-purple vetch species

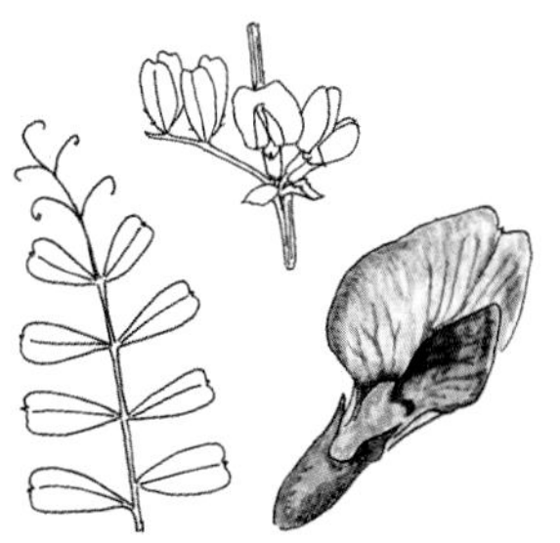

Vicia sativa (Common Vetch)

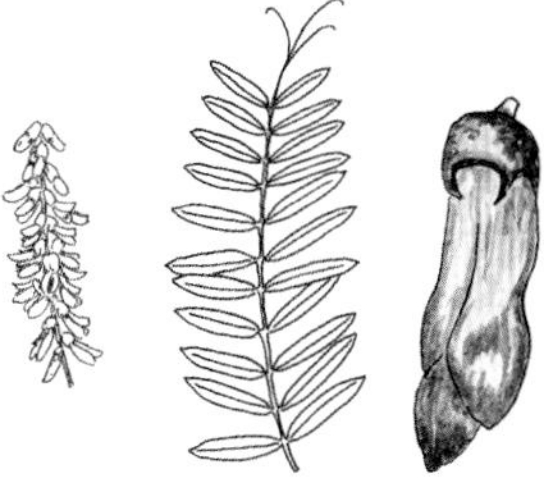

Vicia cracca (Tufted Vetch)

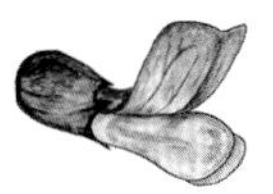

Vicia sepium (Bush Vetch)

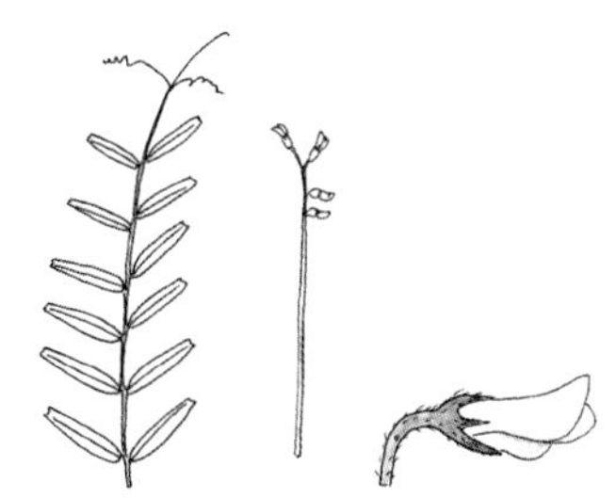

Ervilia hirsuta (Hairy Tare)

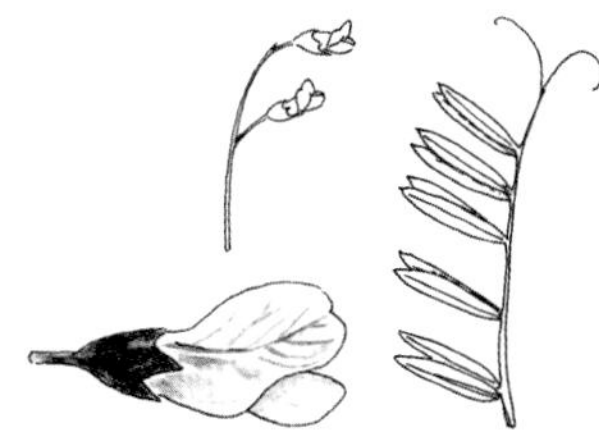

Ervum tetraspermum (Smooth Tare)

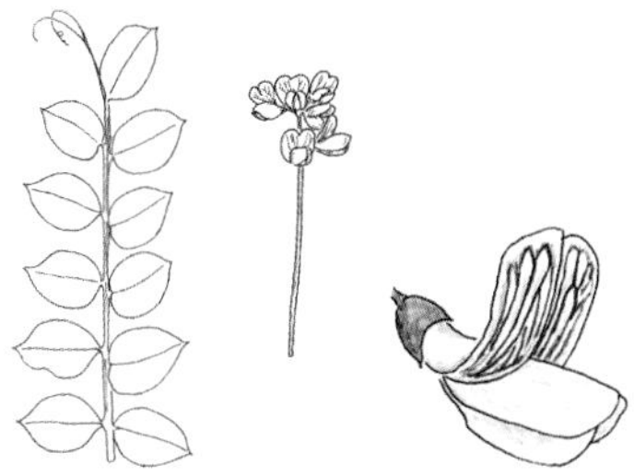

Ervilia sylvatica (Wood Vetch)

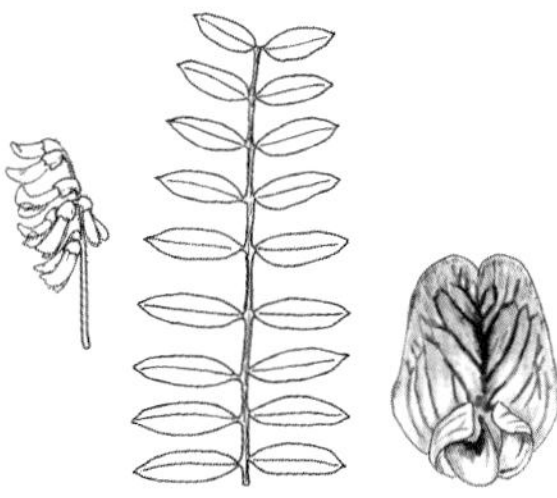

Vicia orobus (Wood Bitter-vetch)

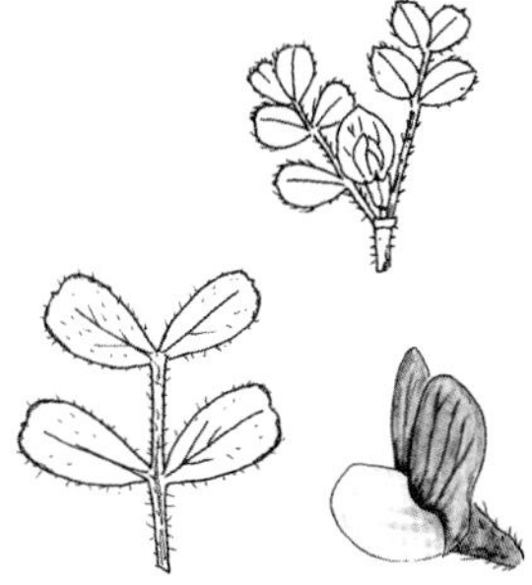

Vicia lathyroides (Spring Vetch)

Leaves, inflorescences and flowers of blue-purple vetch species (*continued*)

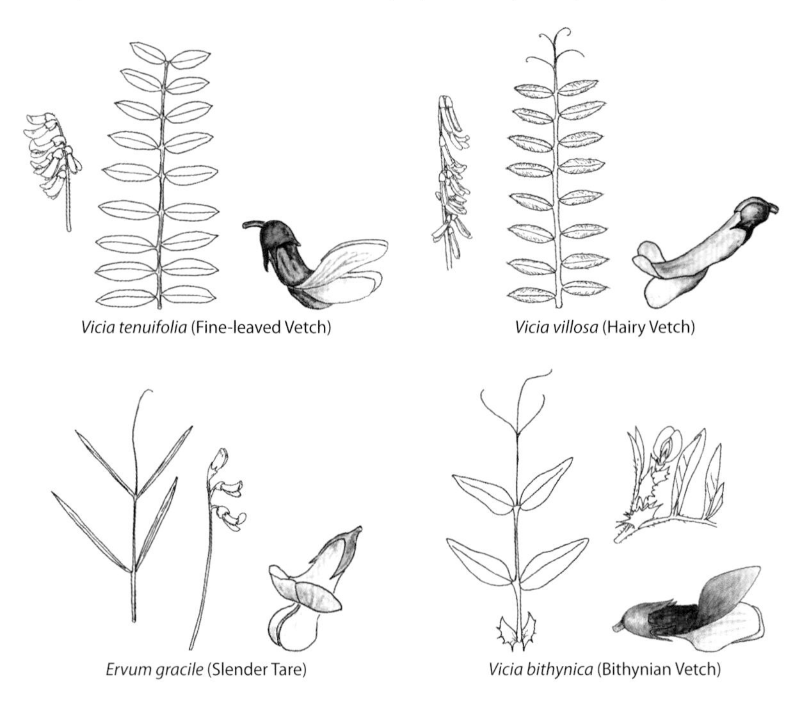

Vicia tenuifolia (Fine-leaved Vetch)

Vicia villosa (Hairy Vetch)

Ervum gracile (Slender Tare)

Vicia bithynica (Bithynian Vetch)

Next let's look at the more common vetchlings and peas

These species usually have fewer, broader leaflets and larger flowers supported by long flower stalks, and they often have distinctly ridged or winged stems.

***Lathyrus linifolius* (Bitter-vetch)**

An upright perennial plant of woodland edges and scrub across Britain and Ireland, although absent from much of eastern England; grows to a height of 40 cm. It has winged stems and leaves with 2 to 4 paired leaflets and no tendrils. The leaflets are elliptical, with a pointed tip. **Its flowers are reddish-purple in colour, turning bluey with age, with darker veining on the upper petals. The flowers occur in groups of 2 to 6.** Each flower measures 1–1.6 cm long. The seedpods measure 3–4 cm and contain 4 to 6 seeds.

***Lathyrus nissolia* (Grass Vetchling)**

An upright annual species of grassy places occasional throughout much of southern Britain, occasional or rare elsewhere; it can grow to 90 cm. The stems are not winged. **The leaves are grass-like blades that clasp the stem.** There are no tendrils. The flowers are an intense bright pink colour. The flowers are solitary or occur in pairs on a long flower stalk. The flowers measure 8–18 mm. The seedpods measure 3–6 cm and **contain 15 to 20 seeds.**

Characteristics of purple- and pink-flowered vetchlings and peas

Species	Tendrils			Number of leaflets		Stems with wings	Flowers per cluster			Seeds per pod	
	none	simple	branched	<2	>2		<2	<5	>5	<5	>5
Lathyrus linifolius (Bitter-vetch)	■				■	■		■		■	■
Lathyrus nissolia (Grass Vetchling)	■			■			■				>10
Lathyrus latifolius (Broad-leaved Everlasting-pea)			■	■		■		■	■		>10
Lathyrus sylvestris (Narrow-leaved Everlasting-pea)			■	■		■		■			■
Lathyrus japonicus (Sea Pea)	■	■	■		■				■		■
Lathyrus palustris (Marsh Pea)			■		■	■		■		■	■
Lathyrus hirsutus (Hairy Vetchling)			■	■		■		■		■	■
Lathyrus tuberosus (Tuberous Pea)			■	■				■		■	■
Lathyrus niger (Black Pea)	■				■				■		■
Your specimen											

Less common species

Lathyrus latifolius (Broad-leaved Everlasting-pea)

This climbing perennial is a widespread garden escape across most of the British Isles, although rare in Ireland; it can grow to 3 m in height. The stems are winged. The leaves comprise **single pair of broad elliptically shaped leaflets,** with a terminal branched tendril. **The flowers are a bright magenta pink,** the lower wing petals are often deeper pink. The flowers occur in clusters of 3 to 12 on long flower stalks, each flower measuring 1.5–3 cm. Seedpods measure 5–8 cm and **contain 16 to 23 seeds.**

Lathyrus sylvestris (Narrow-leaved Everlasting-pea)

Occurs in hedgerows, scrub and rough ground occasionally across most of Britain, almost absent from Ireland. A climbing perennial that can reach 2 m tall. Its stems are winged. The leaves comprise a single pair of **narrow elliptically shaped leaflets,** with a terminal branched tendril. **The flowers are a dull salmon pink,** the lower wing petals often deeper pink. The flowers occur in clusters of 3 to 8 on long flower stalks, each flower measuring 1.2–2 cm. **The seedpods measure 5–7 cm, and have a narrow wing along the top,** they contain 8 to 14 seeds.

There are several less common species of peas and vetchlings to consider

Lathyrus japonicus (Sea Pea)
As the name suggests, this rare creeping perennial grows **in maritime shingles and dunes, with its stronghold being Suffolk; scattered very sparsely around Ireland.** Grows to 90 cm. Its stems are unwinged. The leaves are rather grey in colour and have 2 to 5 paired leaflets. The leaflets are broad and elliptical. Tendrils may be branched or unbranched or absent. The **flowers are pinky-purple in colour (fading to blue)**, often with paler pink wing petals. **The flowers measure 1.5–2.5 cm** and are found in clusters of 2 to 10 on long flower stalks. Seedpods measure 3–5 cm and contain 4 to 8 seeds.

Lathyrus palustris (Marsh Pea)
Occurs in the fenlands and in central Ireland, with a few other scattered locations across England and Wales, where **it grows in damp grasslands**. A climbing perennial that may reach 1.2 m. It has winged stems. Its leaves have 2 or 3 paired leaflets, which are an elongated elliptical shape. The leaves terminate in branched tendrils. **Flower colour is variable – from lilac, through pink, to bluey purple.** The flowers occur in clusters of 2 to 6 and measure 1.2–2 cm. Seedpods measure 3–5 cm and contain 3 to 8 seeds.

Lathyrus hirsutus (Hairy Vetchling)
A rare climbing annual plant, grows in rough grasslands in a few scattered locations across the British Isles – although there are no recent records from Ireland or Scotland, and it has never been found in Wales. May grow to 1 m. Its stems are winged. Its **leaves usually have only a single pair of leaflets**, which are narrowly elliptical in shape. Leaves terminate in branched tendrils. The flowers are pinky-purple in colour, with **pale blue, almost white wing petals**. The flowers are often solitary but may occur in clusters with up to 3 flowers on long flower stalks. The flowers measure 8–15 mm. The seedpods are hairy, they measure 2–4 cm and contain 4 to 8 seeds.

Lathyrus tuberosus (Tuberous Pea)
A climbing perennial plant, that may very occasionally turn up as a casual across much of Britain; not seen recently in Ireland. Grows in road verges and cultivated ground, where it reaches up to 1.2 m. **Has stems without wings.** Its leaves usually have only a single pair of narrow elliptical leaflets. The leaves terminate in branched tendrils. **The flowers are bright crimson, sometimes salmon-pink in colour.** The flowers occur in clusters of 2 to 7 on long flower stalks, each flower measuring 1.2–2 cm. **Seedpods measure 2–4 cm and are almost cylindrical**, they contain 3 to 6 seeds.

Leaves, inflorescences and flowers of vetchlings and peas

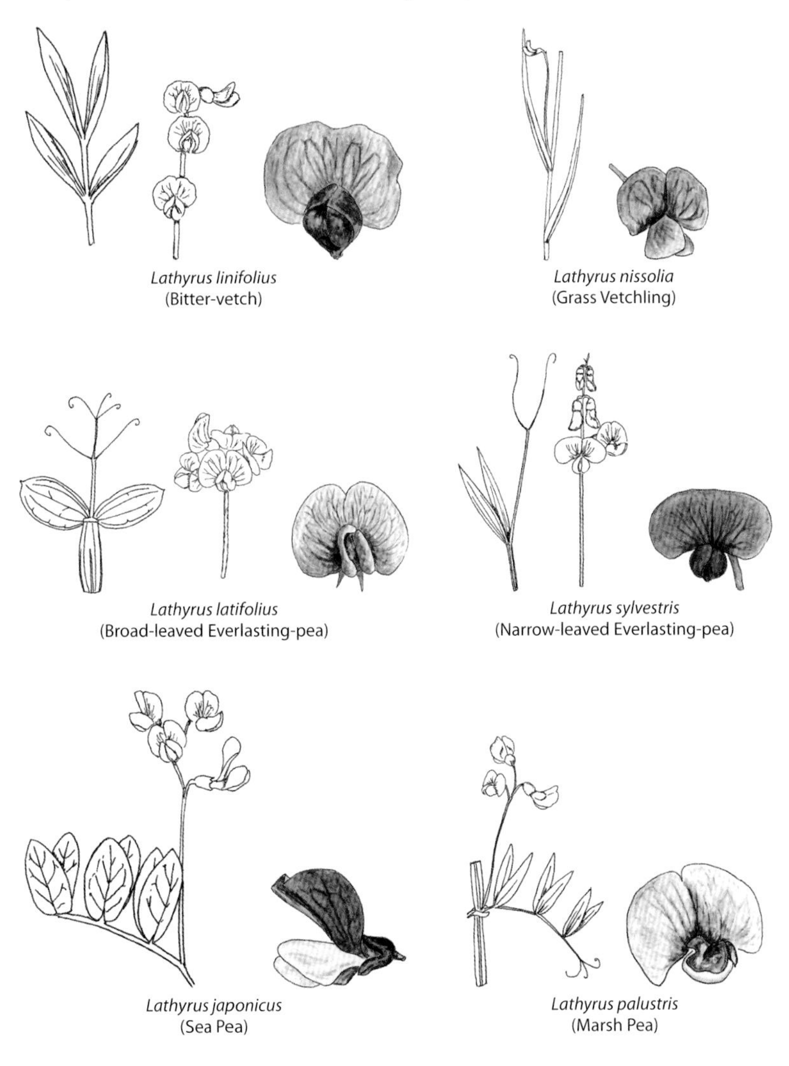

Lathyrus niger (Black Pea)

This rare garden escape is an upright perennial species which occurs at a few scattered sites in England, where it grows in grasslands and scrub. Can get to 80 cm in height. Its stems are not winged. Its leaves have 3 to 6 paired leaflets and lack tendrils. **The flowers are purple (turning bluey with age), often with darker veins.** The **flowers occur in clusters of 2 to 10** on long flower stalks, each flower measuring 1–1.5 cm. Seedpods are black, 4–5 cm long and contain 6 to 8 seeds.

Leaves, inflorescences and flowers of vetchlings and peas (*continued*)

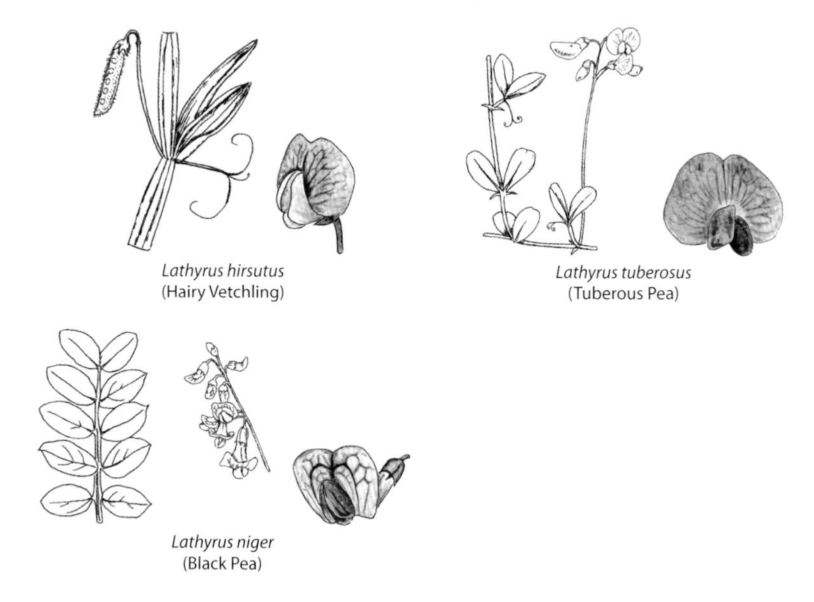

Lathyrus hirsutus (Hairy Vetchling)

Lathyrus tuberosus (Tuberous Pea)

Lathyrus niger (Black Pea)

You may have a forage species that has escaped from cultivation

In addition to their blue-purple pea-like flowers, the members of this rather miscellaneous group of plants are united by being introduced for forage and having escaped from cultivation. Being forage species, they are rather tall and robust plants which have no need of tendrils for support. Their inflorescences contain many, usually quite small flowers.

Galega officinalis (Goat's-rue)

An upright perennial of waste places across much of Britain as an escape, although very rare in Scotland and Ireland. Can grow to 1.5 m. Its leaves have 9 to 17 narrow oblong leaflets, each with a needle point. Has no tendrils. **The flowers are pale lilac to white, occurring in upright clusters of 20 to 30 flowers.** Each flower measures 12–15 mm. Seedpods measure 2–5 cm and contain 2 to 10 seeds.

Onobrychis viciifolia (Sainfoin)

Although possibly native, this semi-erect perennial plant usually occurs as an escape from cultivation, appearing widely across much of southern Britain. Grows in dry grasslands and waste areas, where it may reach 60 cm tall. Its leaves have 6 to 14 paired leaflets. It has no tendrils. Its flowers are pinky-red in colour, often appearing stripy with markedly darker veins. **The flowers occur in a densely packed, upright clusters of up to 50 flowers. There may be many such spikes per plant.** Individual flowers measure 10–12 mm long. The seedpods measure 5–8 mm and contain a single seed which is not released from the pod.

Medicago sativa (Lucerne)

This scrambling perennial is widely cultivated and occurs commonly across most of England but is less frequent elsewhere. Grows in grassy and waste places, where it can reach a height of 90 cm. It is more distantly related to the other species covered here, and its vegetative parts are rather different. **Its leaves are hairy and divided into three leaflets, the ends of which are serrated.** There are no tendrils. The flowers occur in dense clusters of 10 to 20, each measuring 6–11 mm. Flowers are usually dark mauve, although pale lilac, pink, white and even yellow colour forms are also known. **The seedpods are curved or spiralled** and contain 10 to 20 seeds.

Securigera varia (Crown Vetch)

An introduced straggly perennial plant with a scattered distribution across much of southern Britain (rare in Ireland), where it grows in rough ground and grasslands. It can grow to a height of 1.2 m. Its leaves have 9 to 25 paired oval leaflets and lack tendrils. **Its flowers are pastel or lilac-pink, the lower wing petals are almost white and the inner keel petals are tipped with purple.** Each flower measures 8–15 mm. **The flowers occur in circular clusters of 10 to 20 which look rather pompom-like.** The seedpods measure 2–6 cm and are internally divided into 3 to 8 sections.

Finally, you may have a one of the milk-vetches

These three closely related species are rather uncommon or rare. They are low-growing species, usually less than 30 cm tall, that occur in short grasslands.

Characteristics of forage species and milk-vetches

	Flower size mm			Number of leaflets		Flowers per cluster			Seeds per pod	
Species	<10	<15	>15	<10	>10	<20	<30	>30	<5	>5
Galega officinalis (Goat's-rue)										
Onobrychis viciifolia (Sainfoin)									1	
Medicago sativa (Lucerne)				3						
Securigera varia (Crown Vetch)										
Astragalus danicus (Purple Milk-vetch)										
Astragalus alpinus (Alpine Milk-vetch)										
Oxytropis halleri (Mountain Milk-vetch)										
Your specimen										

Milk-vetches

They have leaves with many leaflets and no tendrils. Their flower clusters are dense, with many upward-pointing flowers.

Astragalus danicus (Purple Milk-vetch)

A short perennial plant that grows in calcareous grasslands, particularly in sand dunes in north-east England and Scotland; in Ireland only on the Aran Islands, absent from Wales. Grows to a height of 30 cm. Its leaves are divided into

Leaves, inflorescences and flowers of forage species and milk-vetches

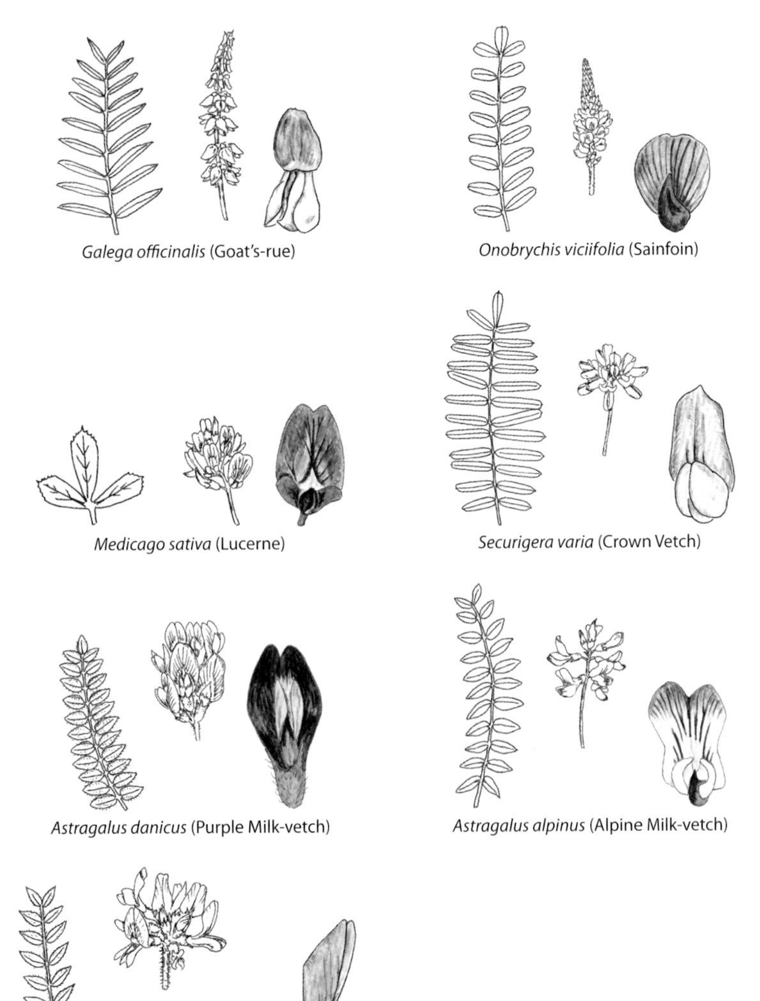

Galega officinalis (Goat's-rue)

Onobrychis viciifolia (Sainfoin)

Medicago sativa (Lucerne)

Securigera varia (Crown Vetch)

Astragalus danicus (Purple Milk-vetch)

Astragalus alpinus (Alpine Milk-vetch)

Oxytropis halleri (Mountain Milk-vetch)

10 to 27 narrow paired leaflets which lack tendrils. **The flowers are an intense deep purple. They occur in dense upright tight clusters of 10 to 20 flowers.** Each flower measures 15–18 mm. The seedpods are hairy and may appear white, they measure 7–10 mm long and contain up to 7 seeds.

Astragalus alpinus **(Alpine Milk-vetch)**
An extremely rare species found at just a few mountainous sites in the Cairngorms, Scotland where it grows to a height of 30 cm. Similar to Purple Milk-vetch, except its flowers are paler. Its leaves are divided into 13 to 27 narrow paired leaflets which lack tendrils. **The flowers are pale blue, fringed with purple.** They occur in dense upright tight clusters of 10 to 20 flowers, each flower measuring 10–14 mm. The seedpods measure 8–12 mm long and contain up to 7 seeds.

Oxytropis halleri **(Mountain Milk-vetch)**
Another rare alpine species of the mountains, scattered over Scotland. Reaches a height of 30 cm. **Its leaves are covered in silky hairs, comprise 10 to 30 paired leaflets and lack tendrils.** Its flowers are pale purple and measure 1.5–2 cm. The flowers occur in round clusters of 10 to 30 flowers on a long stalk. **The seedpods have a distinctive hooked tip**, they measure 1.5–2 cm and appear almost round.

HAVE OTHERS RECOGNISED THIS LEVEL OF VARIATION?

Vetches may have been some of the earliest plants to be cultivated. Many of them have long associations with humans and it is likely that our ancestors were familiar with the species that we recognise today. The etymology of their names also suggest they have been known since antiquity. While a few of our rarer species were only discovered growing here relatively recently, botanists had already described them from elsewhere in their distribution. Thus, although their scientific names may have changed, you will find almost all the species described here in the pages of Victorian floras.

HOW FAR SHOULD I GO?

These species are distinctive enough that with a little familiarity you should have no difficulty in identifying them. However, a degree of caution is required, because many of the commoner species can be variable. Thus, some care is needed if you are to avoid mistaking them for their more unusual relatives or overlooking rarer records. It is worth being aware that the list of introduced species that occasionally turn up is longer than covered here.

CHAPTER 23

Umbellifers: carrots, parsnips, Hemlock etc.

If there is one family of plants that novice botanists find more difficult than the rest, it is probably umbellifers. Psychologically, their intimidating reputation has been exacerbated by their name change from the descriptive Umbelliferae (frequently shortened to the friendly 'umbellifers') to become the more ambiguous Apiaceae. There is of course a good biological reason for the group being scary. While many of our vegetables, herbs and spices are members of this family, so are many toxic species. Therefore, when it was said in the introduction to this book, 'it is not a crime to misidentify plants', if you are unwise enough to test your umbellifer ID skills at the dinner table, then your mistakes may carry the death sentence. DON'T take any chances with these plants. Some people are hypersensitive to the sap of many of them: **Please take great care when handling unknown specimens**.

Umbellifers are generally easy to recognise to family level by their umbrella-like clusters of small flowers, formed by umbels of umbellets. They are herbaceous annuals, biennials and perennials. Their often finely divided leaves appear alternately along the stem. Individual flowers are small, with five lobes, five stamens and two stigmas. Roughly half our species produce separate male and hermaphrodite flowers within their flower clusters.

WHY IS THIS GROUP OF PLANTS COMPLEX?

The answer to this question is simple: there are just so many of them. The Apiaceae is a highly diverse family with more than 3,500 species globally, about 60 of which occur in the British Isles. The situation has been artificially complicated because many more species have been introduced as well. Several members of this family have been introduced as ornamentals, for food or as herbs both medicinal and culinary. In addition, humans have selected lots of different varieties of the cultivated species and some of these have escaped into the wild, with varying degrees of success. For example, cultivated forms of celery have been recorded growing wild, but the fleshy-rooted form of the same species (celeriac) is unknown outside cultivation.

If there were only five species of umbellifers in our flora, you would probably have no difficulty in recognising them. But trying to remember the diagnostic characteristics of more than 60 different species is a lot more challenging. This is especially true because many of these species are rare and you are unlikely to encounter them routinely.

HOW CAN I TELL THEM APART?

This question has frustrated novice botanists in their thousands. To make the task less daunting, it helps to split the family into five, more manageable chunks. First, there is a group of 13 species that are hopefully different enough from the rest that you should be able to recognise them with relative ease. Next, the remaining species are divided in half and then these two subgroups are halved again. The first division is: do they have feathery, fern-like lower leaves? In a few species this may be debateable, but hopefully the unscientificly ranked illustrations should help you develop of feel for leaf-shape variation in this family. You also need to be aware that stem leaves and lower leaves can have different shapes: look at the lower leaves. Stem leaves often have an enlarged membranous base where they attach to the flowering stem. The second division is hopefully less subjective: are bracts present below the umbel (see illustration below)? In some species, bracts may be infrequent or absent – these are listed under no bracts.

Once you have assigned your specimen to one of these five groups, then you should be able to work through the tabular key and determine its ID. Important features to look for include whether the stems are hollow or solid, and if are they hairy or not. The shape of the fruit is often highly distinctive, and so these are fully illustrated. The fruit of these species are dry, often elaborately textured, and they frequently retain the remains of the female stigma at their apex. As they mature, they split in two, releasing the seeds and leaving a fibrous central axis.

The umbellifer flower cluster (an umbel of umbellets)

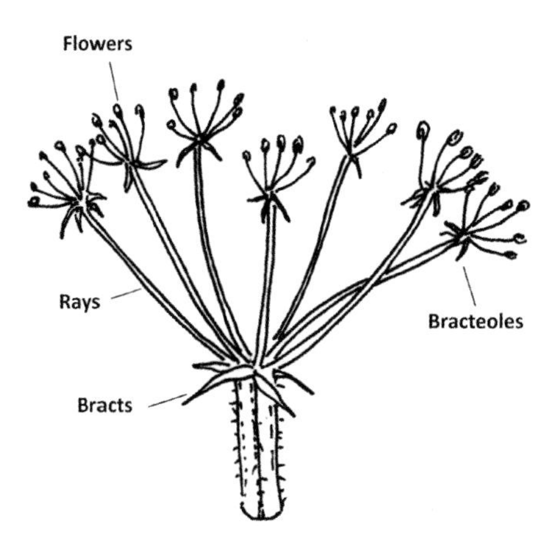

Let's start with the easier, most distinctive species

This first group of 13 species has no overarching biological meaning; rather, they have been selected as species with distinctive features that make them easy to identify with a little effort.

Eryngium maritimum (Sea-holly)
A species of coastal sands and shingles around the coast of Britain and Ireland, but rare in Scotland. Grows to a height of 60 cm. **Its leaves are oval with a distinctively spiny margin. The flowers are pale ice blue** and carried in flowerheads measuring 1.5–3 cm across. The fruit are covered in hooks.

Eryngium campestre (Field Eryngo)
Now a rare species of calcareous grasslands in the south of England. Grows to a height of 75 cm. **Its leaves are long, narrow and spiny; the basal leaves are divided into fingers, while the stem leaves are divided into deep lobes.** The flowers are green and carried in a flowerhead measuring about 1.5 cm across. The fruit are covered in scales and appear almost fluffy.

Bupleurum rotundifolium (Thorow-wax)
Formerly a weed of arable fields but has been extinct in the British Isles since the 1960s. However, it now occasionally turns up as a casual. It is an annual that grows to 30 cm tall. **Its tear-shaped leaves are highly distinctive, with the stems emerging from within the leaf blade.** The yellow flowers and seeds are smaller than the bracteoles (there are no bracts). Fruit measure 3–3.5 mm in length.

Bupleurum tenuissimum (Slender Hare's-ear)
A rare annual species of southern England, where it grows to 50 cm tall. Usually found growing in coastal grasslands, especially the Thames and Severn estuaries and the Solent. **The flowers are yellow with a reddish tinge. Its leaves are narrow, with three main veins. It has 3 to 5 bracts that are longer than the shortest rays.** Fruit measure 1.5–2 mm and are round in shape.

Bupleurum baldense (Small Hare's-ear)
A very rare annual species of the south coast of England and the Channel Islands, where it grows to **just 25 cm tall. Its leaves are narrow, with 3 to 5 main veins. The flowers are yellow. The bracts are longer than the shortest rays.** The fruit measure 1.5–2.5 mm in length, they are oval with narrow grooves.

Crithmum maritimum (Rock Samphire)
A coastal perennial species of the south and west British Isles, rare in Scotland. Grows to a height of 45 cm. Its stems are solid and lack hairs. **The leaves are succulent and usually divided into pairs of lobes.** When crushed it smells of furniture polish. **The flowers are yellow with a hint of green.** Both bracts and bracteoles are present. Fruit are round with prominent ridges and measure 3.5–5 mm.

Smyrnium olusatrum (Alexanders)
Found throughout most of the British Isles, but rarer in Scotland, this species is particularly common around the coast. Usually perennial. A tall, robust plant **growing to height of 1.5 m.** The ridged stems lack hairs and are solid. The leaves are highly lobed, with a toothed margin. The stalks of the stem leaves have a much-expanded base. **The leaves are glossy bright green with yellow veins.** Bracts and bracteoles often do not occur. **The flowers are yellow.** Fruit measure 6.5–8 mm, are black and have distinct ridges. Has a pungent, somewhat unpleasant scent.

Sanicula europaea (Sanicle)
A common perennial species of old woodlands throughout the British Isles, where it grows to about 40 cm. **Most of the leaves are palmate and found at the base of the plant.** Often grows with Wood Anemone, and their leaves can be confused. **The clumps tiny, white-pink flowers are found in loosely clustered umbels.** Male flowers are stalked, while hermaphrodite flowers lack a stalk. The anthers have distinctive long protruding filaments. Fruit measure 2–3 mm and are covered in small hooks.

Astrantia major (Astrantia)
This introduced perennial species is now found scattered across the British Isles. Grows to a height of 80 cm. **Most of the leaves are found at the base of the plant, they have 3 to 5 lobes and long stalks.** Male flowers are stalked, while hermaphrodite flowers lack a stalk. The flowers are white, tinged with pink or green. **The bracteoles are long and distinctive and rather petal-like, they are white with a hint of green or pink.** Fruit are longer than they are wide and measure 5–7 mm.

Foeniculum vulgare (Fennel)
This perennial plant is found across the British Isles but is less common in Scotland. It frequently occurs as a garden escape and on road verges. Can grow to 2.5 m tall. Its stems are hairless and solid at first but the older stems are hollow. **The leaves are filamentous and hairlike and smell strongly of aniseed.** Leaf and stem colour is variable, some plants are green but bronze-coloured individuals are not uncommon. The umbels lack bracts and bracteoles. **The flowers are yellow.** The fruit measure 4–5 mm.

Daucus carota (Wild Carrot)
This is a widespread biennial species in the British Isles, although it tends to be restricted to the coast in Scotland. Plants can grow to a height of 75 cm, but coastal specimens are much shorter. Stems are hairy and solid. Leaves are highly divided and smell of carrots when crushed. There are numerous long-branched bracts and bracteoles. The umbels tend to have a flat top, except in the coastal subspecies (*gummifera*) in which they are almost round. **The flowers are white except the central one which is pink, purple or even black.** Fruit measure 2–3 mm and have rows of bristles along their length.

***Trocdaris verticillata* (Whorled Caraway)**

A perennial plant of marshes and wet grasslands of western Britain and Ireland. May grow to a height of 60 cm. Its stems are hairless and hollow. **Its fine, undivided leaves are unusual in occurring in whorls around the stem.** The flowers are white. There are many oval-shaped bracts and bracteoles. The fruit are smooth and measure 2–3 mm.

***Peucedanum officinale* (Hog's Fennel)**

A very rare perennial plant of coastal wet grasslands in south-east England. Its solid stems grow to a height of 2 m. The stems are hairless, but there is a sheath of thick fibres where the leaves join the stem. **The leaves are long and thin and are unusual in mostly dividing into three.** The flowers are yellow. There are few or no bracts, but many bracteoles. The fruit measure 5–8 mm.

***Hydrocotyle vulgaris* (Marsh Pennywort)**

Older wildflower guides include this species as an umbellifer, but it is now regarded as being in a family of its own. **Occurs in wetlands, where its round leaves can be found growing flat to the ground.** If you look hard you might find a tiny umbel of flowers nestled among the leaves.

Leaves, inflorescences and fruits of distinctive umbellifer species

Eryngium maritimum
(Sea-holly)

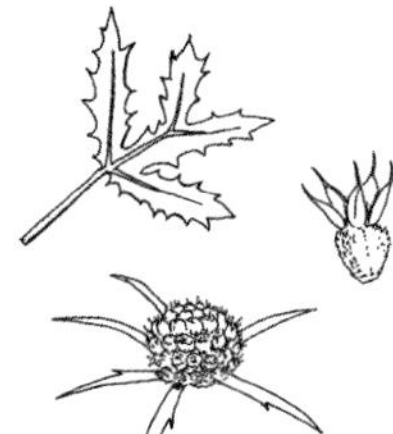

Eryngium campestre
(Field Eryngo)

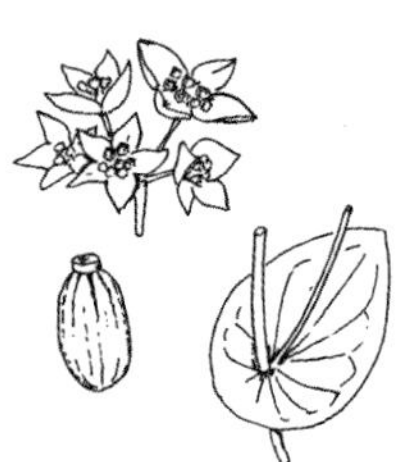

Bupleurum rotundifolium
(Thorow-wax)

Bupleurum tenuissimum
(Slender Hare's-ear)

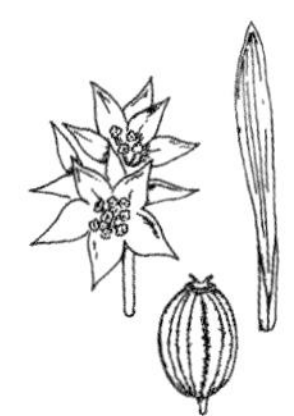

Bupleurum baldense
(Small Hare's-ear)

Crithmum maritimum
(Rock Samphire)

Leaves, inflorescences and fruits of distinctive umbellifer species (*continued*)

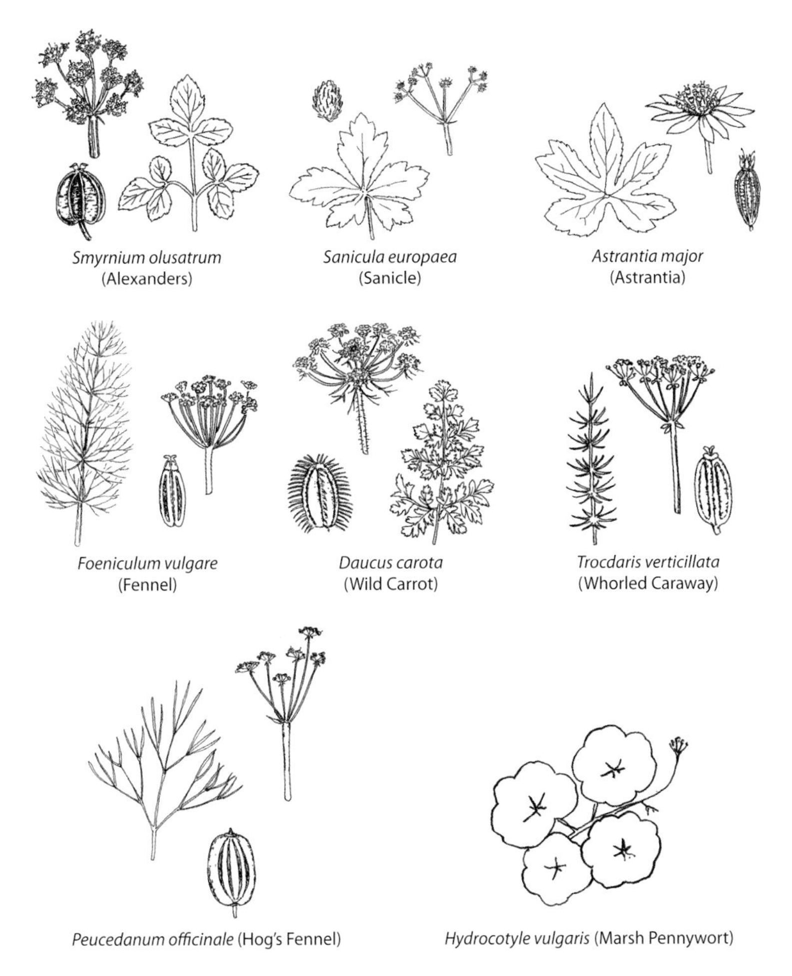

Smyrnium olusatrum (Alexanders)

Sanicula europaea (Sanicle)

Astrantia major (Astrantia)

Foeniculum vulgare (Fennel)

Daucus carota (Wild Carrot)

Trocdaris verticillata (Whorled Caraway)

Peucedanum officinale (Hog's Fennel)

Hydrocotyle vulgaris (Marsh Pennywort)

Next let's look at the species with feathery fern-like leaves and bracts

Defining species as having feathery, fern-like leaves is of course rather subjective. However, hopefully you will develop a feel for which group of species to look at in order to find your specimen. Within each group, species are listed in terms of the featheriness of their leaves. All the species in this group have bracts at the base of their umbels. The number and size of these are highly variable both within and between species.

Bunium bulbocastanum (Great Pignut)

A rare perennial of chalk grasslands in the south of England, where it grows to a height of around 50 cm, sometimes slightly more. **Its leaves are branched and**

hair-like, they usually die before the plant flowers. The stems are solid, ridged and lack hairs. The flowers are white and radially symmetrical. The umbels have many bracts and bracteoles. Fruit measure 3–4.5 mm.

Silaum silaus (Pepper-saxifrage)
This perennial species is found in unimproved damp grasslands across most of England, particularly in the south and west, rare in Wales and southern Scotland, absent from Ireland. May grow to a height of 1 m. **Its leaves are highly divided into narrow linear leaflets.** Its solid stems are hairless. Its yellow flowers are radially symmetrical. The umbels contain few bracts but numerous bracteoles. The fruit are angular and measure 4–5 mm.

Seseli libanotis (Moon Carrot)
An extremely rare short-lived species of chalk grasslands in southern England, where it may grow to a height of 60 cm. Its leaves are divided into numerous narrow lobes. **The solid stems are covered in very short hairs.** Its flowers are white and radially symmetrical. **The umbels are somewhat domed, with many long bracts and bracteoles.** Fruit measure 2.5–3.5 mm.

Thyselium palustre (Milk-parsley)
This biennial species is very rare outside the fens of East Anglia. Grows to 1.5 m in height. Its leaves are highly divided, with the final sections being narrow linear leaflets. Its stems are hollow and usually lack hairs. Its flowers are white and radially symmetrical. **The long bracts curved downwards, and there are many bracteoles.** Fruit measure 3–5 mm.

Torilis arvensis (Spreading Hedge-parsley)
This annual species is a now rare, **an arable weed of southern England. It grows to a height of 50 cm, with spreading branches.** The leaves are hairy and feathery. Its stems are solid and hairy. The white flowers are slightly asymmetric. **The umbel is rather open, with no bracts or just one or two, but many bracteoles. Its fruit are usually covered in tiny hooks and measure 3–4 mm.**

Torilis japonica (Upright Hedge-parsley)
This annual species is found in grassy places across the British Isles, except for the north of Scotland. It grows to a height of 1.2 m. The leaves are hairy and feathery. Its ridged stems are solid, hairy and may have purple spots. **Its white flowers are often tinged with pink and are slightly asymmetric. The umbel is rather open, usually with between 4 and 6 bracts and many bracteoles. The fruit measure 2–2.5 mm and are covered in spines.**

Oenanthe lachenalii (Parsley Water-dropwort)
This perennial species grows in waterlogged ground, mostly in coastal areas around the British Isles, except north-east Scotland. Grows to a height of 1 m. **Its leaves are divided into long and thin, slightly oval leaflets.** **The ridged stems**

may be hollow or solid and lack hairs. Its white flowers are asymmetric, the outer flowers in the umbels appearing larger than those in the centre. The umbels have few bracts and many bracteoles. Fruits measure 2.5–3 mm and are round.

***Oenanthe pimpinelloides* (Corky-fruited Water-dropwort)**

This perennial plant grows in both wet and dry grasslands mostly in the south of England, very rare in south-west Ireland. Grows to a height of 1 m. Its lower leaves are divided into egg-shaped, toothed leaflets. The ridged stems may be hollow or solid and lack hairs. Its white flowers are asymmetric, the outer flowers in the umbels appearing larger than those in the centre. **The robust umbels have thick rays, 1 to 7 bracts** and many bracteoles. Its fruit measure 3–3.5 mm and retain two long stigmas.

Characteristics of species with feathery fern-like leaves and which have bracts

	Plant height			Stems		Flowers				Seeds		
Species	<50 cm	<1 m	>1 m	hollow	hairy	yellow	symmetrical	Bracts	Bracteoles	<5 mm	hooked	hairy
Bunium bulbocastanum (Great Pignut)	■						■	■	■			
Silaum silaus (Pepper-saxifrage)		■				■	■	■	■	■		
Seseli libanotis (Moon Carrot)		■			■		■	■	■			■
Thyselium palustre (Milk-parsley)			■	■			■	■	■	■		
Torilis arvensis (Spreading Hedge-parsley)	■				■			■	■		■	
Torilis japonica (Upright Hedge-parsley)			■		■			■	■		■	
Oenanthe lachenalii (Parsley Water-dropwort)			■	both				few	■			
Oenanthe pimpinelloides (Corky-fruited Water-dropwort)			■	both				■	■			
Conium maculatum (Hemlock)			■	■			■	■	■			
Physospermum cornubiense (Bladderseed)			■				■	■	■			
Sison segetum (Corn Parsley)		■					■	■	■			
Petroselinum crispum (Garden Parsley)		■				■	■	■	■			
Berula erecta (Lesser Water-parsnip)			■	■			■	■	■			
Oenanthe crocata (Hemlock Water-dropwort)			■	■				■	■	■		
Your specimen												

Conium maculatum (Hemlock)
A common biennial species of road verges and damp ground across most of the British Isles except the north of Scotland. Grows to a height of 2.5 m. Its leaves are bright to dark green, slightly shiny and divided into highly serrated leaflets. **Its hollow stems lack hairs are usually covered in distinctive purple blotches.** The flowers are white and radially symmetrical. The umbels have many bracts and bracteoles. Extremely poisonous. Fruit are ridged and measure 2–3.5 mm.

Physospermum cornubiense (Bladderseed)
A rare plant of woodlands in Cornwall and just into Devon. Grows to a height of 1.2 m. **Its leaves are highly divided into long thin-lobed leaflets.** Its stems are solid and mostly hairless. The flowers are white and radially symmetrical. The umbels are rather open, with many bracts and bracteoles. Fruit are smooth and measure 4–5 mm.

Sison segetum (Corn Parsley)
An occasional annual species of grasslands and arable fields in the south of England, rare in Wales, absent from Ireland and Scotland. Grows to a height of 1 m. Its leaves are divided into toothed leaflets. The stems are solid and lack hairs. The flowers are white and radially symmetrical. **The umbel is distinctive, as the delicate rays are unequal in length. There are 2 to 5 long bracts, which are sometimes lobed.** There are also 2 to 5 bracteoles. The fruit are round, deeply grooved and measure 2.3–3 mm.

Petroselinum crispum (Garden Parsley)
This garden escape is a biennial, occurs scattered across the British Isles. Grows to a height of 75 cm. **Its bright green leaves are highly variable with curly and flat-leaved types.** The stems are solid and hairless. The flowers yellow and radially symmetrical. The umbels have 1 to 3 bracts which may be lobed and many bracteoles. Its fruits measure 2–2.5 mm.

Berula erecta (Lesser Water-parsnip)
A common perennial species of waterlogged ground across the British Isles, except Scotland. It has a trailing growth habit but can reach a height of 1 m. **The leaves have paired serrated leaflets, the bases of the leaflets are connected by a distinctive white ring around the stem. The stems are hollow, lack hairs and may root along their length.** The flowers are white and radially symmetrical. **The umbels have many branching bracts and bracteoles. Fruit are small and round, measuring 1.3–2 mm, and retain their stigmas.**

Oenanthe crocata (Hemlock Water-dropwort)
A common perennial species that grows in waterlogged ground across the British Isles, although less common in the east. Grows to a height of 1.5 m. **Its glossy**

Leaves, inflorescences and fruits of species with feathery fern-like leaves and with bracts

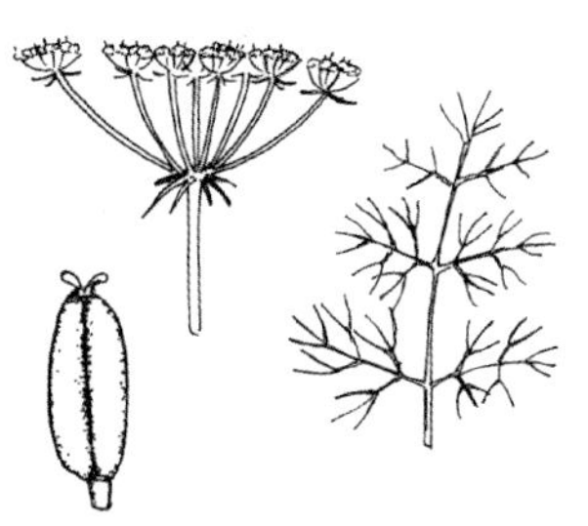

Bunium bulbocastanum
(Great Pignut)

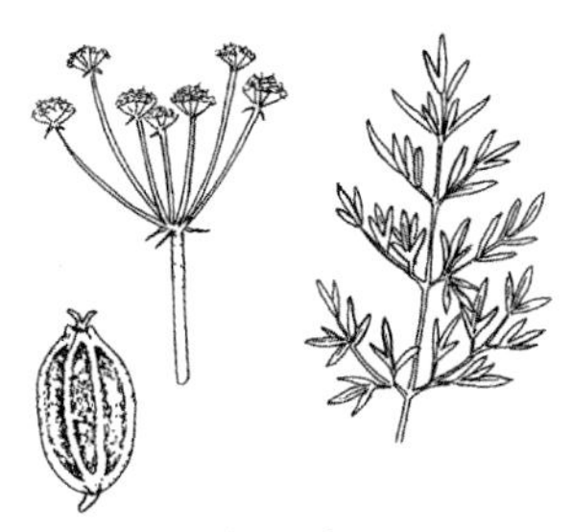

Silaum silaus
(Pepper-saxifrage)

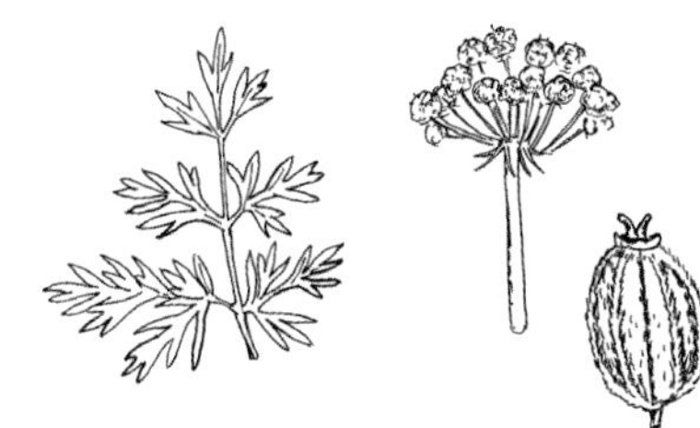

Seseli libanotis (Moon Carrot)

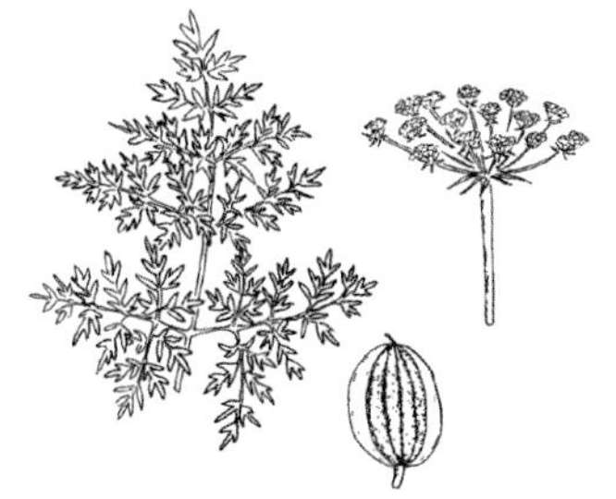

Thyselium palustre (Milk-parsley)

Torilis arvensis
(Spreading Hedge-parsley)

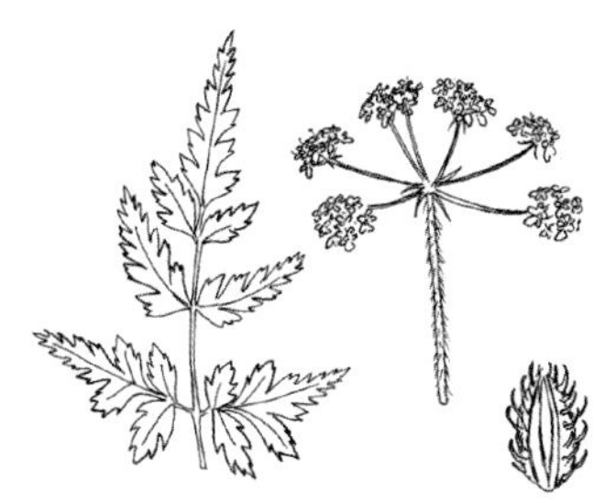

Torilis japonica
(Upright Hedge-parsley)

Oenanthe lachenalii
(Parsley Water-dropwort)

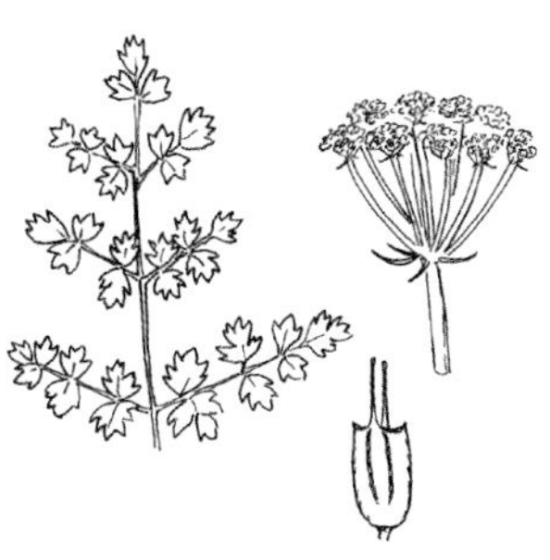

Oenanthe pimpinelloides
(Corky-fruited Water-dropwort)

Leaves, inflorescences and fruits of species with feathery fern-like leaves and with bracts (*continued*)

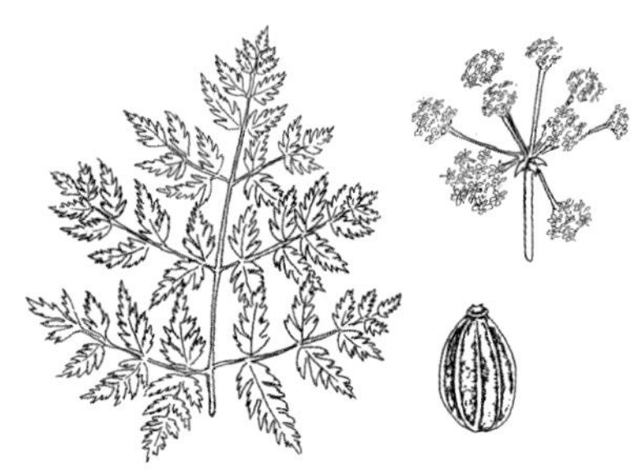

Conium maculatum (Hemlock)

Physospermum cornubiense (Bladderseed)

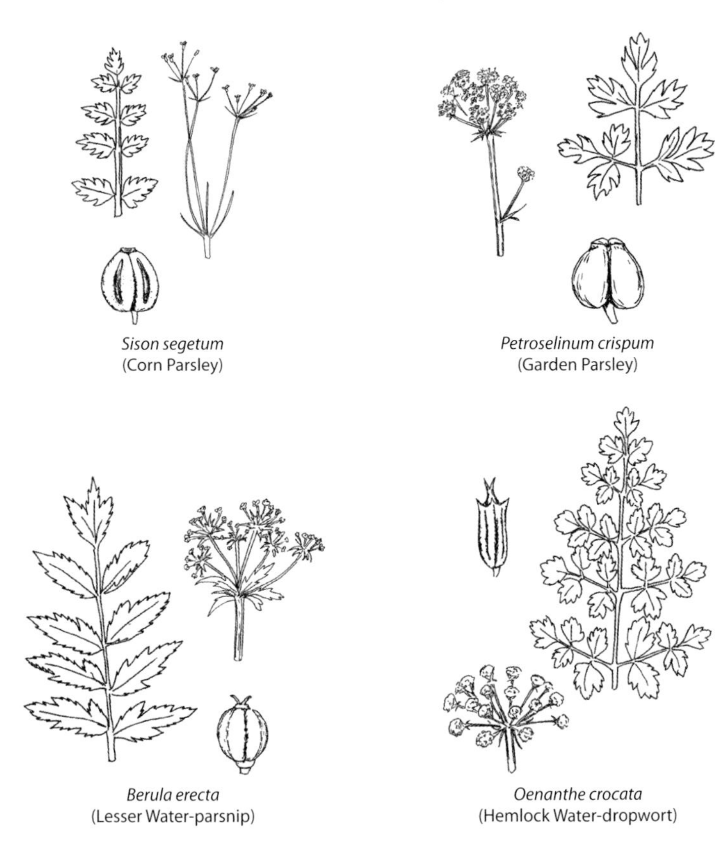

Sison segetum
(Corn Parsley)

Petroselinum crispum
(Garden Parsley)

Berula erecta
(Lesser Water-parsnip)

Oenanthe crocata
(Hemlock Water-dropwort)

green leaves are divided into toothed oval leaflets. It has hollow, ridged, hairless stems. Its white flowers are asymmetric. **The robust umbels are somewhat domed in appearance, they have thick rays, 3 to 6 bracts** and many bracteoles. Extremely poisonous. Its fruit measure 4–5.5 mm and retain their stigma.

Next, let's look at the species with feathery fern-like leaves and NO bracts

Meum athamanticum (Spignel)

A perennial plant of mountain grasslands almost exclusively in Scotland and the Lake District (and very rare in North Wales), where it grows to a height of

Characteristics of species with feathery fern-like leaves and NO bracts

	Plant height			Stems		Flowers				Seeds		
Species	<50 cm	<1 m	>1 m	hollow	hairy	yellow	symmetrical	Bracts	Bracteoles	<5 mm	hooked	hairy
Meum athamanticum (Spignel)												
Carum carvi (Caraway)									few			
Conopodium majus (Pignut)												
Oenanthe silaifolia (Narrow-leaved Water-dropwort)												
Oenanthe fluviatilis (River Water-dropwort)												
Oenanthe aquatica (Fine-leaved Water-dropwort)												
Oenanthe fistulosa (Tubular Water-dropwort)												
Scandix pecten-veneris (Shepherd's-needle)												
Selinum carvifolia (Cambridge Milk-parsley)								rare				
Trinia glauca (Honewort)								rare				
Aethusa cynapium (Fool's Parsley)							vary					
Torilis nodosa (Knotted Hedge-parsley)												
Anthriscus caucalis (Bur Chervil)												
Helosciadium inundatum (Lesser Marshwort)												
Anthriscus sylvestris (Cow Parsley)												
Chaerophyllum aureum (Golden Chervil)												
Myrrhis odorata (Sweet Cicely)												
Chaerophyllum temulum (Rough Chervil)												
Your specimen												

60 cm. Most of its leaves are found at the base of the plant. **The leaves are so filamentous they almost have a cloud-like appearance.** It has solid hairless stems which are surrounded in fibres at their base. The flowers are radially symmetrical, and white sometimes with a hint of pink. There are usually no bracts, but there may be several bracteoles. Fruit are ridged and measure 5–7 mm.

Carum carvi **(Caraway)**
An introduced biennial that can occasionally be found as a garden escape in waste ground scattered across the British Isles. Grows to a height of 60 cm. The leaves are divided into thin leaflets. The stems are hollow and lack hairs. The flowers are white and radially symmetrical. There are usually no bracts but there may be a few, sometimes leaflike, bracteoles. **The seeds have a distinctive smell when crushed.** The fruit are grooved and measure 3–4 mm.

Conopodium majus **(Pignut)**
This perennial species is common in woods and old grasslands across the British Isles, where **it may grow to a height of 40 cm.** It has delicate finely divided leaves. Those at the base of the plant have usually gone by the time of flowering. Its stems are hollow and hairless. The flowers are white and radially symmetrical. There are usually no bracts, but there are typically several bracteoles. **The developing flowering stalk has a distinctive swan-neck appearance.** Fruit are teardrop shaped, retain their stigma and measure 3–4.5 mm.

Oenanthe silaifolia **(Narrow-leaved Water-dropwort)**
This perennial species grows in waterlogged ground such as marshes and ditches mostly in central and south-east England, where it may grow to 1 m tall. Its leaves are divided into narrow leaflets. **Its stems are hollow, with thin walls and lacking hairs.** It has root tubers. The white flowers are radially asymmetric – **the rays are more substantial.** There are no bracts, but bracteoles are usually numerous. Fruit measure 2.5–3.5 mm, **they retain their stigmas which are as long as the fruit.**

Oenanthe fluviatilis **(River Water-dropwort)**
This aquatic perennial grows in slow-flowing rivers scattered across England and Ireland, where it may reach 1 m tall. **It may have two distinct leaf forms: filamentous submerged leaves, and more ridged leaves with fine leaflets above the water.** The stems are hollow and lack hairs. It has root tubers. The white flowers are radially asymmetric. There are no bracts, but bracteoles may be numerous. The fruit measure 5–6.5 mm and have a spiky apex.

Oenanthe aquatica **(Fine-leaved Water-dropwort)**
This short-lived annual or biennial species grows in ponds and ditches across most of the British Isles except Scotland. Grows to a height of 1.5 m. Its leaves are finely divided into leaflets. The stems are hollow and lack hairs. **There are no root tubers.** The white flowers are radially asymmetric. There are no bracts, but bracteoles may be numerous. **The umbels are rather open.** Fruit measure 3–4.5 mm.

Oenanthe fistulosa (Tubular Water-dropwort)
This perennial plant occurs in marshes and other waterlogged sites across the British Isles. Can grow to a height of 80 cm. Its leaves are finely divided into leaflets. The stems are hollow, with thin walls and no hairs. **Some of the stems grow horizontally and root along their length.** The plants produce thin root tubers. The flowers are white and radially asymmetric. There are no bracts, but bracteoles may be numerous. The rays are thick and the umbels rather open. The fruit measure 3–3.5 mm and retain their long stigmas.

Scandix pecten-veneris (Shepherd's-needle)
Now **a rare arable weed in southern England**, this annual may grow to a height of 50 cm. Its leaves are finely divided into narrow leaflets. Its stems are hollow and covered in a few hairs. The flowers are white and radially asymmetric. There are no bracts, but bracteoles are present. **The fruit are distinctively long and needle-like, measuring 3–7 cm.**

Selinum carvifolia (Cambridge Milk-parsley)
A rare perennial species of a few damp meadows in Cambridgeshire, where it may grow to 1 m tall. Its leaves are divided into thin lobes. The stems are solid and hairless. The flowers are white and radially symmetrical. There are occasionally bracts, but these soon fall off. There are many bracteoles. Has many umbellets closely packed within the main umbel. Fruit measure 3–4 mm.

Trinia glauca (Honewort)
A rare biennial species of dry limestone grasslands in the south west of England, where **it grows to a height of only 20 cm.** Its blue-grey leaves are finely divided into leaflets. It has solid hairless stems. The flowers are white and radially symmetrical, **with male and female flowers occurring on separate plants.** There are occasionally one or two lobed bracts, and several unlobed bracteoles. **The rays of the female plants tend to vary in length.** The fruit are round, grooved and measure 2.3–3 mm.

Aethusa cynapium (Fool's Parsley)
A common annual plant of cultivated and waste ground across most of the British Isles except Scotland. May grow to a height of 1 m but often much smaller. Its leaves are finely divided into leaflets. Its stems are solid and lack hairs. Its flowers are white and variably radially asymmetric. It usually has no bracts, **while the bracteoles are long and project out and down beyond the umbel.** The fruit measure 3–4 mm.

Torilis nodosa (Knotted Hedge-parsley)
This annual species is found in cultivated places near the coast, but is increasingly appearing as a weed on rough ground. **It has a rather prostrate habit and grows to only 50 cm tall.** Its leaves are feathery and hairy. Its stems are solid and also very hairy. The flowers are white and radially symmetrical. **The petals are surrounded by hairlike sepals.** There are no bracts but several bracteoles. **Its fruit measure 2.5–3.5 mm and are covered with hooks and spines.**

Leaves, inflorescences and fruits of the species with feathery fern-like leaves and no bracts

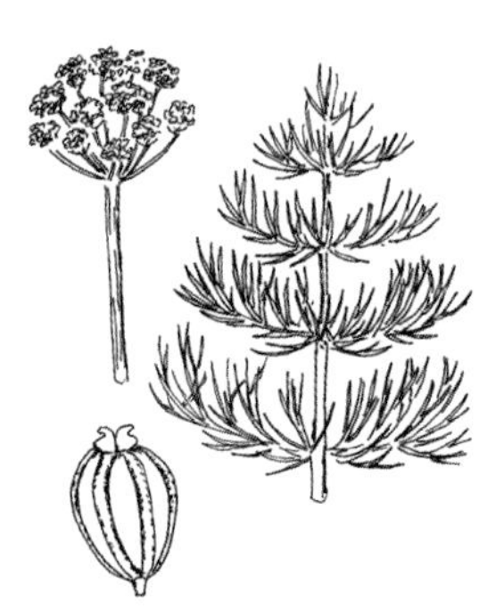

Meum athamanticum (Spignel)

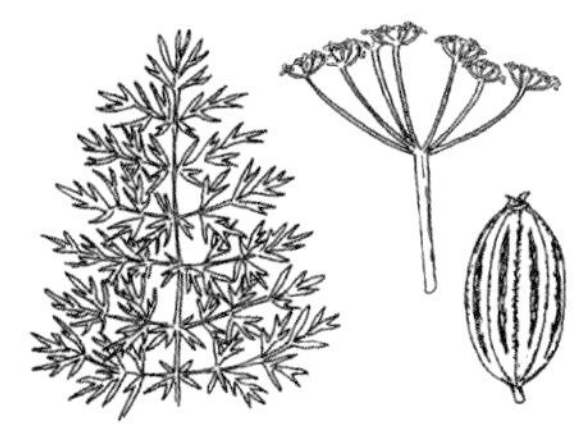

Carum carvi (Caraway)

Conopodium majus (Pignut)

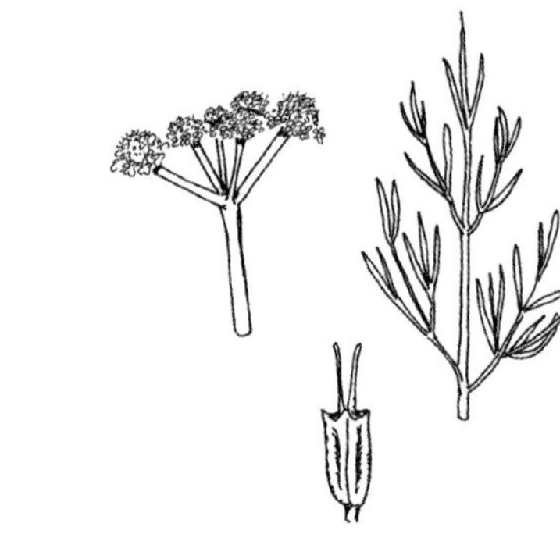

Oenanthe silaifolia
(Narrow-leaved Water-dropwort)

Oenanthe fluviatilis
(River Water-dropwort)

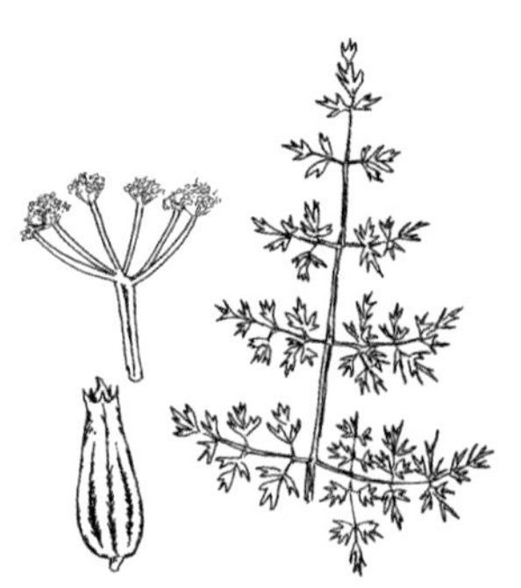

Oenanthe aquatica
(Fine-leaved Water-dropwort)

Leaves, inflorescences and fruits of the species with feathery fern-like leaves and no bracts (*continued*)

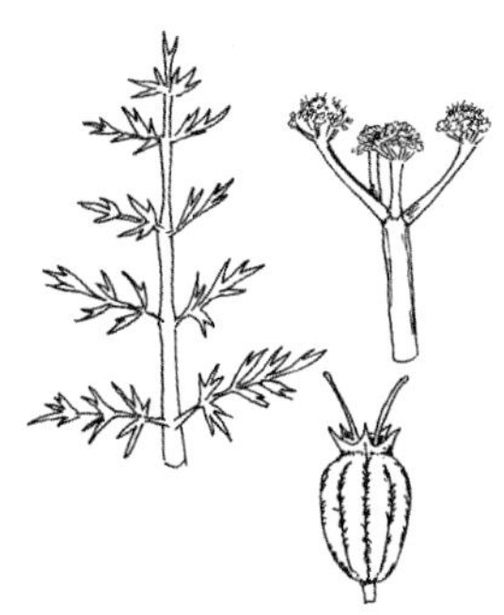

Oenanthe fistulosa
(Tubular Water-dropwort)

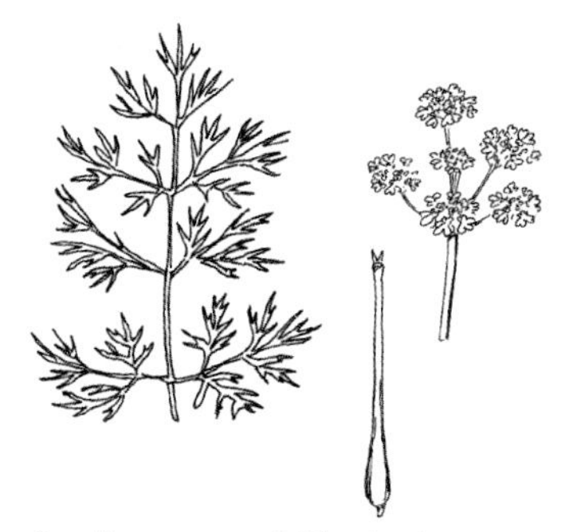

Scandix pecten-veneris (Shepherd's-needle)

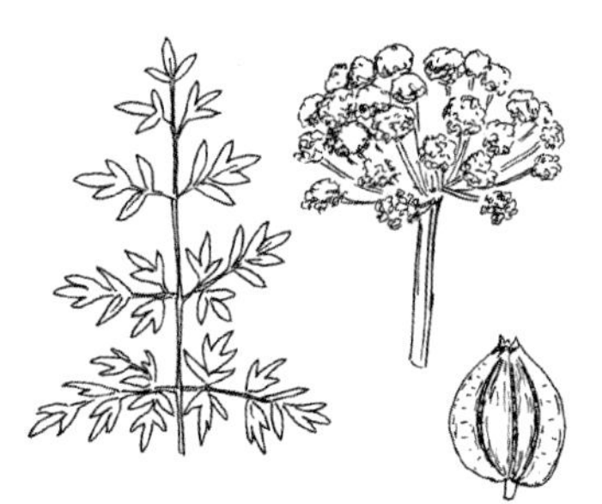

Selinum carvifolia
(Cambridge Milk-parsley)

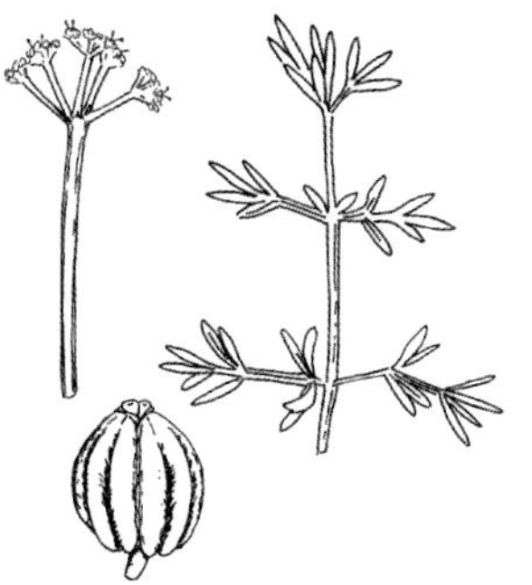

Trinia glauca
(Honewort)

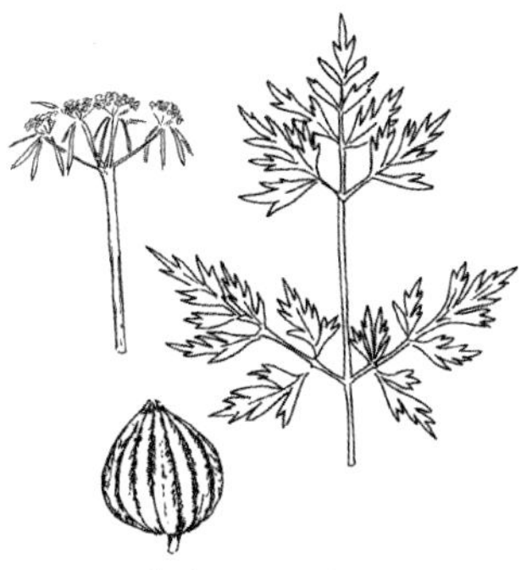

Aethusa cynapium
(Fool's Parsley)

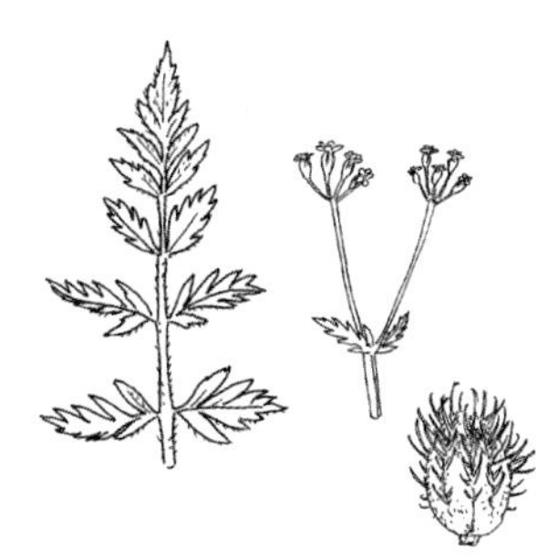

Torilis nodosa
(Knotted Hedge-parsley)

Leaves, inflorescences and fruits of the species with feathery fern-like leaves and no bracts (*continued*)

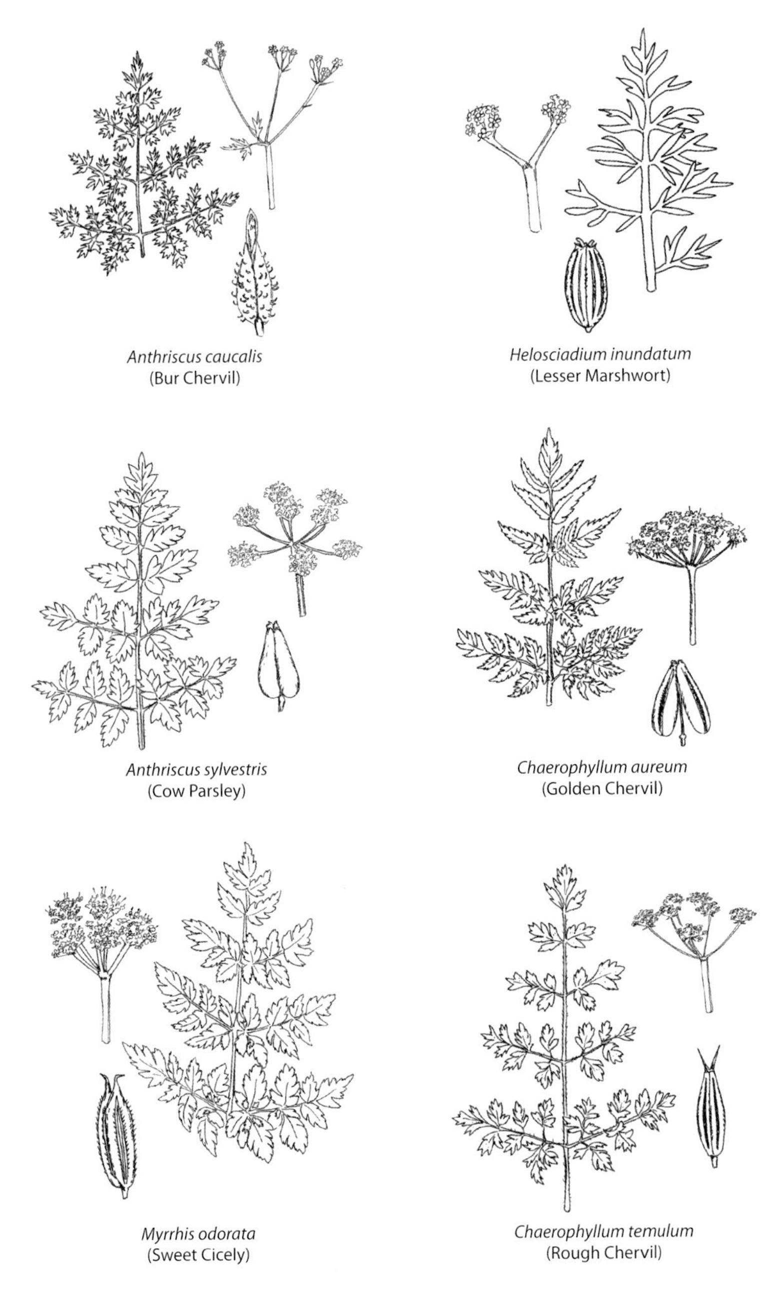

Anthriscus caucalis (Bur Chervil)
This annual plant is found in disturbed ground across the British Isles but more frequently in the east. Grows to a height of 70 cm. **Its leaves are divided into lobes and are slightly hairy.** Its stems are hollow and also hairy. Its white flowers are radially symmetrical. Usually has no bracts, but it does have bracteoles. **The umbel rays are hairless. Fruit measure 2.9–3.2 mm and are covered in hooks.**

Helosciadium inundatum (Lesser Marshwort)
This perennial aquatic plant occurs in shallow water and on mud scattered occasionally across the British Isles. Has a prostrate growth habit, reaching only 50 cm. Its leaves are divided into thin lobes. It has hollow hairless stems that root at the lower nodes. The flowers are white and radially symmetrical. It lacks bracts, but usually has several bracteoles. **The umbels have very few rays, usually just two. The fruit are flattened, deeply longitudinally grooved and measure 2.5–3 mm.**

Anthriscus sylvestris (Cow Parsley)
A very common perennial plant of grasslands and hedgerows across the British Isles, where it may grow to a height of 1.5 m. The feathery leaves are divided into lobes and are hairy. **There is a groove along the leaf stalk. Its stems are hollow and hairy but not at the top. Its stems may appear purple, green or white in colour. Its white flowers are radially asymmetrical. It usually has no bracts but does have bracteoles.** Its rays are hairless. There are many umbellets loosely packed within the main umbel. Fruit are smooth and measure 6–10 mm.

Chaerophyllum aureum (Golden Chervil)
An introduced uncommon perennial plant of grasslands scattered across the British Isles, where it my grow to a height of 1.2 m. The feathery leaves are divided into lobes and are hairy. There is a groove along the leaf stalk. **Its stems are solid and also hairy.** Its white flowers are radially symmetrical. **It usually has no bracts but does have several often purplish bracteoles.** Its rays are hairless. There are many umbellets loosely packed within the main umbel. Fruit taper towards their apex and measure 7–10 mm.

Myrrhis odorata (Sweet Cicely)
This introduced perennial plant is frequent across much of northern Britain and Ireland, but rare in the south, often appearing close to human habitation. Grows to a height of 1.8 m. It has soft feathery, hairy leaves. **There are distinctive small silver-white splodges on the leaves. The plant smells pleasantly of aniseed when crushed.** The stems are hollow and hairless. The flowers are white and slightly asymmetric. There are no bracts, but there are bracteoles. The umbels appear almost fluffy. The fruit are ridged and measure 15–25 mm.

Chaerophyllum temulum (Rough Chervil)
This common biennial grows in grasslands and hedges across the British Isles, although it is less common and possibly introduced in Ireland. Grows to a height

of 1 m. **It has feathery hairy leaflets which arise on their own stalks at 90° from the main leaf stalk.** Its stems are solid and hairy. The flowers are white and radially symmetrical. There are no bracts, but there are bracteoles. Its fruit measure 4–6.5 mm and are tapered towards their apex.

Next let's look at the ones with non-feathery fern-like leaves and bracts

Helosciadium repens (Creeping Marshwort)
A very rare creeping plant of wet ground near Oxford and one site near London. Grows to a height of less than 50 cm. **Its leaves are divided into simple lobes, the lower lobes may have distinctive sub-lobes.** The stems are hollow and lack hairs. The flowers are white and radially symmetric. The umbels have broad bracts and bracteoles. Fruit round and small, measuring 0.7–1 mm.

Sison amomum (Stone Parsley)
This is a short-lived plant of English hedgerows from Yorkshire south, occasional in Wales. Grows to a height of 1 m. Its leaves are divided into simple lobes which are partly subdivided. They often have a distinctive purplish margin. **When crushed the plant smells of petrol.** Its stems are solid and lack hairs. The flowers are white and radially symmetric. The umbels have both bracts and bracteoles. Fruit are round, grooved and measure 1.5–3 mm.

Falcaria vulgaris (Longleaf)
This introduced perennial is found in grassy places scattered across the south of England and at a few sites in Ireland. Grows to a height of 60 cm. **Its leaves are highly distinctive, being divided into long thin lobes with a serrated margin.** The stems are solid and lack hairs. The white flowers are radially symmetric. The umbels are rather delicate and have both bracts and bracteoles. Its fruit measure 2.5–4 mm.

Tordylium maximum (Hartwort)
A very rare species of the lower Thames, where it may grow to a height of 1 m. Its leaves are divided into simple lobes with a roundly serrated margin. Its stems are hollow and hairy, although sometimes the stems are solid. The white flowers are highly asymmetric. **It has a rather stocky umbel with many bracts and bracteoles.** The fruit are flattened, hairy and measure 4.5–6 mm.

Sium latifolium (Greater Water-parsnip)
A rare aquatic perennial plant of the south-west of England and Ireland. **It may grow to 2 m tall. Its aerial leaves have between 3 and 8 simple leaflets with a serrated margin.** Its stems are hollow and lack hairs. The flowers are white and radially symmetric. The umbels have broad bracts and bracteoles. Fruit measure 2.5–4 mm and are round in shape.

Ligusticum scoticum (Scots Lovage)

This perennial plant is found near the Scottish coast and is rare in northern England. May grow to a height of 60 cm. **Its leaves are distinctively bright green and divided into oval leaflets with serrated margins.** Its stems are hollow and lack hairs. It has white, radially symmetric flowers. Has a rather compact umbel with both bracts and bracteoles. The fruit taper towards the apex, are grooved and measure 4–7 mm.

Heracleum mantegazzianum (Giant Hogweed)

This introduced perennial species only flowers once before dying. Found scattered across the British Isles often near rivers, but it is less common in Scotland and southern Ireland. **It may grow to a massive 5.5 m tall. Its leaves are divided into broad leaflets which are sharply toothed.** Its stems are hollow, ridged, covered in purple spots and hairy. The flowers are white and asymmetric. The bracts are short and broad, there are several bracteoles. The sap is caustic. Fruit are flattened and measure 9–11 mm.

Characteristics of species with non-feathery fern-like leaves and bracts

	Plant height			Stems		Flowers				Seeds	
Species	**<50 cm**	**<1 m**	**>1 m**	**hollow**	**hairy**	**yellow**	**symmetrical**	**Bracts**	**Bracteoles**	**<5 mm**	**hairy**
Helosciadium repens (Creeping Marshwort)	■			■			■	■	■		
Sison amomum (Stone Parsley)		■					■	■	■		
Falcaria vulgaris (Longleaf)		■					■	■	■		
Tordylium maximum (Hartwort)		■		vary	■			■	■	■	■
Sium latifolium (Greater Water-parsnip)			■	■			■	■	■		
Ligusticum scoticum (Scots Lovage)		■		■			■			■	
Heracleum mantegazzianum (Giant Hogweed)			■	■	■			■	■	■	
Your specimen											

Next let's look at the ones with non-feathery fern-like leaves and NO bracts

Cicuta virosa (Cowbane)

A rare perennial with a rather scattered distribution. Grows in waterlogged soils and is most common in Ireland. May grow to a height of 1.5 m. **Its leaves are**

Leaves, inflorescences and fruits of the species without feathery fern-like leaves with bracts

Helosciadium repens
(Creeping Marshwort)

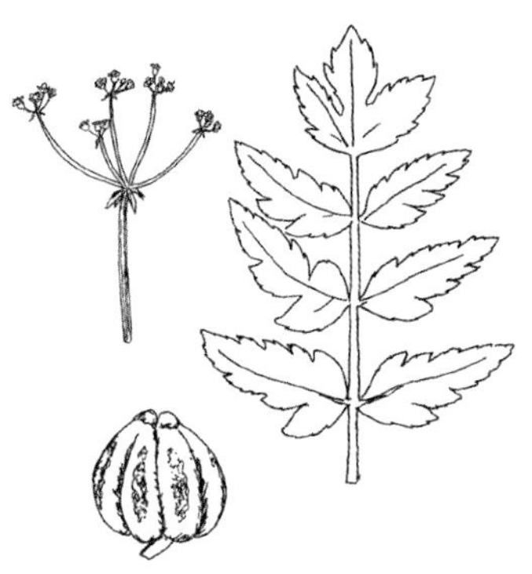

Sison amomum
(Stone Parsley)

Falcaria vulgaris (Longleaf)

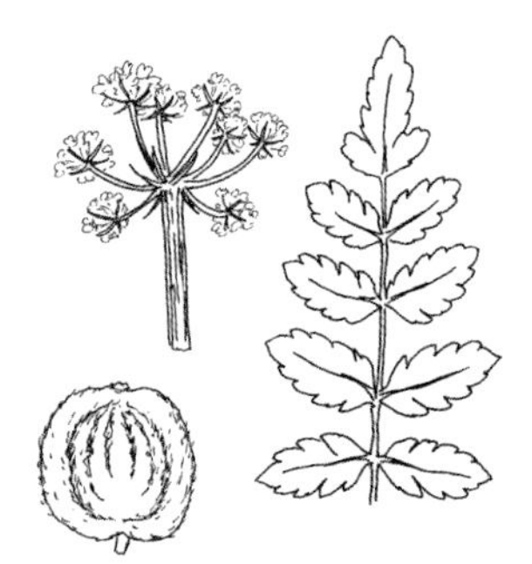

Tordylium maximum
(Hartwort)

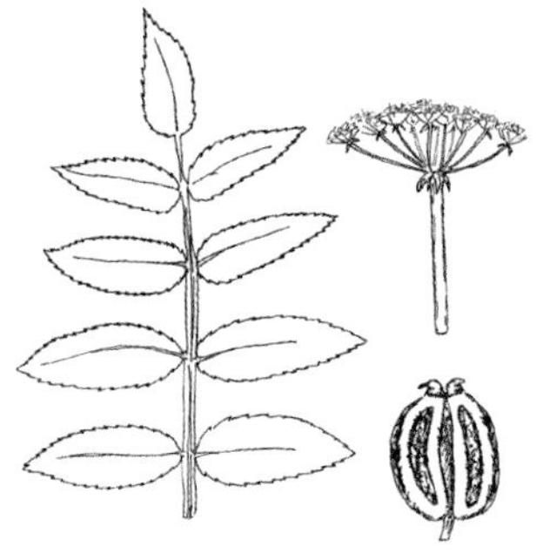

Sium latifolium
(Greater Water-parsnip)

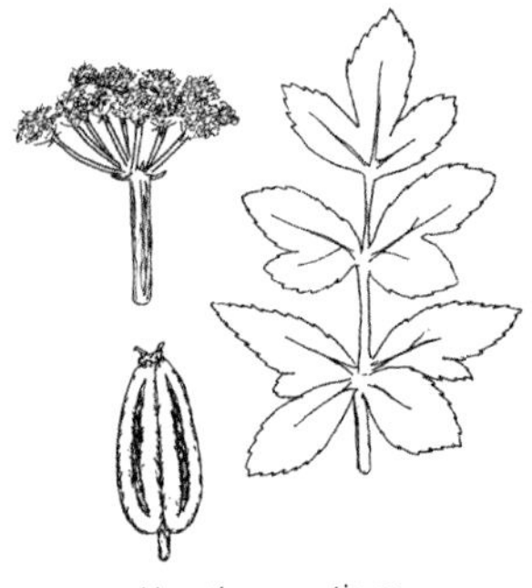

Ligusticum scoticum
(Scots Lovage)

Leaves, inflorescences and fruits of the species without feathery fern-like leaves with bracts (*continued*)

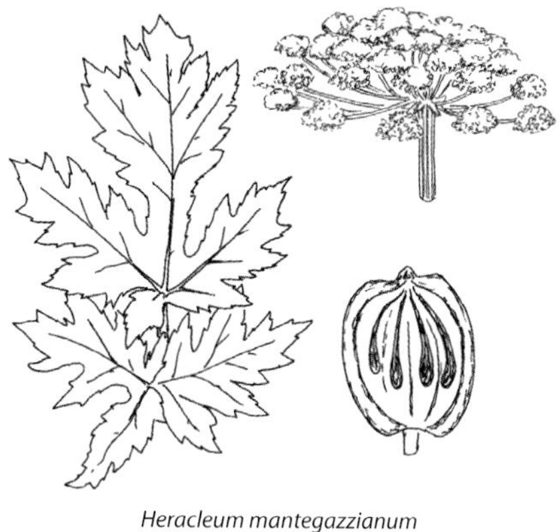

Heracleum mantegazzianum
(Giant Hogweed)

divided into long thin leaflets, often in threes. The stems are hollow and lack hairs. The flowers are white and radially symmetric. There are no bracts but many bracteoles. Highly poisonous. Fruit measure 1.2–2 mm and retain their stigmas.

***Apium graveolens* (Wild Celery)**
In the British Isles this biennial species tends to occur in coastal areas in the south. It grows in wet ground to a height of 1 m. Its leaves are glossy green and divided into lobes. Its stems are solid and lack hairs. **The plant smells of celery.** Its flowers are white and radially symmetrical. Has no bracts or bracteoles. Its fruit are small, measuring only 1–1.5 mm.

***Helosciadium nodiflorum* (Fool's-water-cress)**
A very common plant of drainage ditches and wetlands across most of the British Isles. **May grow horizontally, rooting along its stem, before becoming vertical reaching a maximum height of about 1 m. The leaves are divided into opposite pairs of leaflets along the main leaf stalk.** The stems are hollow and hairless. The flowers are white and radially symmetric. There are usually no bracts but many bracteoles. The fruit are round and measure 1.5–2.5 mm.

***Pimpinella saxifraga* (Burnet-saxifrage)**
This perennial species is common in old grasslands across most of the British Isles, except in the north of Scotland and Ireland. May grow to a height of about 70 cm. **Its lower leaves are divided into pairs of oval leaflets along the main leaf stalk. The leaflets have toothed margins.** The stems are solid and may sometimes be hairy. The flowers are white and radially symmetric. There are no bracts or bracteoles. Its umbels are rather elegant. Fruit measure 2–3 mm.

Characteristics of species with non-feathery fern-like leaves and NO bracts

	Plant height			Stems		Flowers				Seeds	
Species	<50 cm	<1 m	>1 m	hollow	hairy	yellow	symmetrical	Bracts	Bracteoles	<5 mm	hairy
Cicuta virosa (Cowbane)											
Apium graveolens (Wild Celery)											
Helosciadium nodiflorum (Fool's-water-cress)								rare			
Pimpinella saxifraga (Burnet-saxifrage)					vary						
Pimpinella major (Greater Burnet-saxifrage)											
Pastinaca sativa (Wild Parsnip)				vary				rare	rare		
Angelica sylvestris (Wild Angelica)											
Aegopodium podagraria (Ground-elder)											
Heracleum sphondylium (Hogweed)								rare	rare		vary
Imperatoria ostruthium (Masterwort)								rare			
Your specimen											

Pimpinella major (Greater Burnet-saxifrage)

This perennial species grows in woodland margins and grasslands and has a curious, disjointed distribution: occurs in most of England, except the south west, and then again in Cornwall; absent from Wales and Scotland, but occurs in Ireland – mostly in the west. Grows to a height of 1 m. **Its leaves are divided into pairs of leaflets along the main leaf stalk. The leaflets have deeply toothed margins.** Its stems are hollow and ridged, and usually hairless. The flowers are white and radially symmetric. There are no bracts or bracteoles. Fruit measure 3–4 mm.

Pastinaca sativa (Wild Parsnip)

This biennial and variable species is both a native and garden escape. Occurs in grasslands across much of England and Wales but is less common in Scotland and Ireland. May grow to a height of 1.8 m. Its leaves are divided into pairs of leaflets along the main leaf stalk. The lower leaflets may be further divided and have lobes. **The plants have a strong and distinctive smell. The stems are hairy may be solid or hollow. The flowers are yellow** and radially symmetric. There are rarely bracts or bracteoles. There are a number of subspecies. Its fruit are flattened and measure 4–7mm.

Leaves, inflorescences and fruit of the species non-feathery fern-like leaves and NO bracts

Cicuta virosa (Cowbane)

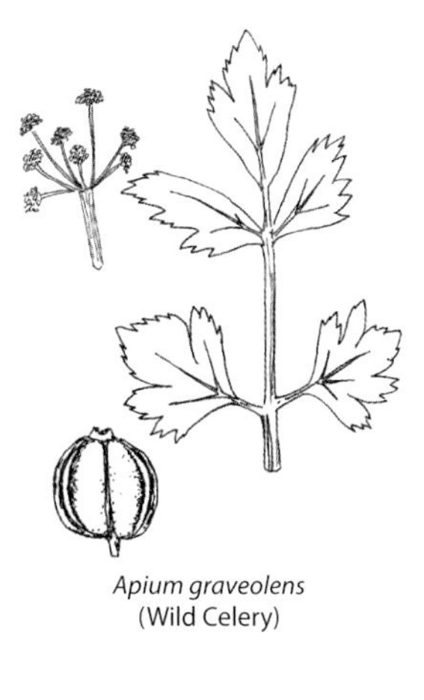

Apium graveolens
(Wild Celery)

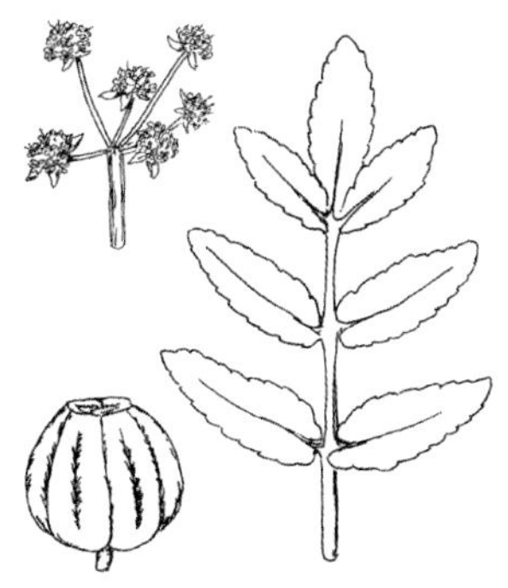

Helosciadium nodiflorum
(Fool's-water-cress)

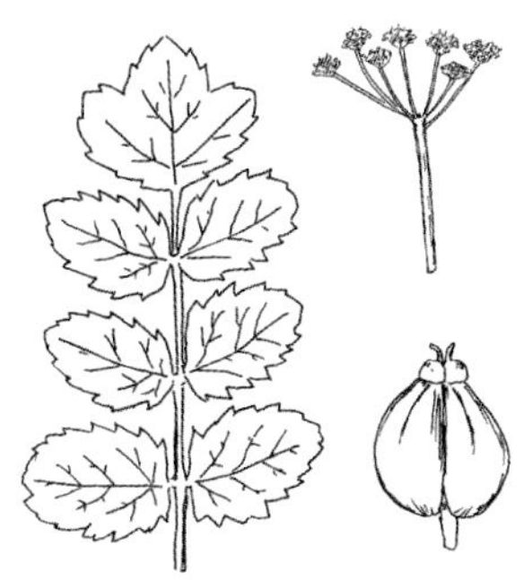

Pimpinella saxifrage
(Burnet-saxifrage)

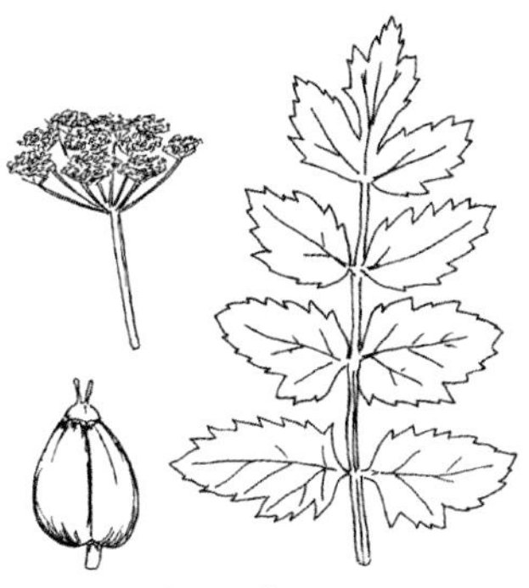

Pimpinella major
(Greater Burnet-saxifrage)

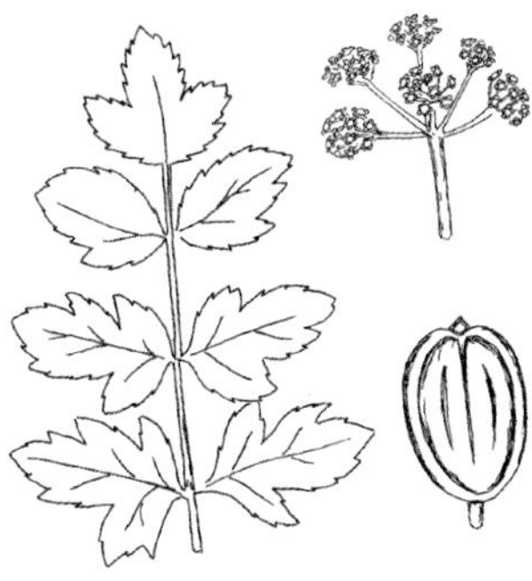

Pastinaca sativa
(Wild Parsnip)

Leaves, inflorescences and fruit of the species non-feathery fern-like leaves and NO bracts (*continued*)

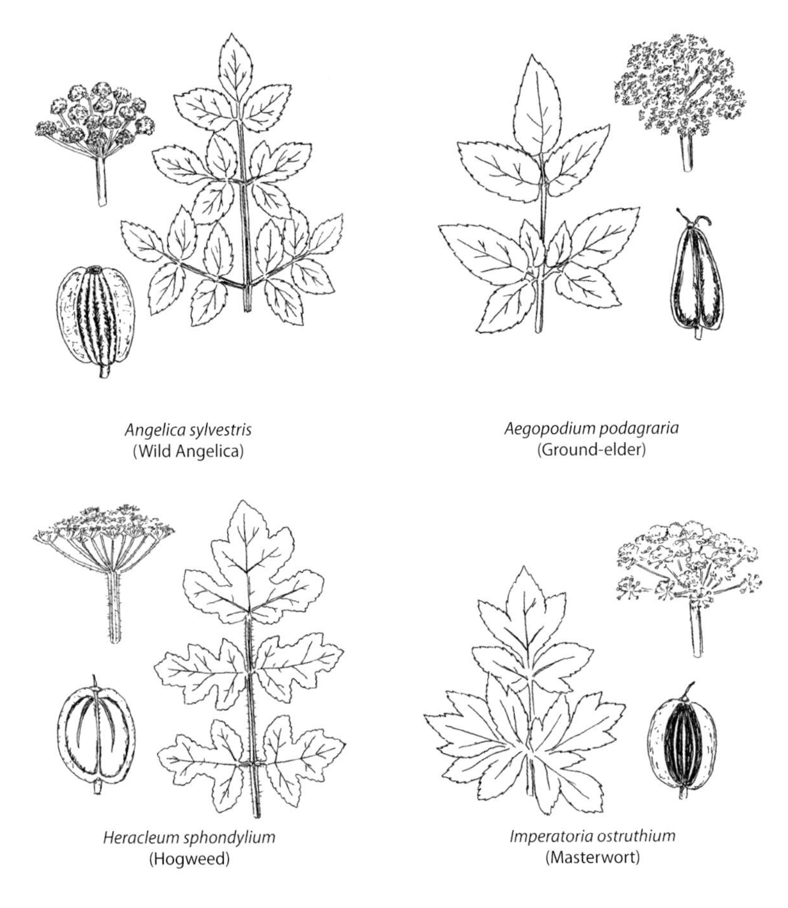

Angelica sylvestris (Wild Angelica)

Aegopodium podagraria (Ground-elder)

Heracleum sphondylium (Hogweed)

Imperatoria ostruthium (Masterwort)

Angelica sylvestris (Wild Angelica)

This wetland plant is short-lived, flowering once before dying. Common throughout the British Isles, it grows to a height of 2.5 m. The leaves are divided into many oval leaflets. **The leaf stalk has a distinctive groove along its upper surface. The stems are hollow, lack hairs and are often purple in colour.** The flowers are white (sometimes pink) and radially symmetric. There are no bracts but several bracteoles. Fruit are flattened, ridged and measure 4–6 mm.

Aegopodium podagraria (Ground-elder)

A common perennial plant of cultivated and waste ground across the British Isles. **It spreads below ground, forming patches.** May grow to a height of 1 m. Its leaves are bright green and divided into several oval leaflets, with small teeth along their margins. Some of the leaflets can often be fused together. The stems are hollow and hairy. The flowers are white and radially symmetric. There are no bracts or bracteoles. The fruit taper towards the apex and measure 3–4 mm.

Heracleum sphondylium **(Hogweed)**
This very common species may be biennial or perennial, it grows across the British Isles in rough ground and in hedgerows, where it may reach 2 m tall or more. **The leaves are divided into several broad leaflets, sometimes the lower ones have distinct stalks of their own. The stems are hollow, ridged and hairy.** The flowers are white (sometimes pink or even purple) and highly asymmetric. There are occasionally bracts and bracteoles. A variable species which has been divided into several subspecies. The fruit are flattened, may be hairy and typically have four lines running longitudinally they measure 9–12 mm.

Imperatoria ostruthium **(Masterwort)**
This scarce perennial plant is most frequently encountered in the north of England, Scotland and Northern Ireland growing in grasslands, where it may reach up to 1 m tall. **Its broad leaves are usually divided into three lobes, which have toothed margins.** The stems are hollow and lack hairs. Its flowers are white and radially symmetric. There are usually no bracts and few bracteoles. Fruit are flattened and measure 3–5 mm.

HAVE OTHERS RECOGNISED THIS LEVEL OF VARIATION?

This section usually reviews when our ancestors first recognised all the species within each complex family. Probably as a consequence of their culinary importance and because of the risks of accidental poisoning, umbellifers have been well documented since ancient times. In fact, these plants were so well described by the Romans that there is a mystery surrounding the identity of one species: Silphium. This Mediterranean umbellifer was so highly prized as a perfume, seasoning, medicine and aphrodisiac that it was worth its weight in gold. Unfortunately, its true identity remains unclear and it is considered to probably be extinct.

In more recent times, Bentham and Hooker's standard Victorian flora of the British Isles included 59 species of *Umbelliferae*, which is close to the 62 described above.

HOW FAR SHOULD I GO?

With a little practice there really is no reason that you should not be able to recognise all the 62 species covered here. Although some hybrids and subspecies are known, these are no more problematic than any other family. Once you overcome any initial anxiety, you'll quickly realise that it is relatively easy to identify the 10 or so most commonly encountered species. Many of the more obscure species have very restricted distributions, so you will need to be lucky or determined to find those.

In addition to the 62 species covered here, there are several others that occasionally escape from cultivation. If you wish to identify these you may need to consult the horticultural literature, or Clive Stace's *New Flora of the British Isles* which includes more ephemeral escapees. There is a BSBI Handbook (No. 2) that covers this family, but it is now rather dated.

Glossary

Agamospermy The production of asexual seeds which genetically are clones of the mother plant.

Agg. (aggregate, species aggregate) A group of related but variable populations that are being treated as a single species – although some may choose to split them into separate closely related species.

Apomixis Asexual reproduction, which may occur vegetatively or through the production of asexual seeds (**agamospermy**).

Bract A modified leafy growth at the base of a flower stalk.

Bracteole A small bract-like growth beneath a branching flower stalk or umbellet.

Capitulum A flowerhead comprising many individual florets, surrounded by bracts. Typically associated with the daisy family.

Druplet The individual unit of a compound fruit such as those found in brambles.

Entire (of a leaf) A leaf that is not divided into lobes.

Facultative apomict A plants that may sometimes reproduce asexually, typically when conditions are less than ideal for sexual reproduction.

Floret A small individual flower, especially one that clusters together with others to form the flowerhead (**Capitulum**) in members of the daisy family.

Founder effect A genetic difference observed in a population that results from the random chance of which individual established the population, rather than from natural selection.

Glands, Glandular hairs Small structures that appear to produce droplets of liquid often found on the end of short hairs.

Lanceolate A long narrow oval shape with a pointed tip, usually used to describe a leaf.

Palmate leaf A leaf that is deeply divided into separate lobes, like fingers on a hand. These lobes may or may not be fused at their base.

Pappus Highly modified sepals that have become feathery; they are attached to the mature seed and help facilitate wind dispersal.

Primocane In brambles, a section of stem that has grown this year.

Recurved A structure that is curved backwards or outwards.

Rosette leaves The basal leaves of some plants, which arise in a circular pattern attached at a central point. They are often very different in form from leaves that are attached to stems.

Semi-deciduous A plant that loses its leaves only for a short time, often associated with a very cold period.

Stem leaves Leaves that are attached to a flowering stem.

Stipule A growth, usually leafy, subtending or surrounding the base of a leaf.

Suberect A plant that is low growing but not prostrate to the ground.

Sub-spikelet A cluster of flowers with each flower held on a short stalk within a larger cluster of flowers comprising several sub-spikelets.

Thorns, spines, prickles and pricklets
These are all sharp protuberances which act as defenses against herbivores. They are derived from different sort of plant tissues, which is important for plant anatomists but less vital for those describing plants for identification purposes.

Thorns These are formed from modified branches or stems.

Spines These are formed from modified leaves or parts of leaves.

Prickles These are formed from the epidermis, the outer layer of the plant's stems. Many species that are commonly regarded as having thorns technically have prickles – these include roses and brambles

Pricklets Structurally similar to prickles, except smaller. These are important in the identification of brambles.

Tubercle A swollen structure on the side of the fruit of many species of docks.

Index

References to figures appear in *italic* type; those in **bold** type refer to tables.